FIDUCIARY DUTY AND THE ATMOSPHERIC TRUST

Law, Ethics and Governance Series

Series Editor: Charles Sampford, Director, Key Centre for Ethics, Law, Justice and Governance, Griffith University, Australia

Recent history has emphasised the potentially devastating effects of governance failures in governments, government agencies, corporations and the institutions of civil society. 'Good governance' is seen as necessary, if not crucial, for economic success and human development. Although the disciplines of law, ethics, politics, economics and management theory can provide insights into the governance of organisations, governance issues can only be dealt with by interdisciplinary studies, combining several (and sometimes all) of those disciplines. This series aims to provide such interdisciplinary studies for students, researchers and relevant practitioners.

Recent titles in the series

Ethics and Socially Responsible Investment
A Philosophical Approach
William Ransome and Charles Sampford
ISBN 978-0-7546-7581-5

Improving Health Care Safety and Quality
Reluctant Regulators
Judith Healy
ISBN 978-0-7546-7644-7

Idealism and the Abuse of Power
Lessons from China's Cultural Revolution
Zhuang Hui-yun
ISBN 978-0-7546-7208-1

Integrity Systems for Occupations
Andrew Alexandra and Seumas Miller
ISBN 978-0-7546-7749-9

Promoting Integrity
Evaluating and Improving Public Institutions
Edited by Brian W. Head, A.J. Brown and Carmel Connors
ISBN 978-0-7546-4986-1

Fiduciary Duty and the Atmospheric Trust

Edited by

KEN COGHILL
Monash University, Australia

CHARLES SAMPFORD
Griffith University, Australia

TIM SMITH
Monash University Australia

Routledge
Taylor & Francis Group

LONDON AND NEW YORK

First published 2012 by Ashgate Publishing

2 Park Square, Milton Park, Abingdon, Oxon OX14 4RN
711 Third Avenue, New York, NY 10017, USA

Routledge is an imprint of the Taylor & Francis Group, an informa business

First issued in paperback 2016

British Library Cataloguing in Publication Data
Fiduciary duty and the atmospheric trust. — (Law, ethics and governance)
  1. Political ethics. 2. Government accountability. 3. Liability for climatic change damages.
  4. Climatic changes—Government policy. 5. Political ethics—Australia. 6. Government accountability—Australia. 7. Liability for climatic change damages—Australia.
  8. Climatic changes—Government policy—Australia. 9. Climatic changes—International cooperation. 10. Climatic changes—Law and legislation.
  I. Series II. Coghill, Ken, 1944– III. Sampford, C. J. G. (Charles J. G.) IV. Smith, Tim.
  344'.046–dc22

Library of Congress Cataloging-in-Publication Data
Fiduciary duty and the atmospheric trust / by Ken Coghill, Charles Sampford and Tim Smith (eds.).
    p. cm. — (Law, Ethics and Governance)
  Includes bibliographical references and index.
  ISBN 978-1-4094-2232-7 (hc : alk. paper)
  1. Climatic changes—Law and legislation. I. Coghill, Ken, 1944– II. Sampford, C. J. G. (Charles J. G.) III. Smith, Tim.
  K3585.5.F53 2011
  344.04'6342–dc23

                                                                                    2011028485

ISBN 978-1-4094-2232-7 (hbk)
ISBN 978-1-138-24553-2 (pbk)

Contents

Notes on Contributors and Editors

Robert Clark, B.Com (Hons), LLB, BA (Melb Uni), is the Member for Box Hill in the Victorian Parliament, Attorney-General and Minister for Finance since December 2010. Formerly:Shadow Attorney-General and Shadow Minister for Finance; Deputy Chair of Law Reform Committee; member Privileges Committee (Assembly); Disputes Resolution Committee; Parliamentary Secretary to Victorian Treasurer (1992–99). He practised as a solicitor in commercial, financial and labour law prior to entering Parliament in 1988.

Ken Coghill is an Associate Professor, Department of Management, Monash University and a founding member of the Accountability Round Table. His research and teaching interests include accountability, integrated governance and professional development for parliamentarians. His PhD concerned ministerial responsibility and accountability in Queensland, Victoria and Western Australia. He has worked in the Victorian public service and served as Councillor (Rural City of Wodonga), Member of Parliament (Legislative Assembly, Victoria), Parliamentary Secretary of the Cabinet and Speaker.

Donald Feaver is an Associate Professor, Graduate School of Business and Law, RMIT University. His research interests include international and transnational law and regulation, including the design of coherent public and private regulatory arrangements, including the international regulation of climate change related issues. He is admitted as a Barrister and Solicitor of the Supreme Court of Victoria and an Attorney and Counsellor at Law of the Supreme Court of New York. He has worked in private legal practice and frequently undertakes advisory and consultancy works for national governments as well as multilateral and non-governmental international organizations.

Paul Finn was appointed a judge of the Federal Court of Australia in 1995. Previously Professor of Law in the Research School of Social Sciences at the Australian National University. Professorial Fellow of the University of Melbourne and Arthur Goodhart Professor of Legal Science at the University of Cambridge in 2010/11. Graduate of the Universities of Queensland, London and Cambridge. Author and editor of a number of books including works dealing with fiduciary duty. He has been a member of the Second and Third Working Groups on Unidroit's Principles of International Commercial Contracts.

Evan Fox-Decent is an Associate Professor, McGill University Faculty of Law and member of the McGill Centre for Human Rights and Legal Pluralism. He has worked on human rights and democratic governance reform in Latin America since 1987, beginning with advocacy and relief work in El Salvador under the auspices of Nobel Peace Prize Nominee Medardo Gómez. He teaches and publishes in legal theory, international law, administrative law, First Nations and the law, the law of fiduciaries, and human rights. He obtained a JD and PhD (philosophy) from the University of Toronto. His forthcoming book in the Oxford Constitutional Theory series, *Sovereign's Promise: The State is Fiduciary*, explores the implications of recognizing the state and its institutions as fiduciaries of their people.

John Glover is a Professor of Law at RMIT University, Graduate School of Business and Law and a barrister practising in Melbourne. Trusts and taxation are his main professional concerns. John has served as a sessional member of VCAT, and on numerous professional committees, including the ATO's National Tax Liaison Group (Trusts). *Texts Equity, Restitution and Fraud* (LexisNexis, Sydney, 2004); and *Commercial Equity: Fiduciary Relationships* (Butterworths, Sydney, 1995) have been authored by him, as well as over 50 book chapters and articles in refereed law journals. Since 2008 John has been a co-author of the leading trusts resource Ford and Lee's Law of Trusts (Thomson Reuters looseleaf and on-line).

Fiona Haines is an Associate Professor in the School of Social and Political Sciences at the University of Melbourne and Adjunct Senior Research Fellow at ANU. Her research interests span globalization, regulation and the control of corporate harm, with publications that include *Corporate Regulation: Beyond Punish or Persuade* (Clarendon Press 1997), *Globalization and Regulatory Character: Regulatory Reform after the Kader Toy Factory Fire* (Ashgate 2005) and *The Paradox of Regulation: What regulation can achieve and what it cannot* (Edward Elgar 2011 forthcoming). She is co-editor of the interdisciplinary journal, *Law and Policy*, with Professors Nancy Reichman (Sociology, Denver) and Colin Scott (Law, UC Dublin).

Rosemary Teele Langford is a Doctoral Candidate and Sessional Lecturer, Faculty of Law, Monash University. Her PhD centres on fiduciary theory and directors' fiduciary duties. She has a first-class honours degree in Law from the University of Melbourne and has been publishing on the fiduciary principle for over 15 years. Rosemary was admitted to practice as a barrister and solicitor of the Supreme Court of Victoria in 1996 and practised with Allens Arthur Robinson before moving into teaching. Her main current teaching area is Corporations Law, and additional research interests include civil and comparative law and equity and trusts.

Will McGoldrick is the Policy and Research Manager at The Climate Institute, an Australian-based think-tank. Will specializes in climate policy, with a particular interest in international climate change law. Before starting at The Climate Institute, Will spent three years in Samoa, as a climate change advisor to the Samoan Government. Will has bachelor degrees in arts and science from Monash

University and a Master of International Law degree from The University of Sydney. Will's Master's thesis assessed the prospects for the post-2012 climate regime providing an improved mechanism to finance climate change adaptation in Pacific island countries.

Andrew Maver's research interests have included biblical and cultural studies and international law, with a recent focus on carbon-trading schemes and climate-change financing. His PhD research focuses on communication and the mass media. He has assisted the Victorian Government with the implementation of the Montreal Process for the Conservation and Sustainable Management of Temperate and Boreal Forests. He has worked on a range of civil and public litigation matters in the energy and resources sectors and in support of Counsel Assisting at the Victorian Bushfires Royal Commission.

Andrew Murray, BA Hons (Rhodes) MA (Oxf), was a Senator for Western Australia from July 1996 to June 2008. He is best known in politics for his work on finance, economic, business, industrial relations and tax issues; on accountability and electoral reform; and for his work on institutionalized children.

Charles Sampford was appointed Foundation Dean of Law at Griffith University in March 1991, Foundation Director of the Key Centre for Ethics, Law, Justice and Governance (the only centre in law or governance to receive centre funding from the Australian Research Council) in 1999 and Foundation Director of the Institute for Ethics, Governance and Law (a joint initiative of the United Nations University, Griffith, QUT and ANU) in 2004. Professor Sampford has written 100 articles and chapters in Australian and foreign journals and collections ranging through public law, legal philosophy, legal education, ethics, economics and governance and has completed 22 books and edited collections for international publishers, including OUP, Blackwell, Routledge and Ashgate. His work on environmental governance commenced when he was invited to open the World Council of Churches colloquium on carbon trading in Saskatoon in May 2000. Foreign fellowships include the Visiting Senior Research Fellow at St John's College Oxford (1997) and a Fulbright Senior Award to Harvard University (2000), following which he was appointed a member of a task force chaired by Secretary Albright on Threats to Democracy.

Tim Smith is an Adjunct Professor, Department of Management, Monash University and Chair of the Accountability Round Table. His interests include democracy, the rule of law, the accountability of government, the governance of the judicial branch of government and sustainability. He has served as a Commissioner of the Australian Law Reform Commission and the Victorian Law Reform Commission and as a County Court and a Supreme Court judge.

Kelvin Thomson (MP) is the Member for Wills, Parliament of Australia. He graduated with first class honours degrees in Arts and Law (winner of the Supreme Court Prize 1987), the University of Melbourne. He worked for the Commonwealth Public Service, as an electorate officer, project officer for the Commonwealth Ombudsman and principal project officer for Australia Post. He served as Coburg councillor, including deputy mayor, until 1988; Member, Parliament of Victoria 1988–1996, Member, Parliament of Australia since 1996. In 1998–2007, he served in the portfolios of Assistant Treasurer, Environment and Heritage, Regional Development, Roads and Housing, Public Accountability, Human Services and Attorney General.

Mary Christina Wood is the Philip H. Knight Professor of Law, Faculty Director for the Environmental and Natural Resources Law Program, and a 2007–2008 Luvaas Faculty Fellow at the University of Oregon School of Law. She teaches property law, natural resources law, public trust law, federal Indian law, public lands law, wildlife law, and hazardous waste law. She is the Founding Director of the school's Environmental and Natural Resources Law Program and is Faculty Leader of the Program's Conservation Trust Project, Sustainable Land Use Project and Native Environmental Sovereignty Project. After graduating from Stanford Law School in 1987, she served as a judicial clerk on the Ninth Circuit Court of Appeals. She then practised in the environmental/natural resources department of Perkins Coie, a Pacific Northwest law firm. In 1994 she received the University's Ersted Award for Distinguished Teaching, and in 2002 she received the Orlando Hollis Faculty Teaching Award. Professor Wood is a co-author of a leading textbook on natural resources law (West, 2006) and has published extensively on climate crisis, natural resources, and native law issues. She is a frequent speaker on global warming issues and has received national and international attention for her sovereign trust approach to global climate policy.

Chapter 1

Rulers' Duties to Our Environment?

Ken Coghill, Charles Sampford and Tim Smith

Long-term policy issues and administration affecting the public interest raise major concerns about how democracy can achieve its ultimate purpose of ensuring 'responsive rule', that is, the 'necessary correspondence between acts of governance and the equally-weighted felt interests of citizens with respect to those acts'.[1] Citizens are the best judges of their *felt* interests, and elections involving universal and equal suffrage are the quintessential means for reflecting them. However, such elections are necessarily crude and blunt instruments of accountability because of their infrequency and the fact that so many overlapping and partially conflicting interests, policies and values have to be reduced to a single decision on which candidate or party to vote for.

While citizens should be able to recognize their longer-term interests in the kind of future they and their children will enjoy, the usual media and electoral political cycles militate against their consideration. Elections cannot make our political institutions responsive to the interests of others who are affected by political and commercial decisions taken today – including future generations, non-citizens, residents of other nations and other species. How can these longer-term interests (and values) of citizens and the interests of various classes of non-voters be taken into account? One possibility may be found in the principles of public trust and fiduciary duty founded in Equity.[2] Such principles seem to offer a basis for arguing that government ministers have a duty to discharge a public trust in respect of the climate.

The implications for accountability of the executive government are considerable. A higher standard of accountability would apply. The executive would be held accountable for the discharge of a public trust over and above simply acting lawfully, ethically and in response to elections.

Here we examine the implications for government in the particular case of climate change. However, note that these implications may not be limited to the political executive. A further question arises as to whether fiduciary duty and/or

---

1   Saward (1996), p. 468.

2   In this volume, Finn defines fiduciary duty as 'standards of conduct properly to be expected of persons occupying fiduciary positions, that is, persons who, by virtue of position, responsibility or function, were expected to act in another's interests and not in their own interests'. For a detailed discussion of the characteristics of fiduciary duty, see Finn in chapter 3.

public trust can be invoked to argue for ordinary members of parliament having a general responsibility to act in a longer-term 'public interest' that is not founded on electoral mandates in the particular case of climate change. Again, fiduciary duty may offer a basis for arguing that MPs have a duty to discharge a public trust in respect of the climate.

If this argument led to a view that members of parliament have a general responsibility to act in this longer-term public interest in the particular case of climate change, there are implications for how parliament would address the issue. For example, parliamentary committee processes similar to scrutiny of legislation might take a greater role than politically charged contests on the 'floor of parliament'.[3]

These issues are highly relevant to concerns of citizens over failures by governments to support effective action on what may be the greatest threat to ever face human civilizations. They may suggest that those governments, like earlier rulers before them, have a fiduciary duty to their long-term interests and the interest of non-voters. They may also suggest that the voters themselves – the ultimate rulers in a democracy – also have such fiduciary duties.

In this work these concerns are addressed by a diverse range of authors, who examine whether the ancient concepts of fiduciary duty and public trust can be revived, re-interpreted and invoked to stimulate governments to more effective action to protect citizens' common interest in the atmosphere that they share with each other and with non-voters. They consider how governments could be held accountable where they neglect to act effectively,[4] asking if accountability can be advanced by invoking the concept of fiduciary duty and/or public trust. Could either or both have particular application to responsibility for policy responses to climate change and/or to accountability for the discharge of those policy responses?

The dawning awareness of the concerning state of the atmosphere has crept up on governments somewhat slowly until they suddenly face *Apocalypse Now!*[5] For more than an entire generation people have seen the evidence slowly emerging that mankind is polluting, depleting and destroying the very environment on which all civilizations depend. To adapt the words of scientist and author Tim Flannery

---

3　Or the media fora in which political debate takes place.

4　A fascinating alternative form of accountability of executive government ministers arose in Iceland at a late stage of writing this book. Former Prime Minister Geir Haarde was referred to the Landsdomur (court of impeachment) for trial for alleged negligence, relating to Iceland's banking crisis, by his government in 2008.

5　The title of a well-known war film. The title has an interesting derivation. According to Hart, it is a '(r)eference to the last book of the New Testament (Revelations of St John the Divine) in which a revelation or knowledge of the future is "revealed"(or uncovered) to John. One of the most important aspects of this revelation is of the "last battle" (Armageddon Rev. 16.16) in which the forces of good and evil are set against each other prior to the final day of judgement'. Hart (undated).

in *The Future Eaters*,[6] the knowledge that mankind is eating its future has been increasingly well known for decades.

Elements of this knowledge were well established in specialist scientific circles long before it entered the popular media with the publication of *The Limits to Growth* in 1972.[7] At that time, any crisis for mankind seemed decades away, but nonetheless actions were begun by some governments. President Jimmy Carter, a trained scientist, came to the White House and, in a highly symbolic measure, had solar hot water units installed on the roof in 1979.[8]

The evidence base expanded and links between the environment's sub-systems became better understood. In 1985, British Antarctic Survey scientists reported alarming rapid thinning of the ozone layer, especially over higher southern latitudes. In summers, it was developing a huge hole extending as far north as southern Australia, exposing people to dangerous levels of ultra-violet radiation.[9] Such dramatic changes egged governments on to action. In this case, the available solution was relatively simple, low cost and did not disturb citizen's life-styles. Within only two and half years, governments had agreed to the Montreal Protocol on Substances that Deplete the Ozone Layer at a meeting in Montreal on 16 September 1987.[10] Within a very few years, the effectiveness of the Protocol would become apparent.

As data mounted confirming risks to climate due to excessive discharges of greenhouse gases (GHG), principally carbon dioxide (CO_2) emissions from combustion of oil and coal, Governments appeared to demonstrate concern for the interests of people they served. In 1992, the United Nations Conference on Environment and Development (UNCED), held in Rio de Janiero and commonly known as the Rio Earth Summit, concluded with what seemed major steps forward. The steps included the *United Nations Framework Convention on Climate Change* (UNFCCC), which was signed by heads of government including then US President George Bush (Senior).[11] It was ratified by the US Senate later that year. Governments began introducing domestic policies to limit CO_2 emissions.

President Bill Clinton and Vice-President Al Gore came to office in 1993. Gore was especially committed to effective action. Clinton's Administration led negotiations which agreed on the 'Kyoto Protocol to the United Nations Framework Convention on Climate Change' (Kyoto Protocol).[12] However, the composition and orientation of the US Senate had changed markedly and it was now strongly opposed to Clinton's initiative, so much so that the Protocol was not

6   Flannery (2002).
7   Meadows (1972).
8   Burdick (2009).
9   Farman, Gardiner and Shanklin (1985).
10 The Ozone Hole (undated).
11 Sessions (1983).
12 United Nations Framework Convention on Climate Change (1997).

presented for ratification (though he did sign it in the dying days of his presidency as a symbolic gesture). His successor, President George W. Bush, who came to office in 2000, repudiated US support for the Protocol.[13]

Nonetheless, many countries ratified the Protocol (see Table 1.1) and a number, including the European Union (EU), made policy changes and implemented measures to put it into effect. The EU established a 'cap and trade' emissions trading scheme (ETS) as provided for in the Protocol.

Despite the readiness to sign, ratify and bring into effect the Protocol by almost all other developed countries responsible for significant CO_2 emissions, the government of two of the worst per-capita offenders, USA and Australia, refused to act. In the case of the USA that opposition was a product of the President's policy position, although it is improbable that the Senate would have ratified it if asked.

In Australia, the Howard Government – a coalition of the conservative Liberal Party and the rural-based Nationals (the 'Coalition') – refused to move in advance of the country's strong ally, the USA. Howard appeared to deny the scientific evidence and justified his action on economic grounds.[14] The issue emerged as a major concern of the Australian public and was a major point of difference between the Coalition and the Australian Labor Party (ALP) at the 2007 national elections. The Coalition had made belated gestures towards an ETS, such as announcing legislation, but it lacked credibility on the issue. The ALP campaigned strongly on the issue and won what seemed to be a strong mandate[15] to implement effective action to curb CO_2 emissions – action which most of the ALP and the commentariat thought involved an ETS.

---

13   Freeland (2001).

14   Howard (2008).

15   The multiplicity of reasons why citizens make the single decision on the party they choose makes any assertion of mandate difficult to establish. Prime Minister Whitlam claimed a mandate for every policy in his manifesto. Prime Minister Howard claimed a mandate for legislation implementing the government's political philosophy even if it was not foreshadowed before the election. Ironically (or hypocritically), Mr Howard was a member of the Opposition that not only opposed legislation implementing Whitlam's explicit election promises but (a) opposed legislation eliminating the pro-country gerrymander which had been the basis of the 1994 Double Dissolution and which was then passed in a joint sitting of parliament, (b) prevented the passage of redistributions required by that legislation and (c) voted to oppose supply in a much maligned (and never repeated) move to force an early election on gerrymandered boundaries. For all these reasons, caution is necessary in asserting mandates.

Table 1.1   Selected signatories to Kyoto Protocol

Selected Participants	Signature	Ratification	Entry into force	% of emissions
European Union	29 Apr 1998	31 May 2002 AA	16 Feb 2005	
Austria*	29 Apr 1998	31 May 2002	16 Feb 2005	0.4
Belgium*	29 Apr 1998	31 May 2002	16 Feb 2005	0.8
Czech Republic*	23 Nov 1998	15 Nov 2001 AA	16 Feb 2005	1.2
Denmark*	29 Apr 1998	31 May 2002 #	16 Feb 2005	0.4
Finland*	29 Apr 1998	31 May 2002	16 Feb 2005	0.4
France	29 Apr 1998	31 May 2002 AA	16 Feb 2005	2.7
Germany*	29 Apr 1998	31 May 2002	16 Feb 2005	7.4
Greece*	29 Apr 1998	31 May 2002	16 Feb 2005	0.6
Hungary*		21 Aug 2002 a	16 Feb 2005	0.5
Ireland*	29 Apr 1998	31 May 2002	16 Feb 2005	0.2
Italy*	29 Apr 1998	31 May 2002	16 Feb 2005	3.1
Netherlands*	29 Apr 1998	31 May 2002 A #	16 Feb 2005	1.2
Poland*	15 Jul 1998	13 Dec 2002	16 Feb 2005	3.0
Portugal*	29 Apr 1998	31 May 2002 AA	16 Feb 2005	0.3
Spain*	29 Apr 1998	31 May 2002	16 Feb 2005	1.9
Sweden*	29 Apr 1998	31 May 2002	16 Feb 2005	0.4
UK*	29 Apr 1998	31 May 2002 #	16 Feb 2005	4.3
Non EU				
Australia*	29 Apr 1998	12 Dec 2007	11 Mar 2008	2.1
Brazil	29 Apr 1998	23 Aug 2002	16 Feb 2005	
Canada*	29 Apr 1998	17 Dec 2002	16 Feb 2005	3.3
Chile	17 Jun 1998	26 Aug 2002	16 Feb 2005	
China	29 May 1998	30 Aug 2002 AA	16 Feb 2005	
Iceland*		23 May 2002 a	16 Feb 2005	0.0
India		26 Aug 2002 a	16 Feb 2005	
Indonesia	13 Jul 1998	3 Dec 2004	3 Mar 2005	
Japan*	28 Apr 1998	4 Jun 2002 A	16 Feb 2005	8.5
Mexico	9 Jun 1998	7 Sep 2000	16 Feb 2005	
New Zealand	22 May 1998	19 Dec 2002 #	16 Feb 2005	0.2
Norway*	29 Apr 1998	30 May 2002	16 Feb 2005	0.3
Russian Federation*	11 Mar 1999	18 Nov 2004	16 Feb 2005	17.4
Switzerland*	16 Mar 1998	9 Jul 2003	16 Feb 2005	0.3
USA*	12 Nov 1998			

* indicates an Annex 1 Party to the United Nations Framework Convention on Climate Change; A indicates acceptance; AA indicates approval; a indicates accession; # indicates Endnote in original source.

Source: Adapted from United Nations Framework Convention on Climate Change (UNFCCC) (2010) 'Status of Ratification of the Kyoto Protocol', available from http://unfccc.int/kyoto_protocol/status_of_ratification/items/2613.php

Ratification of the Kyoto Protocol was one of the very first actions by the incoming ALP Government and the Protocol came into effect in early 2008. A bill for an ETS, the Carbon Pollution Reduction Scheme Bill (CPRS) was introduced. However, it attracted strong criticism from deniers and major business interests on the right of the ideological spectrum and from environmentalists concerned that it was weak and would be ineffective. A minority were concerned that carbon trading schemes were inefficient, inequitable and fatally flawed. The Opposition (Coalition), especially the Liberals, was deeply divided. The ALP Government appeared to play on the divisions amongst the Opposition and made little attempt to explain, defend or promote its climate change policies.

The Government aimed for passage of the CPRS shortly before the 2009 UNFCCC COP15 conference in Copenhagen (COP15). The conference had been planned to reach global agreement on extension of the Kyoto Protocol beyond the initial commitment period.[16]

This, then, was the political context within which the authors met for the workshop that led to this book. The disappointing response to such a fundamental issue facing policy-makers raised accountability for inadequate policy responses as an increasing concern for scholars and others. The Accountability Round Table (ART), already active on various matters affecting the accountability of the political executive[17] shared this concern and was aware of the interesting potential of the fiduciary duty of the political executive as argued by Finn[18] and later work in Canada and Europe by Fox-Decent[19] and by Wood[20] in the USA. ART worked with GovNet (the Australian Research Council's Governance Research Network) to organize the workshop to bring together some of the finest minds to explore that potential. The workshop drew together Australian, Canadian and United States expertise from academia, the judiciary and politics with the objective to 'explore and develop potential application of principles of fiduciary duty and public trust to improve accountability for policy and the administration of long term issues affecting the public interest, such as climate change'. The workshop, conducted 23 November 2009, was supported by the Monash Governance Research Unit (Monash University), IEGL (the Institute for Ethics, Governance and Law, a joint initiative of the United Nations University, Griffith, QUT and ANU that provided the headquarters of GovNet) and RMIT University.

All authors presented papers to the workshop (Wood by video link), except Langford, who was nonetheless a participant. All have had the opportunity to contribute chapters drawing on additional materials and having regard to subsequent developments.

---

16   United Nations Framework Convention on Climate Change (undated).
17   Accountability Round Table. Available from http:// accountabilitytr.org Viewed 10 October 2010.
18   Finn (1995).
19   Fox-Decent (2005).
20   Wood (2009).

Amongst the political developments in that period in late 2009, an agreement between the Government and Hon Malcolm Turnbull, the Leader of the Opposition, produced such tensions within the Parliamentary Liberal Party that Turnbull was replaced by Hon Tony Abbott on 30 November and the agreement repudiated. The CPRS was defeated. Later, COP15 failed to reach any significant agreement, merely noting the Copenhagen Accord reached between a small number of highly influential governments. However, many governments have since agreed to the Accord.[21]

The book is structured to provide a background to the policy issues facing governments, followed by discussion of the possible application of the concept of fiduciary duty and the related concept of public trust.

The next chapter, by Will McGoldrick of Australia's Climate Institute, with Don Feaver and Andrew Maver, introduces the evidence of climate change and the anthropogenic factors fuelling it. This discussion provides a valuable and essential background to the policy responses that many demand of government. The chapter goes on to comment on the potential role of fiduciary duty, discuss the weaknesses of current international processes and arrangements and make some suggestions for reform of the regime for addressing climate change.

Paul Finn, highly regarded as a legal academic at the time of his 1995 work and now as a Judge of the Australian Federal Court, was quite unconsciously the progenitor of this book, as it was that seminal work that alerted us to the potential of a role for fiduciary duty. Finn presents his most recent thinking in chapter 3.

In the following chapter, Sampford, an Australian constitutional lawyer and international governance expert recognizes the value of such concepts but takes issue with Finn on their value in addressing these issues.

John Glover (Australian), Mary Wood (US) and Don Feaver (Canadian Australian) present different, well-argued and complementary perspectives in chapters 5, 6 and 7. Wood introduces the US experience and argues for recognition of an atmospheric trust.

Rosemary Langford's chapter 8 provides a valuable overview of relevant decisions of the Australian High Court.

Political practitioners bring home to us some of the realities faced in the political battleground where even the most profound issues fight for recognition, action and scrutiny, in chapter 9 by Andrew Murray, former Senator (Australian Democrats); chapter 10 by Robert Clark, Member of the Legislative Assembly, Victoria, Australia (Liberal); and chapter 11 by Kelvin Thomson, Member of the Australian House of Representatives (ALP).

Fiona Haines, an academic with a fascinating career history, adds to the range of perspectives when she addresses the intriguing question *A Ponzi Scheme on the Environment?* in chapter 12.

Evan Fox-Decent draws on his scholarly examination of these matters in Canada and the European Union in chapter 13.

---

21   United Nations Framework Convention on Climate Change (undated).

In the concluding chapter, we bring together the arguments advanced by the contributors, gathering common threads and, where they appear irreconcilable, identify the unresolved opinions and perspectives. We suggest directions that the research into these issues, the development of jurisprudence and political practice could take.

References

D. Burdick (2009) 'White House Solar Panels: What Ever Happened To Carter's Solar Thermal Water Heater?' *Journal*, available from http://www.huffingtonpost.com/2009/01/27/white-house-solar-panels_n_160575.html.

J. C. Farman, B. G. Gardiner and J. D. Shanklin (1985) 'Large losses of total ozone in Antarctica reveal seasonal ClOx/NOx interaction', 315 *Nature* 207.

P. D. Finn (1995) *A Sovereign People, a Public Trust*, The Law Book Company.

T. Flannery (2002) *The Future Eaters*, Grove Press

E. Fox-Decent (2005) 'The Fiduciary Nature of State Legal Authority', 31 *Queen's Law Journal*.

S. Freeland (2001) 'The Kyoto Protocol: An Agreement without a Future?', 24(2) *UNSW Law Journal* 532.

D. M. Hart (n.d.) 'Francis Ford Coppola, Apocalypse Now', (1979) 2HRS 35 (LD/WS), available at http://homepage.mac.com/dmhart/WarFilms/OldGuides/ApocalypseNow.html.

J. Howard (2008) 'Howard defends actions on Kyoto protocol', available at http://www.abc.net.au/news/stories/2008/12/08/2440959.htm. Viewed 10 October 2010.

D. H. Meadows (1972) The Limits to growth; a report for the Club of Rome's project on the predicament of mankind, Universe Books.

Michael Saward (1996) 'Democracy and Competing Values', 31(4) *Government and Opposition* 468.

K. Sessions (1983) 'Products of the Earth Summit' (April/June), available at http://www.epa.gov/history/topics/summit/02.htm.

The Ozone Hole (undated) 'The Ozone Hole', available at http://www.theozonehole.com/.

United Nations Framework Convention on Climate Change (1997) 'Kyoto Protocol to the United Nations Framework Convention on Climate Change', available at http://unfccc.int/resource/docs/convkp/kpeng.html. Viewed 9 October 2010.

United Nations Framework Convention on Climate Change (undated) *The United Nations Climate Change Conference in Copenhagen*, 7–19 December 2009, available at http://unfccc.int/meetings/cop_15/items/5257.php. Viewed 10 October 2010.

United Nations Framework Convention on Climate Change (undated) '*Copenhagen Accord*', available at http://unfccc.int/home/items/5262.php.

M. C. Wood (2009) 'Advancing the Sovereign Trust of Government to Safeguard the Environment for Present and Future Generations (part 2), Instilling a fiduciary obligation in governance', *Environmental Law* 93.

Chapter 2

Fiduciary Duty and Climate Governance: Challenges for International Diplomacy and Law

Will McGoldrick, Donald Feaver and Andrew Maver

Introduction

Despite more than two decades of negotiation, policy-makers have yet to conclude a comprehensive and effective global framework to address the threat of climate change. There were high expectations that a breakthrough agreement might emerge from the negotiations held in Copenhagen in late 2009. However, this did not occur. To the extent that limited progress was made at the Copenhagen summit, recent scientific estimates indicate that even if countries abide by undertakings given in Copenhagen, global emissions will continue to rise well beyond 2020, risking global warming of three degrees Celsius or more within this century.[1] Warming of this magnitude could unleash widespread and dangerous changes in the Earth's climatic system, having catastrophic effects.[2]

The failure of the world's nations to conclude an agreement at Copenhagen, yet again, calls into question the current approach towards international climate governance. A growing number of commentators are now declaring that a new approach to international climate change negotiations is required.[3] Yet a new approach is, at best, only a partial solution. Without a more fundamental conceptualization of the nature and gravity of the problem, key barriers to ambitious action will persist. Despite compelling evidence of the need for urgent action and the availability of cost-effective solutions, most national governments have failed to institute effective domestic policy and regulatory frameworks to cut greenhouse gas (GHG) emissions. In fact, the general trend has been to actively encourage economic growth reliant on the expansion of emission-intensive industries and activities.

In an attempt to understand and explain the failure of governments to respond to the threat of climate change, it is routine for commentators to point to a range

---

1   See Project Catalyst (2010); Höhne et al. (2009).
2   For an overview of the potential impacts associated with various global warming scenarios, see: IPCC (2007a).
3   For example, see Nanda and Ris (1975); Sands (2004); Wood (2008).

of barriers. At the domestic level, this includes political influence from vested interests, limited community support and opposition from political foes. The key barriers that exist at the international level include the lack of international consensus on key policy questions and geopolitical tensions. All of these factors are crucially important and deserve the close consideration they generally receive. What is less obvious is the link between the inadequate response from governments and the underlying legal and normative contexts in which governments operate. A more sophisticated and revealing analysis must consider these underlying factors.

In this chapter, it is argued that governments have generally viewed action on climate change as a matter of political discretion, rather than a positive legal or moral duty. Politicians and public servants do not feel compelled to do everything within their power to address this issue. Decisions about whether or not to act are based on political calculations rather than a sense of legal or moral obligation. This approach invites political pressures and influences, such as those listed above, to hold sway in the climate change debate. In this context, it is tempting to seek a solution that short-circuits the usual political tussle that defines public policy formation. One way this is done is by appealing to legal and moral arguments which, if accepted, would compel governments to formulate and implement a policy and regulatory framework that acknowledges the magnitude and urgency of the problem. With this objective in mind, several commentators have argued that the global atmosphere and its associated systems constitute an asset owned equally by all people. According to this view, governments are trustees of these natural assets and, therefore, have a fiduciary duty to protect the asset for the benefit of all people. As one commentator argues, 'government trustees do not have discretion to allow irrevocable damage to the trust'.[4]

According to its proponents, a fiduciary-based approach to climate governance would contribute to the goal of achieving an ambitious global response to climate change in three key ways. First, it would compel governments – based on a legal duty to their citizens – to do everything within their power to reduce greenhouse gas emissions occurring within their jurisdiction. Secondly, since the threat of climate change can only be resolved through international efforts, it would compel governments to intensify diplomatic efforts aimed at achieving an effective, globally coordinated response. This second proposed benefit of the fiduciary approach arises not only because of a government's legal duty to their own citizens, but also an extension of this legal duty to all of the world's people. The third argument, which is the primary focus of this essay, is the notion that a fiduciary duty can be extended beyond national governments to the realm of international organizations.

The first part of this chapter begins with a brief overview of the scale and urgency of the climate change challenge. This is followed by an assessment of the adequacy of the international community's response to date and, assuming no change in the approach to climate governance, the prospects for a more effective

---

4   Wood (2008), p. 10659.

response in the coming decade. It is next discussed how the response to date, both at the domestic and international level, has been wholly inadequate given the scale and urgency of the climate change problem. In the final part of this chapter, a fiduciary-based approach to climate governance, and how it might be used to overcome the inertia that is characteristic of the current paradigm, is discussed.

Scale and Urgency of the Climate Change Challenge

During the twentieth century, the average global surface temperature rose by approximately 0.74 degrees Celsius.[5] This warming has been primarily driven by increased atmospheric concentrations of greenhouse gases that trap the sun's energy in the form of heat.[6] Since the industrial revolution, the atmospheric concentration of carbon dioxide – the most important greenhouse gas – has risen by more than 35 per cent above pre-industrial levels.[7]

It is becoming increasingly clear that the build-up of greenhouse gases in the atmosphere is primarily due to human activities.[8] The biggest contributor is the burning of fossil fuels, but emissions from agriculture and land use change are also large.[9] At the same time, vast areas of forest have been cleared or severely degraded, thus removing or undermining the most important terrestrial carbon sinks.[10] Industrialized nations together are responsible for around 74 per cent of the historical build-up of greenhouse gases in the atmosphere.[11] However, developing nations are rapidly catching up and in 2005 accounted for over half of *annual* global emissions.[12]

Unless action is taken, by the end of this century the global average temperature is projected to rise by as much as seven degrees Celsius.[13] Due to momentum in the earth's climate system, even if global emissions of greenhouse gases ceased today an additional 0.1 degrees Celsius of warming per decade would likely be unavoidable.[14] As the earth warms up, the climate and associated systems are expected to undergo significant, possibly irreversible, changes. Indeed, many changes have already been observed, leading some to conclude that the impacts of climate change are already being felt.[15]

---

5   IPCC (2007c).
6   IPCC (2007c).
7   IPCC (2007c).
8   Forster et al. (2007).
9   Rogner et al. (2007).
10 Nabuurs et al. (2007).
11 Climate Analysis Indicators Tool (CAIT) Version 7.0 (2010).
12 Climate Analysis Indicators Tool (CAIT) Version 7.0 (2010).
13 Allison et al. (2009).
14 IPCC (2007b).
15 Allison et al. (2004) at note 13.

Predicting future climate change and its impacts is not straightforward and there is uncertainty regarding the timing and magnitude of potential changes. Nevertheless, the latest science makes for some sobering reading. It is beyond the scope of this chapter to go into details of the potential impacts of climate change over the coming decades. These potential impacts are well documented elsewhere.[16] Arguably, the most alarming predictions are those associated with sea level rise, which is driven by a combination of melting glaciers and polar ice caps, and thermal expansion of ocean waters.[17] According to the most recent projections, the global average sea level could rise by more than one metre by 2100. Sea level rise of this magnitude would directly affect around 145 million people globally, and would make many existing coastal regions uninhabitable.[18]

Other severe impacts of climate change include those on agricultural production and water supply, which will be driven primarily by changing rainfall patterns. These impacts will vary significantly from one location to the next. For Australia – already the driest inhabited continent – the latest projections are of major concern. While some areas of Northern Australia are projected to become wetter, the country's most important food-producing regions and population centres are projected to see declining rainfall.[19] For the Murray-Darling Basin, which accounts for 40 per cent of Australia's total agricultural output, declining rainfall and other changes could cut production by as much as 92 per cent by the end of this century.[20]

The Challenge for Policy-Makers

The solution to the threat of climate change lies in a coordinated international effort to achieve deep cuts in global greenhouse gas emissions. It is not enough for one country, or even ten countries, to cut emissions if all other nations do nothing. It is the cumulative total of all countries' efforts that counts. It is for this reason that so much time and effort has been invested in multilateral negotiations over the last two decades to develop an international legal regime to tackle climate change.

What would an effective global response aim to achieve? According to the UN Framework Convention on Climate Change (UNFCCC), the aim should be to 'prevent dangerous interference with the climate system'.[21] However, exactly what constitutes dangerous climate change was not defined in the UNFCCC, and has remained a contentious question in international negotiations. While some have resisted defining dangerous climate change, in recent years there has been

---

16   IPCC (2007a).
17   Bindoff et al. (2007).
18   Nicholls et al. (2007).
19   Garnaut (2008).
20   Garnaut (2008).
21   United Nations Framework Convention on Climate Change (NFCCC), Article 2.

a strong push to agree on a more precise and quantifiable goal for international efforts to tackle climate change. The European Union, Australia and others have pushed strongly for the goal of keeping global warming to below two degrees Celsius. Vulnerable developing countries, including many small island developing countries and African countries, have argued that warming needs to be kept below 1.5 degrees Celsius to avoid dangerous climate change. This point of disagreement was not resolved in Copenhagen, but a compromise of sorts was struck. The Copenhagen Accord includes a two-degree benchmark as its overall goal, but also includes a provision to review the need to strengthen its objective to 1.5 degrees Celsius.[22]

The goal of limiting global warming to a certain level can only be achieved if atmospheric GHG concentrations are stabilized at an adequate level, which depends on sufficient cuts in global emissions being achieved within a short enough timeframe. The best available estimates suggest that to provide a reasonable chance of limiting global warming to below two degrees Celsius, no more than an additional 650 gigatonnes of carbon dioxide and other GHGs can be released into the atmosphere over the next 40 years.[23] To ensure this 'carbon budget' is not exceeded, global emissions must peak before 2020 and decline by more than 50 per cent below 2000 levels by 2050.[24] The emissions pathway for the 1.5 degrees Celsius scenario requires even greater emissions cuts in this timeframe. The challenge for the international community, therefore, is to provide an international framework that will deliver these cuts in global emissions.

A key task for government is to decide the magnitude and speed of emissions abatement. This decision should be made on the basis of what is needed to avoid dangerous climate change. However, as explained above, this goal can actually only be achieved if *global* emissions are reduced to safe levels within the timeframe needed. Governments, therefore, must decide what role their country can and should play as part of the overall global effort. This decision is the central function of – and a key sticking point in – international climate change negotiations. Indeed, while very complex and covering a range of topics, ultimately the most important aspect of these negotiations is the division of responsibility amongst countries. The aim is twofold: to ensure each country commits to doing its fair share of the global abatement effort; and making sure these commitments add up to what is needed to avoid dangerous climate change.

Unfortunately, there is no commonly agreed metric or methodology to fairly divide the global abatement task amongst countries. It could be done based on current emissions, historical emissions, population size, wealth, per capita emissions, or a combination of these or any other relevant indicator. The difficulty

---

22   United Nations Framework Convention on Climate Change (UNFCC), Copenhagen Accord (2009), paragraph 12, http://unfccc.int/files/meetings/cop_15/application/pdf/cop15_cph_auv.pdf.

23   Allison et al. (2009); IPCC (2007b).

24   Barker et al. (2007).

arises because for many countries the abatement task varies significantly depending on which indicator is used. Ideally, therefore, governments would negotiate with each other to agree on each other's commitments and to ensure the collective level of ambition is sufficient.

Within climate change negotiations, governments are faced with a decision to either lead, follow or free-ride. A government may choose to set highly ambitious targets, which it hopes will position its country at the forefront of global emission abatement efforts. The aim of this may be to show leadership and encourage other countries to do more, or it may be to gain first mover advantages in the development of new, low emission, technologies. Governments that choose to follow are more likely driven by a desire to ensure their country is doing its fair share of the global effort, but no more. The free-rider strategy aims to benefit from global action to reduce emissions, while taking minimal action to reduce one's own emissions. As outlined in the following section, unfortunately to date most governments have chosen to be followers, with some strong tendencies towards free-riding. There have been too few leaders, with little obvious effect on the position of other countries.

Adequacy of Response

Despite increasing certainty in the science, the international community's collective response to climate change has, to date, been shamefully inadequate. Recent policy commitments and trends in global clean energy investments[25] may offer some signs of hope, but a significant acceleration of mitigation efforts will be required to avoid dangerous climate change. The following paragraphs provide a brief summary of the international community's response to the threat of climate change over the last two decades and an assessment of the prospects for the coming decade.

Progress to Date

In 1992, the international community adopted the UN Framework Convention on Climate Change (UNFCCC), which has the ultimate objective of stabilizing 'greenhouse gas concentrations in the atmosphere at a level that would prevent dangerous anthropogenic interference with the climate system'.[26] By signing up to this treaty, countries committed themselves to (amongst other things) take steps to mitigate climate change by reducing greenhouse gas emissions.[27] In 1997, after intense negotiations, wealthy industrialized countries took this general

---

25   For an overview of recent investment trends: New Energy Finance (2010).

26   United Nations Framework Convention on Climate Change (NFCCC), Article 2.

27   United Nations Framework Convention on Climate Change (NFCCC), Article 4.1(a).

commitment one step further by signing up to the Kyoto Protocol, which contains a commitment from these countries to reduce their collective emissions by 5 per cent below 1990 levels by 2012. To achieve this cut in emissions, each developed country committed to a binding national emission reduction target.

Putting aside questions about the adequacy of the collective goal for developed countries, and their national targets, it is clear that the overall impact of the Kyoto Protocol has been very limited. Indeed a large portion of the emissions abatement claimed under the Kyoto Protocol has been primarily driven by heavy reductions resulting from economic collapse in former Soviet countries. Indeed, by 2005 when the Kyoto Protocol entered into force, the combined emissions of former Soviet countries were already 34 per cent *below* 1990 levels, while emissions in other industrialized economies had risen by 12 per cent over the same period.[28] Several former Soviet countries have subsequently made money out of selling carbon credits to other countries to help them meet their Kyoto Protocol obligations.[29] Japan and European countries have been the largest purchasers of this 'hot air'.[30]

The Kyoto Protocol has also been undermined by the decision the United States (US) to withdraw from the Kyoto Protocol. The US is by far the biggest source of emissions amongst industrialized countries and its national emissions have risen by 16 per cent above 1990 levels.[31] Canada, while not formally withdrawing from the Kyoto Protocol, has made it clear that it will not meet its national target.[32] On paper at least, most other countries appear likely to meet their Kyoto Protocol targets, but on closer inspection there are some worrying trends, with Australia a useful case in point.

Australia has made much of its efforts to rein in GHG emissions, and according to recent estimates is likely to fulfil its Kyoto Protocol target without having to purchase international carbon credits.[33] However, this national trend belies less impressive trends at the sectoral level. Indeed, Australia is likely to meet its Kyoto Protocol commitment almost entirely because of reduced land clearing, particularly in New South Wales and Queensland. Emissions from this source have declined by around 60 per cent since 1990.[34] This is a great achievement and undoubtedly has many other environmental benefits, including the preservation of biodiversity and prevention of land degradation. However, over the same period emissions from electricity generation, transport and industrial processes have recorded increases

---

28   Climate Analysis Indicators Tool (CAIT) Version 7.0 (2010).

29   For an analysis, see note by the secretariat: United Nations (2009).

30   For example see Ukraine's national carbon units registry: http://www.carbonunits registry.gov.ua/en/257.htm.

31   Climate Analysis Indicators Tool (CAIT) Version 7.0 (2010).

32   For example, see: http://m.theglobeandmail.com/news/politics/emissions-reductions-10-times-less-than-governments-projections-report/article1591784/?service=mobile

33   Department of Climate Change (2009a).

34   Department of Climate Change (2009b).

of 54 per cent, 27 per cent and 26 per cent respectively.[35] Reducing land clearing provided a relatively easy way to reduce emissions, but this option has now largely been exhausted (there is only so much vegetation that can be protected). While Australia may meet its Kyoto Protocol target, without much stronger attention to emissions beyond the land sector its national emissions are projected to sky-rocket over the coming decade. Indeed, based on the Australian Government's own estimates, by 2020 Australia's national emissions are projected to rise by an additional 14 per cent above current levels.[36]

The international community's response to climate change cannot be judged solely against achievements under the Kyoto Protocol. The fact is that in absolute terms developing countries now account for over half of annual global greenhouse gas emissions.[37] The vast majority of these emissions occur in emerging economies such as in Asia and Latin America, particularly the likes of China, India, Indonesia, South Korea and Brazil.[38] These countries were not obliged to adopt national targets under the Kyoto Protocol and, to date, have done little to curb emissions growth. This lack of action by developing countries to curb emissions is often justified on equity grounds. Indeed, it is argued that because these countries have contributed so little to the build-up of GHG in the atmosphere they should not be expected to incur costs involved with mitigation efforts. This argument was recognized in both the UNFCCC and the Kyoto Protocol, which have the principle of 'common but differentiated responsibilities and capabilities' at their core. It was for this reason that these international agreements also include provisions obliging developed countries to provide financial and technical assistance to enable mitigation actions in poorer countries. Thus, while it may not be warranted to blame developing countries for failing to curb their emissions, this approach does represent a collective failure of the international community, or at least of developed countries who have failed to provide adequate assistance.

Prospects for the Coming Decade

The 2009 Copenhagen climate summit produced a non-binding political accord, known as the Copenhagen Accord, with the overall aim of keeping global warming below two degrees Celsius. The Accord also calls for a review to be completed by 2015, which, depending on interpretation, should include consideration of the need to strengthen the overall goal from two degrees to 1.5 degrees Celsius.[39]

---

35   Department of Climate Change (2009b).

36   Department of Climate Change (2009a) at note 33.

37   Climate Analysis Indicators Tool (2010) at note 11.

38   Climate Analysis Indicators Tool (2010).

39   The precise implications of the review clause contained in paragraph 12 of the Copenhagen Accord is open to interpretation, particularly its second sentence: 'This would include consideration of strengthening the long-term goal referencing various matters presented by the science, including in relation to temperature rise of 1.5 degrees Celsius.'

While there is some ambiguity, the commitment to an overall goal for global temperature rise can be viewed as an important step forward for the international community.[40] Ultimately what matters, however, is that this global ambition is matched by domestic efforts. Unless each country cuts emissions by a sufficient amount, the two degree goal will remain little more than a symbolic benchmark, and 1.5 degrees will be entirely unachievable.

As part of the Copenhagen Accord all major emitting countries announced targets and plans for reducing their national GHG emissions.[41] Significantly, this includes both developed and developing nations, including China and the US. If all countries live up to their Copenhagen pledges, we can expect global warming of 3.5 degrees Celsius or more by the end of this century.[42] This represents a significant improvement when compared to estimates of what is expected to occur in the absence of these national commitments. However, allowing the global average temperature to rise by 3.5 degrees Celsius would obviously exceed goals set in the Copenhagen Accord of limiting global warming to two degrees Celsius, and would therefore risk the prospects of triggering dangerous interference with the earth's climate.

It is important to note that while, collectively, countries have not made strong enough commitments to avoid dangerous climate change, some countries have made ambitious national pledges. Interestingly, some of the strongest commitments have been made by emerging economies. Indeed, when measured against deviation from business as usual,[43] the pledges made by Brazil, South Africa, South Korea, Indonesia, Mexico and China, are at least comparable to, and in some cases stronger than, those made by the US and EU.[44] Amongst developed countries, Japan's pledge is the strongest. Several analysts have estimated that the collective level of ambition from major emitting developing countries is consistent with their fair share of the global abatement task that is required to provide a reasonable chance of avoiding a two degree Celsius rise in global temperatures.[45] By the same calculations, developed countries as a group have put forward national targets which are inconsistent with this goal.

The current level of international ambition is not consistent with the goal of avoiding dangerous climate change. This will not change unless countries,

---

40   There has never before been such widespread endorsement of a global goal such as this.

41   Details of the commitments made by countries can be viewed online at, http://unfccc.int/home/items/5262.php, accessed 2 June 2010.

42   Höhne et al. (2009).

43   The term 'business as usual' refers to the emissions growth expected to occur in the absence of specific mitigation efforts.

44   This is based on comparing the relative deviation from projected business as usual emissions implied by each country's 2020 target pledge. For analysis, see Jotzo (2010a); Jotzo (2010b).It is important to acknowledge that there are a wide range of plausible BAU reduction estimates. For example, see Howes (2010).

45   Garnaut (2010); Höhne et al.

particularly developed countries, commit to much stronger targets and mechanisms to reduce emissions, and follow up on these commitments with effective action at home. Indeed, this strengthening of domestic ambition will need to occur in the very short term to enable global emissions to peak before 2020, which is crucial to limiting global warming to below two degrees Celsius.

A Shift in Approach?

The Copenhagen Accord should not be solely judged on the strength of the commitments made by countries. Also significant is the possibility that the Accord represents a shift away from a treaty-based approach to international climate change governance. While the Accord may be viewed as a stepping stone to a full international treaty, some view it as vote in favour of a 'pledge and review' model of climate governance, whereby governments make 'politically binding' commitments to reduce emissions and agree to certain levels of international review to ensure these commitments are met. While it is still too early to be sure in which direction we are heading, it is important to contemplate the implications of moving towards a pledge and review model.

Ultimately, it does not really matter what model of international climate governance is applied, as long as the collective effort of countries is ambitious and urgent enough to avoid dangerous climate change. For decision makers and diplomats, the challenge is to determine which model is most likely to deliver this outcome. Given the failure of the treaty-based approach to deliver an effective response, it is tempting to consider the pledge and review approach as a potentially more effective alternative. Under the pledge and review model, countries do not have to sign up to legally binding international commitments. Indeed, it has been suggested that the benefits of the Copenhagen Accord are that it recognizes and reflects 'core geopolitical realities', secures commitments from the major emitters, and can be further developed through bilateral and regional agreements.[46] Furthermore, it offered the US a pathway through its domestic difficulties associated with addressing climate change,[47] particularly those confronted by the Obama administration when attempting to pass emissions trading legislation in the US Senate.[48]

For some, such as the US, this may remove a barrier to their participation, allowing them to pledge an international target, but without having to formally ratify this as a treaty commitment. Given that China and other emerging economies hold a similar antipathy towards the notion of committing themselves to legally binding targets, there is little doubt that the pledge and review model has enabled greater buy-in from a broader cross-section of countries than was achieved through the Kyoto Protocol. Indeed, through the Copenhagen Accord,

---

46   Giddens (2010).
47   Bailey (2010), p. 127.
48   Antholis (2009).

around 62 countries, covering approximately 78 per cent of global emissions, have committed to international emission reduction targets. This is a significant improvement on the Kyoto Protocol which covers 37 countries and around 22 per cent of global emissions.

The key uncertainty for the pledge and review approach is whether it provides for sufficient levels of trust between countries to ensure all of the targets pledged in Copenhagen are actually met, and whether it can pave the way to a more ambitious global response in the future. If it does, this would represent a significant departure for international law, which is built on the premise that global cooperation is the giving of mutually beneficial, but binding, obligations. In reality however, the Copenhagen summit produced an Accord containing no substantive or binding commitments. Only time will tell whether this approach is sufficiently robust, sufficient to build and maintain trust between countries and achieve climate governance objectives, or whether this approach represents a temporary aberration before returning to a treaty-based approach. For the reasons discussed above, there is a very real prospect that the Accord will do little to compel countries to meet commitments in the face of more powerful domestic pressures and therefore not facilitate stronger domestic action, which is needed to avoid dangerous climate change.

Understanding the Barriers to Effective Action – A Failure of Climate Governance?

It is clear that the response from governments to limit GHG emissions has, for the most part, been inadequate. Given what we know about the drivers of climate change and the solutions available, why is it that most governments, and the international community as a whole, have failed to respond effectively? To answer this question, commentators generally focus on two distinct sets of issues: those concerned with policy and those concerned with politics. This part of the chapter focuses on a third level of analysis – the underlying legal and normative contexts in which climate change policies are formed, both domestically and internationally. This approach allows a more revealing appraisal of the adequacy of current approaches to climate governance, and whether there is a case for pursuing alternative approaches, such as one based on a fiduciary duty.

The first set of issues includes the various policy questions that need to be resolved in order to provide an effective response to climate change. The second set of issues includes a range of domestic and international political forces that constrain government action in this area. Without discounting the complexities of some of the policy questions, it would be naïve to believe that the failure of governments to provide an adequate response to climate change is due to their inability to formulate a suitable (and acceptable) policy framework.

Whilst the significance of individual factors influencing policy choice and implementation varies from one country to the next, there are some common themes. Some of the most significant domestic political factors commonly identified include: pressure from vested interests (such as fossil fuel industries); attempts by those who deny there is a problem to undermine community support for action; and opposition from business to the perceived costs of taking action. Internationally, commentators generally point to the role of geopolitical forces, as well as the seemingly entrenched differences between developed and developing countries that featured so prominently at Copenhagen.

In order to understand why these factors have been able to influence governments' response to climate change, one must also consider the underlying legal and normative contexts in which climate change policies are formed. To do so is to consider why political forces (including those listed above) are allowed to shape climate change policies to the extent that they do, and why the responses to date have been allowed to fail to the extent that they have. Despite dire warnings from scientists, governments are not responding in a manner that suggests they feel compelled to do everything in their power to mitigate the threat of climate change.

Without discounting the role of geopolitical factors, it is clear that the main barrier to an effective global response to climate change is the collective lack of political will from national governments, and that this lack of will is an inevitable by-product of the legal and normative frameworks that governments operate within. Striking an effective international deal on such a complex issue as climate change will inevitably involve give and take from all countries, yet this is not occurring to the extent that is required. When countries accept that their own national interests are best served by an agreement that facilitates global efforts to reduce carbon pollution, making small compromises along the way should not be such a concern.

Speaking generally, domestic and international climate governance has to date been shaped by political discretion, rather than a compulsion (either morally or legally) to act. Yet political discretion has clearly failed, having moved progress on climate change policy no further towards an effective global framework to tackle climate change. The failure of the Copenhagen climate summit illustrates the effects of the predominance of political discretion as the key driving force of domestic and international climate governance.

A Fiduciary Conceptualization of Climate Change

If it is accepted that the current approaches are failing to provide an adequate (that is, timely and effective) response to the threat of global warming, the question then arises as to what might compel national policy-makers to become more proactive? Peter Sand argues that the solution to this question 'is simple: The sovereign rights of nation states over certain environmental resources are not proprietary, but *fiduciary*'.[49] The implications of this statement, if the earth's atmosphere is to be regarded as an 'earth resource', is that political action on climate change is not a

---

49   Sand (2004), p. 48.

discretionary decision on the part of politicians, instead, politicians have a positive legal duty to act. The notion of a fiduciary obligation in this context amounts to a 'sort of guardianship for social purposes' whereby governments exercise a 'fiduciary trust' on behalf of their people. Sand further describes what the notion of an environmental trusteeship entails:

> In very simplified language, it means that certain natural resources – e.g., watercourses, wildlife, or wilderness areas – regardless of their allocation to public or private uses are defined as part of an 'inalienable public trust;' certain authorities – e.g., federal agencies, state governments, or indigenous tribal institutions – are designated as 'public trustees' for protection of those resources; every citizen, as 'beneficiary' of the trust, may invoke its terms to hold the trustees accountable and to obtain judicial protection against encroachments or deterioration.[50]

Sources of the Public Fiduciary Obligation

The concept of the shared environment as constituting a 'public trust' is not new. The source of this obligation originates from the longstanding principle of *res communes omnium*, as applied in domestic law, and has been extended in more recent doctrines such as the 'common heritage of mankind' and a fiduciary theory of *jus cogens* as applied in the context of international law. The *res communes omnium* doctrine is derived from the ancient principle that certain classes of cultural and natural resources are *res communes*, or common property, whereby ownership is vested in the state, as the sovereign, in trust for the people.[51] As trustee, the state is deemed to owe a fiduciary duty in respect of those resources. As such, this duty requires that the state safeguard and protect the resource. *Res communes* are considered to be excluded from exclusive private control and a trustee is charged with a positive duty to act to preserve the resources for the benefit of the whole of society.[52] The *res communes* doctrine is said to provide the notional foundation for the more contemporary 'public trust doctrine' that forms part of the domestic environmental law of countries such as the US, Italy, Sweden, India, the Philippines and South Africa.[53]

Although the notion of a public trusteeship of common resources can be found within the domestic law of a number of countries, the question arises as to its status under international law and whether it might be used as a basis to compel multilateral agreement on GHG reduction targets and policy? The common heritage of mankind doctrine (CHM), used as the conceptual foundation for international treaties regulating the law of the sea, sites of historical significance,

---

50   Sand (2004), p. 49.
51   Sax (1970), p. 475; Sax (1980), p. 185.
52   *Illinois Central Railroad v People of the State of Illinois 146 US 387 (1892).*
53   Kameri-Mbote (2007), p. 195.

the Arctic and Antarctica and space is frequently linked to the notion of a public trusteeship – although 'substantial confusion persists over the nature of the concept and its appropriate place in international law'.[54] Because the doctrine is not international customary law, it is used as a conceptual framework underpinning substantive treaty obligations forming part of international regulatory regimes. This conceptual framework provides that under 'a CHM regime all people would be expected to share in the management of a common space area. In other words, States or national governments would be precluded from this legal function, save as the representative agents of all mankind'. Countries, as agents, in failing to protect, preserve and prudentially manage the resource 'would breach the trust and legal obligation implicit in responsibly supervising the earth's heritage for mankind in the future'.[55]

The notion of a public fiduciary responsibility for shared environmental resources, such as a clean and safe atmosphere, has been linked to fundamental human rights. For example, in the *Danube Dams* case, Judge Weeramantry of the International Court of Justice referred to a 'principle of trusteeship for earth resources' based upon the notion that the:

> protection of the environment is … a vital part of contemporary human rights doctrine, for it is a *sine qua non* for numerous human rights such as the right to health and the right to life itself. It is scarcely necessary to elaborate on this, as damage to the environment can impair and undermine all the human rights spoken of in the Universal Declaration and other human rights instruments.[56]

The basis of the connection between human rights and the environment,

> can be found in the preamble of the Universal Declaration of Human Rights, 'the inherent dignity and inalienable rights of all members of the human family'. The right to environment, according to this theory, derives from this dignity. Another important approach bases the environmental right on the right to life and the right to health.[57]

A more recent theory which also provides a juridical basis for a positive duty to act responsibly and proactively in protecting Earth's resources is Criddle and Fox-Decent's 'fiduciary theory of jus cogens'.[58] Although Criddle and Fox-Decent do not explicitly identify the protection of Earth's resources as a peremptory norm, there is only a small step from Judge Weeramantry's view that the protection of the

---

54   Joyner (1986), p. 190.
55   Joyner (1986), p. 195.
56   *Case Concerning the Gabcikovo-Nagymaros Project (Hungary v Slovakia)* (*'Danube Dam Case'*) 37 ILM 162 at p. 217.
57   Fitzmaurice (2002), p. 308.
58   Criddle and Fox-Decent (2009), p. 331.

environment is a vital part of contemporary human rights doctrine to classifying the protection of essential earth resources as a peremptory norm in its own right. According to Criddle and Fox-Decent, if 'the state is a fiduciary of the individual subject to its power ... the international community may act as a surrogate guarantor of jus cogens'.[59] Of particular importance is how the 'fiduciary principle arguably authorizes the international community through the United Nations to establish transnational administrations' to enable the international community to act upon a range of issues such as economic development, disaster and famine relief – and more controversially, even humanitarian intervention.[60] This begs the question whether catastrophic climate change is a serious enough issue to justify international intervention by an appropriately empowered international administration.

Limitations and Challenges

Herein lies the major challenge with this approach. The difficulty is not so much one of finding a substantive basis for a positive duty to act to preserve the object of a public trust obligation, and hence, a positive duty to act to preserve the common property of mankind; more fundamental difficulties arise in relation to the enforcement of this duty in an international context.

The international community is not a single, coherent entity. It comprises numerous state and non-actors, the most powerful of which are nation states.

> Given the decentralized structure of the world community, securing international compromise and consensus often requires adopting vague declarations and ineffective institutional arrangements, thereby avoiding the establishment of the strong enforcement mechanisms needed.[61]

At present, there is no international organization that is granted the broad jurisdictional authority and governance powers required to compel countries to recognize their collective duty to combat climate change. Law, without credible institutional arrangements to implement and enforce it, remains nothing more than just an abstract idea. 'The development of rules of international law concerning protection of the environment is of little significance unless accompanied by effective means for ensuring enforcement and compliance.'

Although not a total solution to the challenge of compelling nation states to negotiate a comprehensive and universal climate change agreement, an important ingredient in achieving this goal involves the creation of international institutions granted appropriate powers and responsibilities to:

---

59 Criddle and Fox-Decent (2009), p. 331.
60 Criddle and Fox-Decent (2009), p. 331.
61 Nanda and Ris (1975), p. 294.

perform a fiduciary role in protecting the environment, in contrast to the more traditional approach towards enforcing international law, in which interstate claims based on the principle of state responsibility, and employing the variety of forms of dispute settlement machinery contemplated in Article 33 of the UN Charter.[62]

Strengthening Institutional Machinery

One step towards providing the means by which countries may be compelled to meet their fiduciary obligation in respect of climate change is to supply institutional machinery designed to promote compliance and, if necessary, enforce those obligations. The UNFCCC and the Kyoto Protocol, together, provide the institutional architecture of a multilateral regulatory regime that is in its very early stages of development (the climate change regime). Climate change regulation, like other forms of multilateral regulation, can be said to comprise a self-contained regime, or a self-contained system of law.[63] A self-contained regulatory system, however, is much more than a body of rules. Regulatory systems not only include special bodies of rules, but institutional/organizational constructs designed to administer and in-built procedural mechanisms to enable the enforcement of those rules. International regulatory regimes can be even more comprehensive, and complex, where the regulatory system is also granted rule-formation and adjudicative powers and machinery.

At present, one of the biggest weaknesses of the current arrangements under the UNFCCC and Kyoto Protocol is that the structure of climate change regime can be described as having an 'instrumental' focus. That means that the regime's *rules* are more developed than the *institutional* structures needed to monitor and, if need be, enforce those rules. Even though robust substantive obligations and widespread consent are critical (both of which are arguably absent in multilateral climate change initiatives to this point), the long-term viability of specialized systems of law, such as the climate change regime, are likely to be undermined in the absence of a well-designed organizational/institutional framework.

The importance of coherent organizational constructs to support the implementation of rule-based regimes has been the focus of much research over the last decade. Organizational constructs play a critical role in institutionalizing

---

62　Boyle (1991), p 229.

63　What is described as the increasingly fragmented structure of international law has caused sufficient concern that it was made the subject of a study prepared at the request of the UN General Assembly by the International Law Commission (ILC). The report, released in 2006, describes how 'what once appeared to be governed by "general international law" has become the field of operation for such specialist systems as "trade law", "human rights law", "law of the sea", "European law"... each possessing their own principles and institutions'. See Koskenniemi (2006), p. 9.

rule regimes in three ways. Coherent organizations play an important role in coordinating a number of activities ranging from policy formation through to regime administration. Secondly, well-designed organizational constructs contribute to the legitimacy of a regime by signalling a commitment to good governance arrangements as well as reinforcing the credibility of the rule frameworks through a commitment to administrative support. Finally, organizational constructs contribute to the authority of a regime by creating a connection between the rules and the clear identification with a body responsible for the monitoring and enforcement of those rules.

At present, the institutional structure of the climate change regime is best described as highly under-developed and fragmented. As a result, this undermines the regime by eroding perceptions of legitimacy as well as the credibility and authority of the regime, in a more practical context. For example, the current climate change regime is referred to by the name of its legal instruments (that is, the UNFCCC or Kyoto Protocol) rather than any organizational body, such as the UN, WTO or IMF. Few people identify with the regime in organizational terms and cannot point to a single organizational body that has responsibility to administer the regime. Furthermore, it is difficult to identify which bodies are jurisdictionally responsible for the several functions the regime is empowered to administer. Finally, it is very difficult to determine where, geographically and organizationally, the administrative locus of regime is based. In short, the regime exists in a conceptual rules-based sense, but lacks an organizational identity.

The UNFCCC and Kyoto Protocol do, however, contain the basic organizational building blocks of an institutional framework even though, at present, it lacks any semblance of a coherent organizational structure. The incoherence of the organizational structure, in part, stems from and is evident in the dysfunctional delegation (and subsequent sub-delegation) of sovereign governance powers (rule-making, executive/administrative, adjudicative authority) to many organizational units (that is, the Conference of the Parties, the Meeting of the Parties, the Clean Development Mechanism, registry body, enforcement bodies etc.). Not only is the nature of the organizational relationship between these constructs poorly defined, but the breadth and scope of the governance powers to be exercised by each of these bodies is vague and, in some cases, non-existent.

One of the underlying causes of the dysfunctional institutional structure is the fact that the regime is derived from two legal instruments. Both instruments contain separate delegations of sovereign power. The creation of organizational structures pursuant to the bifurcated delegations has the effect of creating two management regimes rather than a more uniform and coherent overarching organizational structure (that is, an imperfect relationship between the Conference of the Parties and the Meeting of the Parties is the most obvious example). This approach, in part, can be attributed to the differing scope of application/obligation attaching to the broader UNFCCC and narrower Kyoto Protocol. The consequence of this is an organizational structure made up of semi-autonomous bodies that lack central coordination. Rather than an administratively efficient separation between bodies whose purpose is to exercise discretionary decision-making powers and those primarily responsible for administrative functions, these two aspects are mixed up within the overall organizational structure.[64]

An improved organizational structure would serve the purpose of improving the regime's necessary function of coordinating policy-oriented initiatives and functions as well as support and reinforce the legitimacy and authority of the regime. This result could be achieved by designing an organizational structure that distinguishes between the exercise of governance/discretionary powers, on the one hand, and the exercise of administrative functions, on the other. For example, a better arrangement would consolidate the powers of the UNFCCC Conference of Parties and the Kyoto Protocol's Meeting of Parties within a single plenary body. The purpose of this body would be to perform the substantive rule-making functions as well as to perform an oversight role. The Secretariat should be transformed into a quasi-executive body having greater powers to formulate policy as well as a more direct administrative coordination role. The Secretariat should be responsible for maintaining any schedule of mitigation commitments that might be established. The administrative role would include executive-like discretionary powers. These powers would be exercised in relation to the more functional activities of the sub-bodies. The function-oriented sub-bodies, such as the Clean Development Mechanism, Joint Implementation body and Registry should be consolidated and new functions pertaining to inter-national carbon credit management oversight should be established. Finally, all compliance related functions, including monitoring of commitments as well as enforcement functions, should be consolidated within a single body that is granted appropriate powers.

In the absence of a delegation of adequate powers, re-designing and strengthening the international organizational machinery charged with the responsibility for administering and enforcing international climate change obligations is not a complete solution to the issues raised in this chapter. Nevertheless, a coherent and appropriately empowered international organization provides a more credible institutional foundation from which international obligations of a collective nature, such as a clean and safe atmosphere, can be asserted and, where possible, enforced against those nation states that do not meet their fiduciary obligations.

Conclusion

Despite more than two decades of negotiation, policy-makers have yet to conclude a comprehensive and effective global framework to address the threat of climate change. The failure of the world's nations to conclude an agreement at Copenhagen, yet again, calls into question the current approach towards international climate governance. A growing number of commentators are now declaring that a new approach to international climate change negotiations is required. In this chapter,

---

64   Feaver and Durrant (2008), p. 394.

it is argued that governments have generally viewed action on climate change as a matter of political discretion, rather than a positive legal or moral duty. Politicians and public servants do not feel compelled to do everything within their power to address this issue. Accordingly, without a more fundamental conceptualization of the nature and gravity of the problem, key barriers to ambitious action will remain. The response to date, both at the domestic and international level, has been wholly inadequate given the scale and urgency of the climate change problem. Rather than viewing climate change as a matter of political discretion, this chapter considers how a fiduciary-based approach to climate governance might be used to overcome the inertia that is characteristic of the current paradigm.

The implications of this statement, if the earth's atmosphere is to be regarded as an 'earth resource', is that political action on climate change is not a discretionary decision on the part of politicians. Instead, politicians forming Government have a positive legal duty to act. The notion of a fiduciary obligation in this context amounts to a 'sort of guardianship for social purposes' whereby governments exercise a 'fiduciary trust' on behalf of their people. The source of this obligation is the longstanding principle of *res communes omnium*, as applied in domestic law, and has been extended in more recent doctrines such as the 'common heritage of mankind' and a fiduciary theory of *jus cogens* as applied in the context of international law. The difficulty is not so much one of finding a substantive basis for a positive duty to act to preserve the object of a public trust obligation, and hence, a positive duty to act to preserve the common property of mankind; more fundamental difficulties arise in relation to the enforcement of this duty in an international context.

At present, there is no international organization that is granted the broad jurisdictional authority and governance powers required to compel countries to recognize their collective duty to combat climate change. Law, without credible institutional arrangements to implement and enforce it, remains nothing more than just an abstract idea. One step towards providing the means by which countries may be compelled to meet their fiduciary obligation in respect of climate change is to supply institutional machinery designed to promote compliance and, if necessary, enforce those obligations. The UNFCCC and the Kyoto Protocol, together, provide the institutional architecture of a multilateral regulatory regime that is in its very early stages of development (the climate change regime). An improved organizational structure would serve the purpose of improving the regime's necessary function of coordinating policy-oriented initiatives and functions as well as support and reinforce the legitimacy and authority of the regime. Finally, an appropriately empowered and coherently constituted international organization would be much better placed to do so.

References

I. Allison et al. (2009) *The Copenhagen Diagnosis: Updating the world on the Latest Climate Science*, The University of New South Wales Climate Change Research Centre (CCRC), Sydney.

W. Antholis (2009) *Toward a successful climate agreement: Building Trust and Ambition*, available at http://www.brookings.edu/~/media/Files/rc/papers/2009/09_climate_change_poverty/09_climate_change_poverty_antholis.ashx, 10 July 2010.

I. Bailey (2010) 'Copenhagen and the new political geographies of climate change', 29(3) *Political Geography* 127.

T. Barker et al. (2007) 'Technical Summary', in B. Metz et al. (eds), *Climate Change 2007: Mitigation. Contribution of Working Group III to the Fourth Assessment Report of the Intergovernmental Panel on Climate Change*, Cambridge University Press.

N. L. Bindoff et al. (2007) 'Observations: Oceanic Climate Change and Sea Level', in S. Solomon et al. (eds), *Climate Change 2007: The Physical Science Basis. Contribution of Working Group I to the Fourth Assessment Report of the Intergovernmental Panel on Climate Change*, Cambridge University Press.

A. Boyle (1991) 'Saving the World? Implementation and Enforcement of International Environmental Law Through International Institutions', 3(2) *Journal of Environmental Law* 229–245.

World Resource Institute (2010) *Climate Analysis Indicators Tool (CAIT) Version 7.0*, Washington, DC.

E. Criddle and E. Fox-Decent (2009) 'A Fiduciary Theory of Jus Cogens', 34 *The Yale Journal of International Law* 331.

Department of Climate Change (2009a) *Tracking to Kyoto and 2020: Australia's Greenhouse Emission Trends*, Canberra.

Department of Climate Change (2009b) *National Greenhouse Gas Inventory*, Canberra.

D. Feaver and N. Durrant (2008) 'A Regulatory Analysis of International Climate Change Regulation', 30(4) *Law and Policy* 394.

M. Fitzmaurice (2002) 'International Protection of the Environment', Kluwer Law International, Hague Academy of International Law.

P. Forster et al. (2007) 'Changes in Atmospheric Constituents and in Radiative Forcing', in S. Solomon et al. (eds), *Climate Change 2007: The Physical Science Basis. Contribution of Working Group I to the Fourth Assessment Report of the Intergovernmental Panel on Climate Change*, Cambridge University Press.

R. Garnaut (2010) 'Global warming after the Obama Accord', Keynote Address to the Annual Conference of Supreme and Federal Court Judges, available at http://www.rossgarnaut.com.au/Documents/Global%20Warming%20After%20The%20Obama%20Accord%20250110.pdf.

R. Garnaut (2008) *The Garnaut Climate Change Review*, Cambridge University Press.

A. Giddens (2010) 'Big players, a positive accord', available at http://www.policy-network.net/uploadedFiles/Articles/Big%20players,%20a%20positive%20 Accord.pdf.

N. Höhne et al. (2009) 'Copenhagen Climate Deal – How to close the gap?', *Climate Analytics*, http://www.climateactiontracker.org/briefing_paper.pdf.

S. Howes (2010) 'China's energy intensity target: On-track or off?', available at http://www.eastasiaforum.org/2010/03/31/chinas-energy-intensity-target-on track-or-off/

IPCC (2007a) 'Contribution of Working Group II to the Fourth Assessment Report of the Intergovernmental Panel on Climate Change', in M. L. Parry, O. F. Canziani, J. P. Palutikof, P. J. van der Linden and C. E. Hanson (eds), *Fourth Assessment Report of the Intergovernmental Panel on Climate Change*, Cambridge University Press.

IPCC (2007b) 'Summary for Policy Makers', *Climate Change 2007: Synthesis Report,* Cambridge University Press.

IPCC (2007c) 'Contribution of Working Group I to the Fourth Assessment Report of the Intergovernmental Panel on Climate Change Summary for Policymakers', in S. Solomon et al. (eds), *Fourth Assessment Report of the Intergovernmental Panel on Climate Change 2007: The Physical Science Basis*, Cambridge University Press.

F. Jotzo (2010a) 'Comparing the Copenhagen climate targets', *ANU Crawford Policy Forum,* available at http://www.crawford.anu.edu.au/pdf/events/2010/ policy_forum/Frank_Jotzo.pdf.

F. Jotzo (2010b) 'Copenhagen implications for Australia', *ANU Crawford Policy Forum*, available at http://www.crawford.anu.edu.au/pdf/events/2010/policy_ forum/Frank_Jotzo.pdf,

C. Joyner (1986) 'Legal Implications of the Concept of the Common Heritage of Mankind', 35(1) *The International and Comparative Law Quarterly* 190–199.

P. Kameri-Mbote (2007) 'The Use of the Public Trust Doctrine in Environmental Law', 3(2) *Law, Environment and Development Journal* 195.

Martii Koskenniemi (2006) 'Fragmentation of International Law: Difficulties Arising from the Diversification and Expansion of International Law', *Report of the Study Group of the International Law Commission,* International Law Commission, Fifty-eighth session, Geneva, 1 May–9 June and 3 July–11 August 2006 at 9.

G. J. Nabuurs et al. (2007) 'Forestry', in B. Metz et al. (eds), *Climate Change 2007: Mitigation. Contribution of Working Group III to the Fourth Assessment Report of the Intergovernmental Panel on Climate Change,* Cambridge University Press.

V. P. Nanda and W. Ris (1975) 'The Public Trust Doctrine: A Viable Approach to International Environmental Protection', 5 *Ecology Law Quarterly* 291.

New Energy Finance (2010) 'Green Investing 2010', *Policy Mechanisms to Bridge the Financing Gap*, World Economic Forum, available at http://www.newenergymatters.com/UserFiles/File/Presentations/Green_inv_report_2010%20FINAL.pdf.

R. J. Nicholls et al. (2007) 'Coastal Systems and Low-lying Areas', in M. L. Parry et al. (eds), *Climate Change 2007: Impacts, Adaptation and Vulnerability. Contribution of Working Group II to the Fourth Assessment Report of the Intergovernmental Panel on Climate Change*, Cambridge University Press.

Project Catalyst (2010) *Taking Stock – The emission levels implied by the pledges to the Copenhagen Accord*, available at http://www.project-catalyst.info/images/publications/project_catalyst_taking_stock_february22_2010.pdf.

United Nations (2009) *Annual report of the administrator of the international transaction log under the Kyoto Protocol*, UN Doc FCCC/KP/CMP/2009/19.

H-H. Rogner et al. (2007) 'Introduction', in B. Metz et al. (eds), *Climate Change 2007: Mitigation. Contribution of Working Group III to the Fourth Assessment Report of the Intergovernmental Panel on Climate Change*, Cambridge University Press.

P. H. Sand (2004) 'Sovereignty Bounded: Public Trusteeship for Common Pool Resources', 4(1) *Global Environmental Politics* 47–71.

J. L. Sax (1970) 'The Public Trust Doctrine in Natural Resources Law: Effective Judicial Intervention', 68 *Michigan Law Review* 471–556.

J. L. Sax (1980) 'Liberating the Public Trust Doctrine from its Historical Shackles', 14 *University of California-Davis Law Review* 185–194.

M. C. Wood (2008) 'Law and Climate Change: Government's Atmospheric Trust Responsibility', 10 *Environmental Law Reporter*.

Cases

Illinois Central Railroad v People of the State of Illinois 146 US 387 (1892)
Case Concerning the Gabcikovo-Nagymaros Project (Hungary v Slovakia) (*'Danube Dam Case'*), 37 ILM 162 (1998).

Legislation

United Nations Framework Convention on Climate Change (UNFCCC), Copenhagen Accord, Decision – /CP. 15 (19 December 2009).

Chapter 3

Public Trusts and Fiduciary Relations

Paul Finn

Over 30 years ago I wrote a book on fiduciary obligations. It was probably the first general treatment of this subject in the common law world. There are two things I should say about it. The first is that it had two quite unequal parts. The first – and much shorter – part dealt with what I described as fiduciary powers. Put shortly, this was about proper fiduciary decision-making; that is, decision-making in another's interests. The second dealt with what I called duties of good faith. Its concern was with the standards of conduct properly to be expected of persons occupying fiduciary positions; that is, persons who, by virtue of position, responsibility or function, were expected to act in another's interests and not in their own interests.

My second general observation is this. What became apparent to me as I wrote the book – and my concern was simply with the fiduciary obligations as they related to non-governmental actors – was that there were clear and obvious parallels between what I was writing about and the requirements of proper decision-making by, and the standards of conduct imposed on, persons in the Public Sector. I later began writing on the Public Sector – and I will talk about that in a little detail in a moment. What I want to emphasize at the outset is that in the Public Sector no less so than in the private there is the same divide between fiduciary law as it contrives and confines proper decision-making and fiduciary law as it imposes appropriate standards of conduct.

In 1995 I wrote a short piece which I entitled 'The Forgotten "Trust": The People and the State'.[1] I would like to be able to say to you that this immediately sparked the Australian legal imagination and that we now talk of little else but public trustees and public fiduciaries. We do not. It is important to appreciate this point at the outset. I consider the present Australian propensity is not likely to change much, at least when we are talking about the ends of, and constraints upon, governmental decision-making.

Let me describe briefly the outline of the argument I put in 'The Forgotten Trust'. It was, in essence, that much the most fundamental of fiduciary relations in our society is that which exists between the community (the people) and the State and its agencies that serve the community. In some parts of the common law world and particularly in the United States, this idea acquired real purchase particularly after the Revolution. The Americans after all were then called upon to explain the

---

1 Finn (1995).

architecture of their system of government which had been shorn of the King. The American historian, E. S. Morgan, has brilliantly explained their embrace of the notion of popular sovereignty in a book entitled *Inventing the People: The Rise of Popular Sovereignty in England and America*.[2] Popular sovereignty was and remains a potent fiction and one which people in effectively operating democratic societies are in varying degrees prepared to embrace and to do so with a willing suspension of disbelief. What popular sovereignty can contrive are such potent political and legal ideas as those expressed in the *Pennsylvania Declaration of Rights (1776)*: 'all power being ... derived from the people: therefore all officers of Government, whether legislative or executive, are their trustees and servants and at all times accountable to them'.[3]

The idea of trusteeship, of fiduciary responsibility, that informed this declaration was transformed into legal doctrine in the United States in at least two forms. One simply involved an adaptation of a long-standing idea in the common law. From medieval times officials who exercised power they derived from the King were characterized as being in a relationship of trust with the King. It was nonetheless acknowledged that the power and position of officials could be used both for the oppression of members of the community and for serving the self-interest of the official. From that time a growing web of laws sought to control and contain the use and abuse of office. So began what I will describe as the imposition of standards of conduct on public officials. These laws held the officials personally accountable in the courts for their improper conduct, for the most part through the criminal law. The second transformation related to State property in which the public had rights. The State was perceived as holding such property for the public. I refer particularly to public rights to use tidal and navigable waters. Here in an infant form lay the seeds of the United States' 'Public Trust' doctrine.

After the tumultuous constitutional events of the seventeenth century in England, the language of English common law began to change. The 'King's officer' became the 'public officer' and from that time one starts to find both in popular and scholarly writings the assertion that public power, legislative and otherwise, was 'fiduciary'. It was of course from this time that the ideas of a constitutional monarch and a parliamentary government were inaugurated. Nonetheless, public offices in England and later in Australia were as a rule formally still held under the Crown, and the Judges in England were unable to draw the treasonable conclusion that public power came directly from the people. That was the expedient that the now republican United States could embrace – hence the trusteeship of the Pennsylvanian Declaration.

Over time, a very considerable body of law developed particularly in the United States dealing with what we today well accept as controls upon the abuse of public office by officials of all stations, from Congressmen or Members of

---

2   Morgan (1988).

3   *Pennsylvania Constitution of 1776, Declaration of Rights Congress January 1, 1776*, IV, http://press-pubs.uchicago.edu/founders/documents/bill_of_rightss5.html.

Parliament down to the most menial, whether the abuse was to further an official's own interests or to oppress others.

For much of our history in this country that body of law was ignored, at least insofar as it could result in the imposition of sanctions upon officials personally. Also forgotten was the language of the public fiduciary or trustee. It is fair to say that events in the second half of the twentieth century compelled us to rediscover and expand upon laws designed to correct abuse of office by public officials and to promote official probity. I need only refer to what are known colloquially as the Fitzgerald Enquiry[4] and the WA Inc Royal Commission[5] and to the changes these wrought in relation to official conduct. What is notable about our present concerns with setting standards of conduct is that once again the idea of the public trust is alive and well. To give a very simple example, *The Independent Commission Against Corruption Act 1988* (NSW) in New South Wales describes 'corrupt conduct' as, amongst other things, 'conduct of a public official that constitutes or involves a breach of public trust.[6] Relatively similar provisions are to be found in, for example, *The Crime and Misconduct Act 2001* (Qld) and the *Corruption and Crime Commission Act 2003* (WA). We equally see notions consonant with the standard-setting function of the public trust in the codes of conduct for Members of Parliament and Ministers, in Public Service Rules, Regulations, statements of values and the like. To this extent, and it is not an unimportant one, my 'Forgotten Trust' has been remembered. But this is not the trust, not the fiduciary idea, with which we are concerned here. Our concern is with that first and difficult part of the fiduciary idea which concerns fiduciary power.

Before I deal with that directly, I should for the sake of non-lawyers explain both what a trust and a fiduciary relationship are. To put it crudely, a trust can be described as a relationship to property in which a person has that property vested in him or her to be held and/or used in some way for the benefit of another person or persons or for a purpose recognized by law. The trustee characteristically will have powers conferred by the terms of the trust to be exercised in the interests of those persons (the beneficiaries) or for that purpose. For their part, the beneficiaries or the Attorney-General in the case of a purpose trust, have a legal right to enforce the trust both by compelling the trustee to perform his or her duties and by challenging decisions taken by the trustee on the basis that they have not been made in good faith, or within the limits of the authority given by the terms of the trustee's powers, or for proper purposes. Fiduciary relationships generally can be described as ones in which parties are so circumstanced relative to each other for some purpose, as to give one the right reasonably to expect that the other will act in his or her interests or in their joint interests in discharging that purpose and not in his self-interest. The paradigm examples are trustees and beneficiaries, company directors and their company, agents and principals and many types of adviser and client.

---

4   Fitzgerald (1989).

5   Government (Western Australia) (1990).

6   *The Independent Commission Against Corruption Act 1988* (NSW), s 8(1) (c).

Some, but by no means all, fiduciaries have discretionary powers given to them, and again the beneficiaries of the fiduciary relationship can challenge the exercise by a fiduciary of such powers on similar grounds to those which would justify a beneficiary challenging a trustee's decision.

While the courts have recognized the possibility that the State or public agencies might be brought into a strict trust relationship with a designated person, or group or class of persons by virtue of powers (invariably statutory in origin) possessed over that person's or group's property or property interests, this is a course they have been very reluctant to take even in circumstances where the relevant Statute which created the parties relationship with a State agency and conferred powers on the agency in question, actually uses the term 'trust'. The same can be said of finding fiduciary relationships save where legislation in question bespeaks unmistakably such a relationship applying well-accepted private law criteria as, for example, in State-child guardianship relationships. I will come back to this phenomenon.

By the mid-nineteenth century it had become characteristic of English law and now even more so of Australian law that, when the language of trust was used to describe the responsibility of government and its agencies, it was seen as a political metaphor and so imposing only a moral or political obligation. Such a trust, such a fiduciary relationship, was not one which the court would enforce as such. Chief Justice Gleeson speaking in 2000 was clearly speaking in this sense when he said of the judiciary that 'judicial power is held on trust. It is an express trust, the conditions of which are stated in the Commission of a Judge or Magistrate',[7] and later 'the characterisation of the High Court as the agent of the Australian people, entrusted with the responsibility of ensuring observations of the Federal compact, signifies the fiduciary capacity in which it exercises its power'.[8]

Even the former Justice Michael Kirby, who was most sympathetic to the fiduciary idea in the public sector, was quick to accept its ready collapse into metaphor.[9]

I earlier noted that over 300 years ago the language of trusteeship and fiduciary responsibility were part of popular and political discourse. While that persisted in the United States, it all but evaporated in England and consequently in Australia from the first half of the nineteenth century. Save for a brief period following the First World War when State Members of Parliament were found to have fiduciary responsibilities such as would preclude them from using their positions for private advantage, the language of public trusteeship and the like were a notable absence from Australian law reports and statutes as, I would add, was the language of popular sovereignty.

Popular sovereignty, though, was resuscitated with a fanfare in the early 1990s in a number of decisions of what I will describe as the Mason Court. So, for

---

7　Gleeson (2000a), p. 5.
8　Gleeson (2000b).
9　Kirby (2008).

example, in *Nationwide News Pty Ltd v Wills* (1992) 177 CLR 1 at 22, Justices Deane and Toohey observed:

> the central thesis of the doctrine [of representative government] is that the powers of government belong to, and are derived from, the governed, that is to say, the people of the Commonwealth. The repositories of government power under the Constitution hold them as representatives of the people under a relationship, between representatives and represented, which is a continuing one.

Like judicial observations were made in several other High Court cases around that time. If it was thought that this was the herald of a new dawn and a new conceptualisation of our constitutional architecture – and a number of scholars at the time enthusiastically embraced the idea – it proved to be a false one. Australian constitutional agnosticism triumphed again. Without there being the embrace of a fiction such as popular sovereignty as is the case in the United States, it seems to me that the use of the idea of public trusteeship or of a public fiduciary obligation as devices to signify a legally enforceable, confining force on at least some governmental powers is likely to remain limp indeed. There are three reasons.

The first is itself a reflection of our agnosticism. Apart from unhelpful slogans in our public discourse, we lack sustaining and widely accepted fictions which describe the ends of our system of government. I would instance the quite misleading formula of 'parliamentary sovereignty' or the contrast (invariably pejorative) between elected and accountable politicians and unelected and unaccountable judges. I need not enlarge upon this.

The second is a peculiarly Australian phenomenon. Unlike in the other common law countries, the balance between statute and the common law has always heavily favoured statute. It is fair to say that from the time of white settlement, Australians were born to statutes with the role of the common law correspondingly diminished. A consequence of this today is that, for the courts, much their major functions are the interpretation and enforcement of statutes and, in the public sector, judicial review of official action taken under statutes. The legal imagination is no longer concerned with grand themes in the common law. It is concerned rather with more prosaic statutory matters, albeit matters in which confining and constraining the exercise of power and ensuring regularity in its exercise are constant preoccupations. The significance of this will become apparent later.

The third reason is that if the State or its agencies is to be found in a trust or fiduciary relationship with a class or group of persons other than with the public at large a potentially very significant problem presents itself. There are many segments of the community which rightly could be said to have claims upon the State for support, protection or assistance of varying sorts. Most obvious are the Aboriginal people of this country but one can instance many other disadvantaged groups within the community. I would instance the disabled, homeless, war veterans, refugees and so on. I would also note in passing, though, that the Supreme Court of Canada has acknowledged there could be a fiduciary relationship, for example, between the Canadian Crown and disabled veterans.[10] If the courts are to be asked to mediate governmental decision-making in relation to such groups relative to the rest of the community by reference, for example, to such traditional fiduciary criteria as fair dealing as between beneficiaries having differing rights and interests, then very obvious issues not only of competence but also of legitimacy arise.

There are, nonetheless, three areas across the common law world in which, to varying degrees, the fiduciary and trust concepts have been invoked so as to enable challenges to be made to the decisions of an alleged trustee or fiduciary. The first of these relates to Local Government and to the relationship of ratepayers to their Council. This appears to be an almost peculiarly English innovation designed to check expenditure on unauthorized purposes, for example, free bus tickets. Our High Court has indicated the unlikelihood of its being seduced by the trust idea in that arena. I need not enlarge upon it.

The second area of trusteeship relates to the public trust of natural resources. The common law and Roman law before it recognized that there were certain things which were ownerless or in which the public collectively had rights. These were a narrow and motley collection – tidal waters, fishing and wild animals and the like. As is well known to many readers, from this flimsy foundation American courts in the nineteenth century developed the public trust doctrine which was limited initially to the State's ownership of navigable and tidal waters. From that it expanded to afford protection to government held resources dedicated to public uses (or public trust lands and waters). You will hear more of this today I venture, and of Joseph Sax's seminal article in the United States, 'The Public Trust Doctrine in Natural Resource Law: Effective Judicial Intervention'.[11]

Whatever the allure of this doctrine in the United States and notwithstanding the apparent recent flirtation with it in Canada,[12] it has had almost no discernible impact in Australian law. I rather suspect our legislative legal history has a deal to do with that, as also does the different way in which we tend to conceptualize the ancient public rights, that is, of fishing and navigating in tidal waters, from the way the courts in the United States felt able to do. Equally, we display little propensity to develop or recognize new public rights in the environment.

Having said this, there is a distinctive Australian variant on the public trust doctrine that warrants brief note. Land is regularly vested in the Crown or in a local authority under statute for 'public purposes' or even on trust for 'public purposes'. While the courts for reasons I gave earlier have not treated such conferral as giving rise to a trust in the strict sense, they have nonetheless used the fact of such a grant under legislation as imposing a restriction upon how the land can be used. So in a

---

10   *Authorson v Canada (Attorney General)* [2003] 2 SCR 40.
11   Sax (1970), p. 471.
12   *British Columbia v Canadian Forest Products Ltd* (2004) 240 DLR (4th) 1.

New South Wales Local Government case where land was vested in a council on trust for public purposes, the High Court commented:

> The term 'trust' in cl 6(2)(b) of Sch 7 is apt to include those governmental responsibilities which, whilst not imposing a trust obligation as understood in private law, may fairly be described as a 'statutory trust' which bound the land and controlled what otherwise would have been the freedom of disposition enjoyed by the registered proprietor of an estate in fee simple. The trust was 'not a trust for persons but for statutory purposes'. It would be no answer to the existence of such a constraint that there was lacking a beneficial owner of the nominated lots with standing in a court of equity to enforce observance by the Council of the dedication of the nominated lots to the provision of parking spaces. It had been within the competence of the Attorney-General to seek to restrain action incompatible with 'the due exercise of the powers of the [C]ouncil or the due discharge of its duties'.[13]

However, there are two important methodological techniques used in the United States public trust case law which are significant for us given, as I suggested earlier, the concern the judiciary in this country has with statutory interpretation and judicial review.

It has been a long-accepted principle of statutory interpretation in this country that parliament should not be taken as overthrowing fundamental principles, infringing basic rights or departing from the general system of law without expressing its intention to do so with unmistakable clarity. Consequently, the courts will not give such an effect to general words in a statute simply because they have such a meaning in their widest or usual or natural sense. What is clear from the American trust cases is that a like principle of interpretation is used where the general language of a statute would seem to abrogate a public right recognized by the public trust doctrine itself. The technique used in the United States courts is to read down the statute so as to protect that right on the basis that if the legislature wished to abrogate the right it needed to do so in a way that demonstrated an explicit legislative awareness that it was so doing. Equally, in conducting a judicial review of administrative decision-making under statutes, a ground for the invalidation of a decision in our law is that the decision-maker has failed to take into account a relevant consideration, that is, a consideration that the statute expressly or impliedly required to be taken into account. Again, a like technique is used in United States public trust case law so as to oblige agencies to take into account public trust uses when an administrative decision is being made which will impact upon such uses. What differentiates Australian law from case law in the United States on the public trust doctrine is that we seem, for the moment, far less ready to regard the types of rights and interests protected in the

---

13   *Bathurst City Council Respondent v PWC Properties Pty Limited* (1998) 195 CLR 566 at 592.

United States as being ones which would justify reading down legislation or in making them significant relevant considerations in decisions which might affect such rights and interests.

The characteristic approach of Australian courts is to leave it to the legislature to identify what are the considerations a decision-maker is obliged to take into account. So, for example, under planning legislation, the requirement of 'ecologically sustainable development' is now a consideration of which account is often required to be taken but is so because of statutory prescription and not judicial innovation.

If we are to reach the point where we change our course in this country in either of these respects, it does not seem to me that it will be the product of our embrace of a public trust or of a fiduciary obligation of some sort. Rather, it will be because of changes in our recognition in the law of new rights and values that should be given the protection of the principles of interpretation and of judicial review to which I have referred.

The third area of trusteeship is that relating to Australia's aboriginal peoples. I put to one side that class of case illustrated in the Stolen Generation litigation in which it was accepted in the Federal Court that a statutory regime of wardship, according to criteria applied in private law cases, could be such as to create a fiduciary relationship between the State as guardian and aboriginal children as wards. My concern is with whether the State owes fiduciary responsibilities to aboriginal people as such.

The conclusion I expressed in my 'Forgotten Trust' piece is that advocacy for an enhanced, but distinct, fiduciary relationship of this variety in this country 'is likely to flounder'. Nothing that has occurred in Australia in the intervening fifteen years would suggest I should qualify what I said in any way. On the contrary. As is well known, in the 1830s in the United States a fiduciary relationship was found to exist between the United States government and various Indian tribes based upon the guardian and ward analogy. While a large body of law has developed in the United States from that time, the relationship itself is becoming increasingly regulated by statutes both specific to the relationship and those of more general character. In consequence, the contemporary issues which are now emerging owe less to the common law than to what are the appropriate principles of statutory construction to be applied, as also to the issue when, in the exercise of general statutory powers, should the interests of a particular tribe be taken specifically into account or otherwise.

For their part, the Canadians have recognized fiduciary duties owed by government to indigenous people, though on a *sui generis* basis. I mean no disrespect when I say that it is unlikely that Canadian jurisprudence in this area will have much influence on Australian law for the moment at least.

Against the background of the principles of the Treaty of Waitangi, the New Zealand courts in turn have found a relationship of a fiduciary nature akin to a partnership between the State and the Maori such as would require the Crown to act fairly and reasonably to the Maori. What is to be achieved by reliance upon fiduciary law here, which might not otherwise be better achieved by other means, is by no means clear and is becoming a subject of contemporary debate. The one theme in Canadian jurisprudence which is, I think, of some present significance in Australia does not derive from any principle of fiduciary or trust law as such but rather from the status of the Crown in the Canadian polity. The Canadians have referred in more recent case law to the 'honour of the Crown' as a source of fair dealing obligations to indigenous people affected by some decision or action of the Crown. That concept is alive and well in Australian law and is reflected in the observation of Chief Justice Griffith in 1912, when he referred to 'the old fashioned traditional, and almost instinctive, standard of fair play to be observed by the Crown in dealing with subjects, which I learned a very long time ago to regard as elementary'.[14]

I mention this idea for this reason. It may, as in Canada, be able to be employed by indigenous peoples against the State in respect of decisions which affect their rights and interests so as to author duties to advise and to disclose. This said, I consider the pressure in this country is not great for the evolution of a fiduciary obligation owed to Aborigines. Such pressure as may otherwise have developed has been dissipated in considerable degree by the *Native Title Act 1993* (Cth) and the *Racial Discrimination Act 1975* (Cth).

Where does this all leave us? I suppose that, subject to their important role in setting standards of conduct for public officials, the concepts of the public trustee or public fiduciary will remain for us very much the metaphor that they have been for so long now. To say that though is not to end the story. What I have tried nonetheless to indicate is that in our principles of interpretation, in our grounds of judicial review and in the standards of fair play and fair dealing we expect of the State itself, we have the tools to achieve a deal of what has been achieved elsewhere in the common law world by direct resort to the notions of trusteeship and fiduciary responsibility. Moreover, these tools are ones which are consistent with our legal history and methodology. They do not involve the judicial usurpation of the decision-making powers of Parliament or the Executive which is a recognized hazard of the public trust/fiduciary obligation ideas. Rather, they impose on those institutions a level of accountability to the public by requiring a more open acceptance by them of responsibility for the consequences of their decisions.

All that remains to be done – and it is a large 'all' – is that the courts breathe further life into those principles by acknowledging that there are emerging public interests and values which warrant protection from legislative or executive encroachment and which should be protected in the same way that we now protect fundamental rights and interests.

The distance we have to go in this is reflected in the recent decision of the High Court in *Griffiths v The Minister for Lands, Planning and Environment* (2008) 246 ALR 218. In that case, the Northern Territory Government compulsorily acquired

---

14   *Melbourne Steamship Co. Ltd v Moorehead* (1912) 15 CLR 333 at 342.

all rights and interests, including native title rights and interests (if any), in certain unalienated Crown land. The land was acquired for the purpose of its use by interested third parties who were later to be granted freehold title in the property. In other words, the acquisition was merely to clear land of native title interests so as to permit land grants to be made for private purposes. The relevant Lands Acquisition legislation[15] permitted the nominated Minister to acquire compulsorily land 'for any purpose whatsoever'. The land itself had been the subject of a native title application under the *Native Title Act*.[16] By a majority of five to two, the High Court considered the compulsory acquisition unimpeachable. Two members of the High Court, Justices Kirby and Kiefel, in separate judgments did not. For slightly differing reasons, both judges invoked the principle of statutory interpretation I earlier referred. As Justice Kirby said:

> [i]f the legislature of the Northern Territory means to empower the Minister ... to acquire native title interests of Aboriginal communities ... in order to extinguish them in favour of private interests ... [the Act] must make this expressly clear. Then only would the territory legislature assume responsibility, and accept electoral accountability, for taking such a course.[17]

Justice Kiefel read down the power so as to require a governmental purpose for its exercise. Merely clearing land of native title interests further to effectuate private purposes was not a public purpose.[18]

That native title interests could be dealt with so differently in such a case perhaps illustrates why I said that the 'all' that remains for us to do is a big 'all'.

References

Paul Finn (1995) 'The Forgotten "Trust": The People and the State', in Malcolm Cope (ed.), *Equity Issues and Trends*, Federation Press.

G. Fitzgerald (1989) *Report of a Commission of Inquiry Pursuant to Orders in Council*, Queensland Government.

Murray Gleeson (2000a) 'Judicial Legitimacy', 20(4) *Australian Bar Review* 5.

Murray Gleeson (2000b) 'Judicial Legitimacy', *Australian Bar Association Conference New York*, 2 July, available at http://www.hcourt.gov.au/assets/publications/speeches/former-justices/gleesoncj/cj_aba_conf.htm.

---

15   *Lands Acquisition Act 1978* (NT).

16   *1993* (Cth).

17   *Griffith v Minister for Lands, Planning and Environment* (2008) 246 ALR 218 at 250.

18   *Griffith v Minister for Lands, Planning and Environment* (2008) 246 ALR 218 at 251–9.

Government (Western Australia) (1990) *Royal Commission into Commercial Activities of Government and Other Matters.*

Michael Kirby (2008) 'Equity Australia's Isolationism', *W. A. Lee Lecture*, Queensland University of technology, 19 November.

Edmund S. Morgan (1988) Inventing the people: the rise of popular sovereignty in England and America, Norton.

Joseph Sax (1970) 'The Public Trust Doctrine in Natural Resource Law: Effective Judicial Intervention', 68 *Mich. L. Rev* 471.

Cases

Authorson v Canada (Attorney General) [2003] 2 SCR 40

Bathurst City Council Respondent v PWC Properties Pty Limited (1998) 195 CLR 566

British Columbia v Canadian Forest Products Ltd [2004] 240 DLR (4th) 1

Griffiths v The Minister for Lands, Planning and Environment [2008] 246 ALR 218

Melbourne Steamship Co. Ltd v Moorehead [1912] 15 CLR 333

Nationwide News Pty Ltd v Wills [1992] 177 CLR 1

Legislation

Corruption and Crime Commission Act 2003 (WA)

Lands Acquisition Act 1978 (NT)

Native Title Act 1993 (Cth)

Pennsylvania Constitution of 1776, Declaration of Rights Congress January 1, 1776.

Racial Discrimination Act 1975 (Cth)

The Crime and Misconduct Act 2001 (Qld)

The Independent Commission Against Corruption Act 1988 (NSW)

Chapter 4

Trust, Governance and the Good Life

Charles Sampford

Introduction

While he was still a Professor of Law, Paul Finn made a major contribution to the reinvigoration of Equity in Australia and on the approach taken to it by the High Court. Equity's historic role as the conscience of the law was given new life in ruling unconscionable some sharp practices in private transactions. His chapter indicates some disappointment that more use was not made of it in Australian Public Law – regretting the much more limited use of the equitable concepts of Trust and Fiduciary Duty than in US, Canada and even New Zealand Public Law. My admiration for Paul Finn's work and especially the first-mentioned achievement does not translate into sharing his regret on the latter. I once made the point in a jocular remark in late 1992 when he was visiting me at Griffith University during my second year as Foundation Dean of Law. I congratulated him on reviving equity but suggested that if he were completely successful in colonizing Australian law with trust ideas from eighteenth-century England we would need a new *Mabo*[1] to free us of the intellectual yoke of trust concepts.[2]

In this chapter, I will make the serious arguments that underlay that friendly (and respectful) quip. Equity provided an inspiration for some really important developments during the 'great leap forward' of the North Atlantic enlightenment. That inspiration has continued in the development of modern corporate governance and public governance (especially administrative law) in which the responsibility of office holders to use their power for the benefit of those they claim to serve is absolutely central. However, I would also argue that concepts of trust and fiduciary responsibility suffered inherent limitations in their origins which modern versions of corporate and public law and governance have managed to transcend (and in so doing move towards a form of 'institutional law'[3] and institutional governance

---

1 *Mabo and Others v Queensland* (No. 2) (1992) – decided on 3 June 1992.

2 I have to admit that I had not always thought that way. When Paul Finn was penning his landmark 1977 book, I was studying both equity and constitutional law as an undergraduate studying third year law. We had the opportunity of suggesting a topic for our constitutional law essay and I proposed writing a speculative piece on 'Constitutional Equity'. Gareth Evans, my lecturer, dismissed the idea as pure fantasy and I wrote on the fundamental problems of 'sovereignty' in public law.

3 Sampford (1990) later published in Sampford (1992).

that emphasizes the common issues of governing corporate and governmental institutions). The debt owed to trust concepts is so great and so widely recognized that this might seem to be a semantic debate about whether the relationship between equity and modern institutional law is one of inspiration or continuity – and a semantic debate that generates little difference in policy prescription.[4] However, I will argue that an institutional approach can ultimately take us further – not least in our approach to the governance of climate change.

Trust and the Eighteenth Century's Great Leap Forward

The eighteenth-century transatlantic enlightenment[5] fundamentally transformed governance by what I have called a 'Feuerbachian'[6] reversal of the way rulers and ruled related to each other. Until then, 'subjects' had to demonstrate their allegiance and loyalty to their 'sovereign'. Afterwards, governments of states had to justify their existence to 'citizens' who chose them. This is powerfully subversive notion, reinforced by the generally mendacious claim by sovereigns that they ruled for the benefit of the people. I have likened this reversal to those looking at Jastrow's diagram of a duck-rabbit[7] where an image of a duck suddenly appears to be a rabbit – except that once it is seen in the new way, it does not make sense to see it in the old way.

I have used the term 'great leap forward' to describe this shift because the great strides in governance that followed also generated two great sets of miseries. The first set of miseries was suffered by the colonial peoples, who were not generally considered eligible for such citizenship and sovereignty. The second set of miseries were those of the military and civilian victims of the particularly vicious forms of warfare that nation states have since honed as citizens became much more willing to die, and kill, for the states that were now, at least in theory, their own. But unlike Mao's great leap forward, great good came of it – not least when, three hundred years later, a majority of states were counted as democracies.[8]

Locke's application of trust to constitutional theory was critical in this leap. Hobbes had described a 'state of nature' that was a hell on earth. The only way to

---

4   In particular, in governance reform to build integrity and combat corruption (issues on which he focussed in his WA Inc report and which I found myself involved in when I came to Queensland and on which I have spent much of the last 20 years of research).

5   I have added the 'North Atlantic' prefix to indicate the importance of the ideological and, especially, institutional contributions made by Americans.

6   Feuerbach (2008 [1841]) said that Christians imagined that God created man in His own image but that it was rather more likely that man had created God in his own image. See Sampford (2000) later published in Zifcak (2005).

7   Jastrow (1899).

8   I should emphasise that I do not subscribe to a Whig view of history in which change is unidirectional and positive. History is a 'long game', sometimes going 'backwards'.

avoid this hell on earth was for individuals to freely contract with each other to subject themselves to this sovereign power to enforce order. The Sovereign was not a party but was created by the contract. For Locke, the sovereign was not given power by the social contract but was entrusted with it. If he broke that trust, his subjects had a right to take it back. Initially, this power was by the right to revolt[9] – something that was institutionalized in the less drastic forms or methods of impeachment and losing the numbers in the House of Commons or the electorate.

Locke was particularly popular with the American revolutionaries because his constitutional theory legitimated the revolution in which they were engaged. He was also popular because his theory of property gave the American colonials property in any part of nature with which they had mixed their labour (and that of the labour they had purchased from others through wage labour or slavery).[10] As this title is conceptually prior to the social contract, it was not alterable by the sovereign and any attempt to deprive free men of that property gave citizens a right to revolt.[11] Locke's view of property was, indeed, inspired by his image of North America as a state of nature more benign than Hobbes', in which free men would go into the wilderness and turn it into productive land. This view failed to acknowledge the rights of the American Indians – a position that was more congenial to the slave-owning sons of liberty who chafed at the British government's refusal to allow them to move west into Indian territories that they recognized.

The Contribution of Trust to Constitutional Theory and Public Law

Despite the necessary caution that the above reminders should engender, trust terminology is an extremely fruitful way of thinking about constitutional theory and public law. It makes three very important contributions to the relevant theory and law. The first contribution is that it addresses the fundamental problem of Social Contract theory – that it never happened and that it could not happen. I am fond of saying that 'human beings are social animals descended from social primates without interruption', using one sentence to dismiss all social contract theory as not only wrong but pointless and fundamentally misguided. Trust terminology does not require original citizens to have entered the contractual bargain or to deem current citizens to have adhered to it. The citizens are the

<hr />

9   Locke (1689) Second treatise para. 149 and 221–2.

10   Locke (1689) para. 55 not only justified slavery but participated in through shareholding in the Royal African Company and explicitly condoned in the Constitution of Carolina which he assisted in drafting.

11   Locke (1689) para. 222. The priority of property over democracy is one of the reasons why Locke remains so popular with neo-liberals and so called 'neo-cons' (a strange oxymoron that is 'rescued' by the fact that what they say is not conservative and not particularly new).

beneficiaries of a trust and, like most trusts, it is not created *by* the beneficiaries but *for* the beneficiaries.[12] Furthermore, trusts can create classes of beneficiaries – something that is useful in climate change where one must not only take into account the electorate but future generations and other people, other communities, who are affected by decisions. Again, like many trusts, it is implied on the basis of the conduct or office of the trustee.

The second contribution is the idea that power is held for the purpose of benefiting the community and provides an antidote to the tendency of constitutional lawyers to see the purpose of constitution law as limiting government power or preventing its abuse. If that is your principal goal, there is a very simple and effective way of achieving it more completely, perfectly and instantly than the most perfect of governance systems – just do not have a government. While this goal has been traditionally attractive to some sections of the left and, more recently, the right, most people will recoil at the suggestion. Why? Because there are things that they want government to do and which they believe cannot be adequately provided without government and/or can only be provided by governments. There is ideological debate about what those things might be – security, education, health and other services – and whether governments should be the sole or residual provider. However, there is general agreement that we do need a government to do some of these things and that the things that should be done should be the subject of debate and decision within a constitutional democracy.

This is where Trusts can provide a useful way of thinking about governments. I acknowledge that trusts were originated to prevent abuses to the inheritance of minors (as well as the avoidance of obligations from feudal dues to death duties and income tax). However, Trust law is primarily positive – a trustee is given powers in order to advance the interests of the beneficiaries. A trustee who does not use his powers for the purposes for which they were given fails in his duty just as did the servant in the parable[13] who buried the talent given to him by his master to keep it safe.

The American revolutionaries can be seen as a prime example of this phenomenon. They thought that they had had a very bad experience of government under the British colonial regime – bad enough to trigger a right to revolution.[14]

---

12   In the classic Trust, a settler transfers property to a Trustee on behalf of the beneficiaries.

13   Parable of the Talents: Matthew 25:14–30.

14   Whether or not it was so bad is not for discussion here and I will leave to the footnotes quips about whether the success of the British government in freeing the settlers of French and Indian threats might have had some influence on their finding that government was an excessive burden. I will also consign quips about the deft way in which the British lifted the threat from the French and the French lifted the threat from the English and the combined effects of both allowed them to deny the rights of life, liberty and the pursuit of happiness to the Indians who had been allied to the French and protected by the British (who, like the Colonial Office in the nineteenth century were more interested in the rights of indigenous peoples than colonists). The Republic that was born in a declaration of the

However, the colonists did not decide to eschew government. While the famous second sentence of the Declaration of Independence[15] refers to inalienable rights to life, liberty and the pursuit of happiness, the next sentence claims: 'that to secure these rights, governments are instituted among men'. The drafters of the Declaration and later the Constitution were determined to create a new government, not just to abolish the old one. However, following their bad experience, they were going into the exercise with their eyes open, recognizing that government power that should be used for the benefit of the community could be abused and used against them. Accordingly, they developed a risk management strategy, devising ways in which it would be difficult for governments to abuse power by using it against those who gave it to them.

The third contribution is to Administrative Law, in which the public power is seen as delegated to officials who must exercise it for the purposes for which the power is delegated. Any purported exercise of power for an improper purpose or that does not take into account all relevant considerations is rendered void. This approach ties in with the most common definition of corruption and the one preferred by Transparency International – the abuse of entrusted power for personal ends.[16]

Accordingly, trusts theory (including fiduciary duty) is enormously important in public law, from its Lockean inspired basic concept of holding power on trust, to the very important principles of administrative law, which actually determine and limit the ways in which power is used. They also have a profound influence on corporate law – the law that creates and seeks to regulate the other set of extremely powerful institutions created by modern law.

Limitations of Trust Terminology

Despite this vital contribution, there are limitations to the use of trust terminology. The first is that Trust Law is, in its origins and in its heart, about individuals. Equity in the Court of Chancery involved an individual trustee with responsibilities to individual beneficiaries. When Locke transferred it into constitutional thinking, the King was very much an individual. Sovereignty was found in a sovereign individual. More generally, the eighteenth century was not a time of huge institutions. There were few corporations (with few of them large), there were no unions, and agriculture, the major industry, was in the hands of individuals who were lifetime beneficiaries of long-standing trusts. Of course some households were very large – including the householder's family, servants and, especially in the Americas, slaves. But these other members of the household were not

---

unalienable rights of man and a complaint that they had been violated set about denying those rights to other residents of the continent.

15   (1776) http://www.archives.gov/exhibits/charters/declaration_transcript.html
16   TI (2000).

particularly visible in the Lockean world of the American Founding fathers. The male householder was the core constituent of the state, and the limitations of the enlightenment are very apparent in the following quote from Locke: 'Day labourers and tradesman, the spinsters and dairy maids' must be told what to believe, 'the greatest part cannot know and therefore they must believe'.[17]

The second limitation was the relationship to the environment built into Locke and Trust thinking at the time. When Locke was writing and when the American revolutionaries were revolting, we were going through a period of global warming following on from the little Ice Age. The British were doing very well out of this event. The expansion of agriculture provided increased food supply and generated wealth that could fund industrial development and feed an industrial labour force. I occasionally quip that 'global warming is conspiracy by Canadians to make their country habitable'. Global warming in the seventeenth and eighteenth centuries made England a lot more habitable as well as profitable.

Trustees of estates had to exercise their powers solely for the benefit of the human beneficiaries and not for the environment. Indeed, the eighteenth century was at the centre of a short 300 year interlude when land was thought to be so unimportant as to be subject to the dominion of a single person. This period started with the enclosures and ended when pollution and lack of planning led the state to insist that there were limitations in the rights of owners and beneficiaries to the land. Before that time, feudal property was central to society and the Common Law recognized a range of interests, including rights to wander, to gather wood, to graze animals, to use streams for power. With the enclosures, the rights of any other than the landlord were ignored and the irrelevancy of feudal dues meant most dues were translated into cash entitlements that quickly eroded into irrelevance.

This leads on to a third limitation – the origins of trust law in the handling of property. The classic trust involves a settlement of property on a trustee for the benefit of the beneficiaries. In a sense, this was an important limitation, given the growing absolutism of 'private property' involving complete dominion over a part of nature that had been appropriated by an individual. This fitted well the concept of sovereign power being asserted by monarchs to which English equity law was a neat riposte. Their response could be rephrased as follows: 'Yes, you have this power, but it is held in trust for someone else – the people you have so long claimed to protect. God may well have ordained that you have this power, but the great settlor in the sky has entrusted it to you to use for our benefit and if you abuse that trust and do not use it for that purpose, we have the right of beneficiaries to collectively dissolve the trust and take it for ourselves'. This was a great riposte to the argument of the time, especially when property was not only the source of much power but was, in the estates of aristocrats, fundamentally linked to it. However, power is not best seen as a form of property and modern theories of power are more likely to be institutional or relationship based.

---

17   Locke (1695) para. [279].

While the eighteenth century started with trust-based notions of government, it ended with a more institutionalist approach by the American Federalists. While the notion of an implied trust was a good model where state institutions had not been developed by the beneficiaries, it did not fit so well when the beneficiaries have dissolved the trust and 'institute a new government laying its foundation on such principles and organizing its powers in such form, as to them shall seem most likely to affect their safety and happiness'.[18] The model of a trust involves a settlor who is not a beneficiary and modern trust deeds established for tax minimisation and the avoidance of potential creditors still have to go to some lengths to ensure that those who establish trusts for their own benefit can do so. But there is no need for such indirect methods in establishing a government (or for that matter a corporation[19]). The American founders certainly went about the business of thoughtfully creating a set of institutions designed to carry out the will of the citizens and check each other if they did not. While the design had many valuable features, it has been developed in the country of its origin and subsequent attempts have sought to build on its experience and that of other countries.[20]

Some might describe the allocation as a matter of agency and apply fiduciary duties to this. However, this is still over-personalized and sees power in the hands of individuals rather than a set of institutions with a set of interlocking functions comprising individuals with particular roles, powers and duties. It puts a great deal of emphasis on the head of state or the head of government rather than the structure of management (a fault that is all too evident in corporate governance – a matter on which I have commented).[21]

Another limitation related to the property basis of Trust law is that Locke saw the taking of property as the first listed, and apparently most serious, abuse of sovereign power. The right to revolution was triggered by the 'endeavour to invade the property of the subject'[22] or the 'endeavour to take away, and destroy the property of the people'.[23] A fourth limitation is that the basis of removal is seen as a breach of trust. The right to revolution was triggered by a significant failure of governance. As Locke put it: '*revolutions happen* not upon every little mismanagement in public affairs. *Great mistakes* in the ruling part, many wrong and inconvenient Laws, and all the *slips* of human frailty will be *born by the*

---

18   The Declaration of Independence of the United States of America (1776). Though not, of course, the safety and happiness of non-citizens and territorial neighbours.

19   This is a view that I took with respect to the family companies with which I am associated. On my return from Oxford in 1983, I persuaded my father to ditch the complex trust structure, returning to the corporate form and only retaining trusts where family members were under age or potentially under unjustified financial threat.

20   UK for responsible parliamentary government, Sweden for ombudsmen, Hong Kong for anti-corruption agencies and Queensland Australia for the 'integrity systems' approach to combine all the elements.

21   Sampford (1992).

22   Locke (1689) para. 221.

23   Locke (1689) para. 222.

People, without mutiny or murmur'.[24] This is reflected in the approaches to the removal from office of government ministers in England at the time and reflected in the impeachment process incorporated into the US constitution. The point of the democratic revolutions of the eighteenth century was that new, more peaceful, mechanisms were introduced for the removal of governments and the threshold for their invocation could be reduced. It was not necessary for governments to breach their trust but merely to be less preferred than an alternative, with the decision being made by representatives of the people. While the US Constitution set these things in stone, the British Parliament responded to the loss of the American colonies by demanding Lord North's resignation. When the King did not want to lose his first minister, Lord North famously wrote to George III to tell him that he had no option. Thus, parliamentary government was finally crystallized in Great Britain.

This is connected to a more fundamental change related to the fundamental shift in sovereign power from monarchical to democratic government. Where the former held sovereign power, it might be argued that it must be exercised for the benefit of the community, as the Trustee of a constructive trust must act for the benefit of the unrepresented and otherwise powerless beneficiaries. However, a democratically chosen government acts as representative/agent of those who appointed it – the citizenry. Indeed, as the ultimate power is now held by the citizens, it might be argued that they are the ones with fiduciary duties over the exercise of public power with respect to those who do not have a vote (such as non-voters, future generations, members of affected foreign communities, other species and/or the environment).

The final limitation of trust law is that it tends toward originalism. The trust-based approach looks naturally to the instrument which establishes the trust and the trustees' performance is judged entirely according to that Deed and any subsequent alterations, which are generally difficult to secure. However, institutions grow, change and develop in response to the changes in the environment, to new thinking and, centrally, the wishes of the electorate. Think of the purposes for which State power was used in 1788 (the date of disembarkation from the 'First Fleet' that brought convicts, soldiers and a few settlers to Australia and the year in which the US Constitution was ratified by sufficient states to come into force the following year). Think of the purposes of state power now. This is true of all long-standing institutions. States emerged in the seventeenth century (with the 1648 Treaty of Westphalia a convenient date for a process that continued long afterwards). Joint stock companies in modern form emerged in the mid-nineteenth century and European Universities preceded each by six and nine centuries respectively.

---

24 Locke (1689) para. 225. The Declaration of Independence put it: 'Prudence, indeed, will dictate that governments long established should not be changed for light and transient causes; and accordingly all experience hath shown that mankind are more disposed to suffer, while evils are sufferable, than to right themselves by abolishing the forms to which they are accustomed.'

One of the essential elements of institutions is that they reinvent themselves, not only finding new ways to do what they did before but new purposes, justifications and, if they find new ways of serving communities of which they are a part, a long-term future. This is the dynamic of institutions and trust thinking may not pick this up. If universities were still operating as they were in the eleventh century, they would be almost completely irrelevant. Universities started as places where young men could be educated with a remarkably limited degree of freedom but much more than they could in the monasteries in an environment dominated by Latin, religion, monkish pupils and argument unsustained by empirical work. They have now changed beyond all recognition.

The institutions of governance have changed just as radically – often with a violence that makes impossible the degree of continuity that universities can attain. The institutions of governance have developed, challenged, and finally transcended the personal sovereignty of pre-enlightenment polities. They have become in theory (and, not infrequently, in practice) instruments of citizens rather than rulers. However, the institutional core of governance is not always appreciated when the need for its reform is greatest.

More than Tough Laws and Ardent Prosecutors

Similar to Paul Finn, my approach to governance reform was formed in response to the business and political scandals of the late 1980s – in which Queensland and Western Australia were exemplars of bad practice and exposed as such in the 'WA Inc' and Fitzgerald Inquiries. Paul Finn took the leading role in the former. My Queensland role was much more limited and came later – as an advisor on the subsequent reforms and as someone who tried to analyse the new approaches to governance reform emerging at opposite corners of Australia that was aimed at combating corruption and building integrity. Paul Finn in the West and Tony Fitzgerald in the north recognized that the response had to go beyond the Hong Kong approach of strong laws and a strong ICAC (Independent Commission Against Corruption). Paul Finn has focussed on the addition of public trust and the fiduciary duty of officials. I have focussed on ethics and institutional governance.[25]

The need for anything more than tough laws and ardent prosecutors is not always obvious. In the first stages of a crisis caused by governmental or business wrongdoing, the initial public reaction is to seek out the culprits and punish them. However, those who think that increasing the penalties and catching the bad apples is, or even can be, the answer remind me of H. L. Mencken's great comment:

---

25 Having said all that, I doubt that Paul Finn and I differ too much on the kind of reforms that are desirable. Indeed, his report on WA Inc suggested similar reforms to those developed in more detail in Queensland.

There is always an easy solution to every human problem – neat, plausible and wrong.[26]

At first sight, tough law enforcement seems deceptively obvious:

Bad things have happened.
There are plenty of bad people.
Therefore, the bad people caused the problem.
Therefore, the problem can be fixed by catching the bad people.

Prosecutions do have a cathartic effect and may help to mobilize reform. Laws can support other reforms. But they are not the key part of the answer.

First, prosecutions take a long time and are frequently inconclusive. Even if successful, they will not bring back the destroyed shareholder wealth, the stolen money, the uncollected revenue or even a significant proportion of it. Even for the few who are brought to justice, most of the wealth that has been destroyed or stolen will be irrecoverable. This is not just because it cannot be traced but often because it no longer exists. Second, as we all know, laws whose purposes are not internalized are rarely effective. This is the reason why many emphasize the importance of ethics. Third, laws do not address the key institutional questions of why the 'bad apples' advanced to such positions of power and were tempted to abuse that power for their own ends. If there are more 'crooked' politicians or CEOs, it is not because there are more bad people in a particular country. It is because its corporate, bureaucratic and/or political institutions generate a lot of temptations and opportunities for corruption and tend to promote those who will give in to those temptations.

The point is that many of the problems are essentially institutional rather than individual and you cannot fix institutional problems by punishing individuals. Much of this is appreciated. In fact, there are almost as many zealous proponents of ethics and institutional reform as single solutions to governance problems. Thus, *pace* Mencken, there is not one simple solution – there are three. After law reform has failed – as it always does if tried in isolation – the other solutions are preached from a range of soapboxes.

Those pressing for essentially ethical solutions emphasize that law is ineffective if not backed up by the values of those they are supposed to govern. This approach leads to attempts to create codes of conduct and to persuade relevant players to abide by them. Some enthusiasts (not including myself) push for a form of 'bare ethics' as a singular solution involving voluntary codes and 'all regulation short of law'. Yet, ethics without the sanction of law to back it up is a 'knaves' charter' – a guide for the good and a dead letter for the bad.

Those pressing for institutional solutions are attuned to the institutional nature of many of these problems. They recognize that much of the problem lies

---

26   Mencken (1949), p. 443.

in the opportunities and temptations for corrupt and unethical behaviour and the difficulty in detecting it. The solution becomes the creation of new agencies and the reform of existing ones – ticking every box on the list of institutions that have worked in other countries.[27]

Each of these three solutions is inadequate and bound to fail if tried in isolation. What is needed is a combination of the three components – ethical standard setting, legal regulation and institutional design. None are sufficient by themselves but together they provide a powerful trinity – what I have called an 'ethics and integrity regime',[28] what the OECD called an 'ethics infrastructure'[29] and what Transparency International and the World Bank called a 'national integrity system'[30] – the latter proselytized internationally.

Institutional Ethics and Values Based Governance

My approach to governance reform was born in response to the ethical meltdowns of the 1980s. It has emphasized the importance of ethics and institutions – especially 'institutional ethics' and 'values-based governance'.

Accordingly, I have long argued[31] for a values-based approach to governance of institutions – be they corporations,[32] government agencies[33] or professional groups.[34] Such an approach uses a form of 'institutional ethics' to integrate ethical standard-setting, legal regulation and institutional design and utilize the insights of the four main governance disciplines in looking for potential norms. This methodology starts with Peter Singer's basic ethical question – how should we live our lives?[35] Answering that question involves asking yourself hard questions about your values, giving honest and public answers, and trying to live by those answers. If you do, you have integrity in the sense you are true to your values, and true to yourself. In fact, if you do not live up to the answers you give, the first person you cheat is yourself.

---

27 We need to build institutions that make it easy to do the right thing, hard to do the wrong thing and make the risk of discovery too great for rational persons to choose. (If this leaves us with irrational wrongdoers, I am not particularly perturbed. I believe we can deal with them!)

28 For further discussion of this idea see, Sampford (1994b), p. 114.

29 OECD (1999).

30 The term used by Transparency International's first CEO (Jeremy Pope) to describe the comprehensive integrated approach pursued by Queensland's Electoral and Administrative Reform Commission established following the recommendations of the Fitzgerald Inquiry see Pope (2000); (2008).

31 Sampford (1990).

32 Sampford and Wood (1993).

33 Sampford (1994b).

34 Sampford and Parker (1995).

35 Singer (1993).

Institutional ethics applies the same approach to institutions (be they public agencies, political parties, professions, corporations or NGOs). It involves an institution asking hard questions about its values, giving honest and public answers and living by them. Doing so for an institution is more complex than for an individual: but it is both possible and necessary. It requires leadership in posing questions and seeking answers from members. This process starts with the vital questions that must be asked of any institution or organization. What is it for? Why should it exist? What justifies the organization to the community in which it operates, given that the community provides privileges such as powers, immunities, funding, monopolies (professions), and the privileges of incorporation from the licence to operate to limited liability? Why is the community within which it operates better for the existence of the government/corporation? Asking those questions involves an institutional and collective effort under its own formal and informal constitutional processes (including getting acceptance from relevant outsiders – including shareholders and or relevant regulators). This process does not make the institution a charity – some of the most effective institutions in the long term are those that find profitable ways in which to serve the public (as opposed to those who find unprofitable ways to serve or profitable ways that do not serve the public interest). This is not an exercise that should be resented. Public bodies are always expected to so justify themselves and the search for new ways in which institutions can serve the community is one of the great dynamics of change. Even corporations should not resent the challenge to justify themselves. Very few believe that they are there, or would long remain, if they were doing harm to the community. Most believe that a system in which people, ideas and resources can be accumulated in joint stock companies operating in more or less free markets is better for the community than other alternatives. Political parties and the profession of politicians who lead them are used to justifying themselves in terms of how they can benefit the community – it lies at the heart of their activity. While political parties in particular and politicians in general should be the ones to set out their values and put these values on public display, it is almost certainly going to be on the basis that they co-ordinate proposals for the use of public power in ways that benefit those who live within the community.[36] The parties propose and package alternative principles and policies about how public power can be deployed for the benefit of the electorate. These proposals will often reflect different values or different versions of the 'public good' that institutions should pursue. These ideas are presented to the electorate to justify choosing one group of politicians over another.

---

36 Some might limit the justification to citizens rather than members of the community living within the borders of the sovereign entity of which they are a part – with special responsibility for the electors of the constituency they represent. However, most would see a responsibility to those of the wider group for reasons of prudence, humanity or the acceptance of the human rights obligations that all sovereign states have endorsed through international human rights treaties.

An institution has integrity if it lives by its answers. However, it does so in a different way. It cannot merely be a personal commitment but an institutional commitment that involves creating mechanisms which make it more likely that the organization keeps to the values it has publicly declared and to which it is publicly committed. These mechanisms are collectively called an 'institutional integrity system' – a reflection of the larger 'national integrity system'.

Leaders of any organization under challenge should initiate this process and consider the justification for their existence, for the concentration of resources within them and the privileges accorded them. Why is the community better off for their existence? Is it better off? These are questions that should always be asked. In some cases, there is a demand for answers from outside as well as a need to provide them internally. While others may be seen in more urgent need of this process than politicians (for example, financial institutions, ratings agencies and any economic organization that has built the assumption of 'efficient markets hypothesis' into the way they do business), there are always good reasons to do so and the public can attack politicians more quickly than others.

In all of this, long-standing notions of fiduciary duty will play an important role – especially in the ethical and legal dimensions of governance reform. However, more recent Enlightenment and post-Enlightenment values will also play a part – particularly democracy, citizenship and environmental values so that Singer and Suzuki may be as important as various Supreme Court judgements. Ethics will play a particularly important role in addressing the tensions and conflicts between those values – such as the potential conflicts between representational and fiduciary elements of parliamentary ethics and the line management and fiduciary elements of public service ethics. The values underlying 'values-based governance' will not be 'just' those of equity lawyers from the Lords Chancellor in the fourteenth century to Paul Finn and Anthony Mason in the twentieth.

The institutionalization of these values will not merely be through the adoption of improved ideals of equity by Australian courts recognizing that their debt to, and continuity with, the Courts of Chancery. It will be through the range of new institutions established since then and particularly in the range of 'integrity institutions' established, strengthened and co-ordinated by the reforms that followed WA Inc and Fitzgerald – including ethics offices, ICACs and their variants, merits review tribunals, parliamentary committees, ombudsmen and integrity commissioners. It will also require a wide range of non-governmental, international and transnational bodies as well as corporations led by 'ethical entrepreneurs' who see the future of their companies in providing low carbon products and services that enable sustainable good lives for their employees and their customers (see below).

Just as the ethical meltdowns of the 1980s indicated the urgency of governance reform in corporations and state governments, and the requirement of a co-ordinated institutional reform process for the recent global finance's near death experience (the 'GFC'), so, too, the emerging 'Global Carbon Crisis' (or GCC) requires governance reform in the gamut of institutions that have generated it.

These changes will require values based reform of the kind advocated above rather than adaptation of trust ideas.

Why Problems like Climate Change are so Difficult

The most challenging governance issues of today are complex for two reasons – one of which was highly relevant to the institutional problems arising from the 1980s scandals and one of which has become relevant to the issues of today.

Governance and Interdisciplinary Rivalry

The importance of good institutional governance is recognized by many disciplines that might make a contribution to institutional governance and reform. The problem is not that it is ignored: the problem is that each discipline has a strongly theorized but limited conception of institutions, which colours and structures their view of the nature of institutional problems and the best means for addressing them. For example, lawyers look at institutions and see sets of formal norms, ethicists see informal norms and the values the institution claims to further, economists see incentives and disincentives, political scientists see power relations, social psychologists see complex webs of interpersonal and group relationships, and management theorists see structures and systems. Accordingly, the problems are seen in the deficiency of laws, ethical standards, incentives, power relations, systems, and so on, and the solutions are seen as lying in remedying those specific deficiencies.

All these partial insights into institutions and their problems are important and any solution that ignores them is likely to fail. However, as proffered solutions tend to be developed from only one disciplinary perspective, they are necessarily limited, perhaps over-emphasizing legislative solutions or the impact of economic incentives. This approach was not a problem in the eighteenth century before the disciplines of philosophy, politics, economics were separated.[37] However, the explosion of literature within each of the relevant disciplines means that we need strong interdisciplinary teams with mutual understanding and respect for what their disciplines can contribute if we are going to provide insights into how these institutional reforms may be achieved.

Globalization and Governance

Over the last twenty years, the flow of money, goods, people and ideas across borders has threatened to overwhelm the system of sovereign states. Much activity has moved outside the control of nation states at the same time as nation states

---

37   As I have argued elsewhere Sampford (2010a) Smith and Bentham would not have considered such specialisation as feasible.

have 'deregulated' and 'privatized'.[38] Such policies have transferred power from those exercising governmental power at the nominal behest of the majority of its citizens to those with greater wealth or greater knowledge in markets in which knowledge is typically asymmetric – and in which power is distributed on a very different basis of one dollar one value rather than one vote one value.

It is now recognized that many governance problems have arisen because of globalization and can only be addressed by global solutions.[39] It must also be recognized that governance problems at the national level contribute to governance problems at the global level and vice versa. This is true of current issues, from the melting Greenland glaciers to the ethical and financial meltdown of Wall Street. In both cases, there are glaring and mutually reinforcing weaknesses in global governance institutions, national governance institutions, and corporations. In the case of the financial crisis, there have been significant failures of professions and those whose advice is trusted. From the ratings agencies, to corporations, to superannuation funds, to banks, to governments and multilateral agencies, institutions must be redesigned to increase the probability that they will use the power entrusted in them to serve the public interest in the way they claim. With climate change, there have been serial and mutually reinforcing failures in global governance (as seen in Copenhagen), national governance (with failures to agree on the extent of the problem and the means for addressing it), and corporate governance (from short termism to green-wash). However, if we are going to demand that institutions are to serve our interests and values, it is critical that we are clear to ourselves what our values are and how those values are integrated into our view of the good life and that of our actions as citizens, consumers and investors.

Recognizing the Multiple Roles of Individuals

While the architecture of sustainable global governance and sustainable globalization is largely institutional, we should never ignore the individual dimension. We should identify our own actions that can further stated good governance values. We must recognize that we can act at three levels: as citizens, as investors, and as consumers. When we act, we have responsibility for the consequences of our actions. The fact that we are acting as consumers and investors does not excuse us from that responsibility. However, between our actions and the achievement of intended consequences lie a number of institutions: as citizens we rely on political parties, parliaments and bureaucracies to implement our collective choices; as investors, we rely on advisors, trust funds, fund managers and corporations to connect our values with our investments; and, as consumers we rely on manufacturers, service providers, retailers and advertisers to inform our

---

38 Deregulation has generally preceded 'privatisation' – though the attempted privatisation of natural monopolies has required a high degree of regulation which the former movement rubbished as ineffective.

39 Stiglitz (2008).

choices and deliver them. We empower these institutions by voting, investing and consuming. We must recognize that those institutions may well abuse that power. Accordingly, we should demand institutional changes to limit the ability of those institutions to abuse the power entrusted to them.

We should recognize that action on one front can affect action on other fronts and campaigns should press for action on all three fronts. We should especially seek to harness the ultimate owners of most corporations – superannuants. The latter have been actively discouraged from thinking of themselves as having any interests or values – effectively, and insultingly, required to be 'economic man'. Their interests, however, are long term and not confined to the market return on their investments. They have other economic interests as employees, taxpayers and parents. An action that marginally increases the return on their investments but raises unemployment or requires taxpayer-funded clean-ups or bailouts are against their overall economic interests. The best entrepreneurs are those who build sustainable businesses; but a large part of the problem has been that the financial intermediaries who handle superannuants' money are driven by short-term incentives.

Investors also have values that go beyond economic interests. They are not only entitled to seek to further these values through their investments but are responsible for their choices. Shareholders' values may vary, but this merely means that funds should differentiate themselves on the basis of the values they seek to further. As most superannuation funds aim for diverse investments and align shareholdings with stock market indices, superannuants are becoming 'universal investors'. Any attempt by businesses to externalize their costs hurts another one of the superannuant's investments – and often the superannuants themselves. Accordingly, the externalization of costs is not a game that superannuants can afford and neither they, nor the funds who invest their money, should be willing to play. There is a direct line between ethical and socially responsible investment by individuals, funds adopting and implementing the UN Principles of Responsible Investment, and corporate social responsibility initiatives such as the Global Compact.[40]

Reconceiving the Good Life – The Key to Environmental Governance?

It is just over a decade since I was first invited to think about the ethical and governance problems surrounding the issues of global warming – giving the opening keynote to a world council colloquium on carbon trading. The commonly perceived ethical issue at the time was that carbon trading would allow developed countries to avoid their responsibility for fixing the problem they had caused. My views on carbon trading were informed by recognizing:

---

40   For a discussion on these issues, see Ransome and Sampford (2011).

a. The value of well-governed markets and the effectiveness of clear price signals. Putting a price on carbon could have dramatic effects on the decisions of consumers, investors and providers of goods and services. However, through direct experience of currency markets during the 1980s, I recognized that market players could profit from generating volatility in which they profited at the expense of those investing in the provision of goods and services.[41]

b. That markets involve the trading of property rights and the 'initial distribution' of those property rights was irrelevant in some theories of market operation but vitally important as a matter of ethics and justice.

c. Economic incentives are an important governance tool but, like all governance tools, they are most effective as part of a package of ethical standard-setting, legal regulation and institutional reform, and such packages are necessary for major reforms to succeed (Sampford, 1990, 1992). The primary ethical question is not about how the economic incentives are activated but the overall values major reforms seek to secure.

The conclusion that I drew was that the fundamental problem was an unsustainable 'high carbon' version of the 'good life' that had developed in the West and was increasingly sought by the rest, for example:

a. The wealthiest countries were pressing for carbon trading schemes because all such schemes allocated more per capita rights to emit to themselves than to others. This proposal effectively created property rights in unsustainable activities (emitting carbon) and allocated most of those to the countries which had already contributed most to the problem. The outcome of which was neither fair nor likely to be agreed by the less-developed countries.

b. The wealthiest countries wanted to buy some of these limited rights from less-developed countries. However, if the latter still sought high carbon lifestyles the extra resources would be expended on high carbon activities.

c. Accordingly, low-carbon versions of the good life that both the West and the rest wanted to live were essential to solving global warming. This could be assisted by putting a price on carbon (preferably through carbon taxes).

d. Grandfathered trading schemes encouraged investment in unsustainable activities.

e. Part of the 'good life' involved meaningful and rewarding work and we should look to stimulate low-carbon or no carbon industries that would provide such work. Carbon taxes would help promote low-carbon industries.

---

41   In promoting volatility, traders benefited at the expense of those who needed to exchange currencies to conduct business through a combination of asymmetric knowledge and outright manipulation. Although some risks could be hedged (generating major profits for the market players), long term risk was uninsurable. By increasing the risk of doing business, it discouraged what would be otherwise profitable investment.

In the intervening decade, I have not returned to this theme in the same holistic way although some elements have been included in other papers and publications and new arguments made about the advantages of a 'carbon added tax' (Sampford, 2008a) and similarities between the problems of water trading and carbon trading (Sampford, 2008b). However, I have continued to suggest the centrality of conceptions of the good life to addressing climate change within the United Nations University and Griffith University. As a result, the World Institute for Development Economics Research (UNU-WIDER) initiated discussions on the theme and the Griffith Institute for Social and Behavioural Research (GISBR) chose this theme for its launch conference.[42] In setting out these ideas and their interaction with globalization and the global financial crisis (GFC), the discussion of each will necessarily be brief.

Carbon, Climate Change and Unsustainable Versions of the Good Life

Unlike the increasing flows of money, goods, people and ideas across national borders that constitute the heart of globalization, carbon flows across borders independently of human action. It is a headline issue because all of the above-mentioned global flows have exacerbated climate change, and because solutions involve global agreement on goals and the creation of untried institutional mechanisms. If global warming is to be halted this century, total emissions have to be capped and cut and all states will have to participate in securing that outcome.

The fundamental driver of our climate problems, however, is arguably the incipient spread of an unsustainable Western version of the 'good life'. Resource-intensive, high-carbon, Western lifestyles are frequently criticized as unsustainable and deeply unsatisfying, and yet it would appear that their ethical legitimacy has been established by the adoption of a bowdlerized version of utilitarianism that its most famous exponents would have derided. Jeremy Bentham himself believed in a form of utilitarianism that maximized, but which applied to everyone equally, and which included a very important principle – the principle of diminishing marginal utility.[43] The first loaf of bread makes one happy, the second loaf of bread does not add to one's happiness nearly as much as the first and the third may be positively unhealthy. Of course, it is hard to measure utility directly, so many decide to measure dollars (which, until derivatives, were easy to count) rather than utility or happiness,[44] and ignore the equalizing role of diminishing marginal utility. This approach arguably leads to a 'dollarized' version of the good life that is

---

42   My conference paper was later published as the lead essay in a special edition of the *Australian Journal of Social Issues*. These arguments were developed further in an invited paper to the Fourteenth International Anti Corruption Conference.

43   Bentham (1962 [1843]), p. 228.

44   Demand and supply curves may recognize that the rich man does not value the third loaf of bread but does not recognize that the fact that the poor values the first loaf even more and certainly much more than the rich man's next transient treat.

not 'good' and may not be much of a life. However, whether by good marketing or bad habits, these lifestyles are still attractive to the majority of Westerners and to a high proportion of the developing[45] world's middle classes. In so doing, northern profligacy has become southern aspiration. Even if confined to the West, such lifestyles are unsustainable: their extension to the rest of the world increases the downward spiral to ecological catastrophe. Since the 1970s, there have been many pleas for Western nations to desist from unsustainable aspects of their lifestyle and more ascetic lifestyles have been advocated. While some will choose less energy-intensive and environmentally damaging versions of the goods and services they desire, self-denial has rarely been widely popular among those who can indulge themselves, and the numbers pursuing unsustainable lifestyles has increased over the last thirty years rather than decreased. In sum, the key problem is that the West has invented and proselytized an unsustainable version of the 'good life' that other countries seek to emulate.[46]

There can be no solution to climate change until sustainable conceptions of the good life are developed that Westerners want to live and which others might want to adopt. A dialogue between East and West might be very instructive in imagining such conceptions of the good life. Fortunately, many of the things that human beings value most do not require huge investments of energy and an unsustainable use of resources – for example: companionship, conviviality, conversation. None of Martha Nussbaum's extensive list of human values – to take one prominent example of the emerging broader and deeper approach to these questions – need break the ecological bank.[47] Other alternatives based on maintaining the unsustainable Western lifestyles (the evaluative status quo), including coercing low-emitting countries to cap their carbon emissions (which is not possible even if it were morally acceptable) and paying those countries to cap their emissions (which is self defeating while unsustainable images of the good life prevail, because one way or another, those being paid to live more sustainable lifestyles will seek the unsustainable 'good life') lack plausibility. A third possibility is that elites in less-developed countries will be induced to commit their countries to cap their emissions. While there is a long tradition of such corrupt deals, they should not be contemplated here because they are unsustainable for both parties to such deals.

---

45 I use the term 'developing' rather than 'less developed', 'low income' or 'very low income' despite what is sometimes seen as a neo-liberal bias in the term. First, the term predates neo-liberalism. Second, and more importantly, I still maintain the view that we need to develop the economies and polities of the world to allow individuals to take part in the good life through the development of their capabilities and through delivery of material and non-material goods.

46 Sampford (2000); (2008).

47 Nussbaum (1997).

Concerns about Carbon Trading Schemes

While the ultimate solution to climate change is the development of low carbon lifestyles, it is important that economic incentives support and stimulate that search. The sustainable versions of the good life provide an ethical pull. The incentives provide an economic push. The currently favoured approach is to set a cap and then cut total emissions with the trading of emission rights to provide incentives to those who can most efficiently cut their carbon and minimize the cost. This approach is unsurprisingly popular in states emitting the most carbon because it effectively gives them a property right to emit. However, where an activity is shown to be harmful and unsustainable, it is not immediately obvious that the appropriate response is to create property rights to continue the harmful activity and to give the greatest property rights to those countries or corporations who have done the most harm and have been externalizing the costs on others who have suffered and continue to suffer from the harm done. This idea is popular in the West and with those who would profit by the operation of those markets. It is unpopular with non-Western countries which would be given less rights. Indeed, why would they agree?

Wherever large amounts of assets are found, so will there be attempts to appropriate them. When a very large body of assets is created, the temptation/corruption risk is correspondingly very large. Most proposals suggest giving away most of these valuable assets. If these permits go to the major polluting corporations and companies, it will be the greatest private appropriation of public assets since the Russian privatizations of the early 1990s – except they will be global in scope.

Faith in such markets may be misplaced in this case, just as it is being sorely tested in the current global economic crisis. The relevant commodity – carbon – is not well understood, and knowledge will thus be asymmetric, allowing market players many opportunities for arbitrage and taking advantage of the ignorance of those who need to access the market to continue their businesses. This process leads to increased costs and risks of doing business (as the market for carbon can fluctuate wildly),[48] so that much of the extra cost of doing business will end up in the hands of market players rather than consumers or producers who have to pay higher prices. There is also the possibility that everybody loses as new markets get it wrong even more often than established markets. Overall, I find the ideological embrace of markets as a solution to every problem not only profoundly unattractive but rather silly. After the GFC, it amounts to either stupidity, cupidity

---

48   I recognize that the European carbon market has settled down so that the maximum prices are only about double the minimum prices and that these make it less volatile than many commodity prices. However, there are many artificial aspects of the European carbon market, including the fact that most of the carbon permits required by those who need them are given them for free.

or both (and neither is made more attractive by the self-serving ideological basis of so much of the debate and the debaters).

Revisiting Carbon Taxes in a New Form: 'Carbon Added Tax'

The clearest alternative approach to carbon trading involves the taxation of unsustainable activity rather than granting rights to it. I would suggest a 'carbon added tax' (CAT) to operate like a value added tax (VAT). If a CAT operates like a VAT, carbon taxes will be 'passed on up the line' until they are ultimately paid by the consumer of the relevant goods and services. The VAT treatment of imports means that those who keep outside the system of carbon taxes would still face the CAT when the goods are imported into a market within the system.[49] It also means that the burden is on those countries which consume high carbon goods and services rather than those who produce them.[50] Finally, substituting CAT for VAT/GST means that there is no effect on inflation or the percentage of GDP going to revenue.

This strategy involves the harnessing of the power of markets – though by using a direct and controlled price signal which inhibits high-carbon industries and stimulates low-carbon industries and provides clear signals to where future entrepreneurs can make their fortunes. It closes off two other ways in which individuals and corporations can make money – through lobbying (and worse) to get free carbon credits or through leveraging asymmetric knowledge and resources to profit from an immature market.

Volatility is good for traders and not for those engaging in long-term business decisions. A decision to adopt a carbon tax over a carbon trading scheme provides incentives to channel entrepreneurial talent into the new industries without which we cannot grow our economies in ways that provide a sustainable good life for this planet's peoples. The next group of great entrepreneurs are those who will have new ideas of how sustainably to provide goods and services that consumers want in ways that provide a decent living for those who work in the industries those entrepreneurs create. This approach will involve new ways of providing old goods

---

49 Some might question how the carbon emitted in producing imports is calculated. This is a reasonable question – though the question of measurement is an issue for goods produced locally and for carbon trading systems as well. The answer for carbon taxes is a simple one. The carbon emitted by producing particular classes of goods would be estimated on the basis of traditional practice and it would be open to any importer (or manufacturer) to demonstrate that they emit less carbon than that standard. If the cost of proving the lower carbon emissions is greater than the tax benefits to be gained, then they will run with the estimate.

50 It is a concern is that countries which produce high carbon goods or components are treated as just as much of the problem as those who consume them. Much of the manufacturing, mining and smelting that was once done in the West is now done in China, Australia etc. A carbon tax will address both consumption and production but the burden for the latter should be on the ultimate consumers not the producers.

and services and new goods and services that meet human needs. A decision to adopt carbon trading makes investment in low-carbon industries more risky by reducing the certainty of the price advantage that sustainable goods and services can provide. In so doing, it increases the required rate of return for the investment to be made.

Institutional Reform

If we are going to secure sustainable low carbon lifestyles, we need more than the ethical pull and the economic push. Each needs to be institutionalized – built into the governance of global, regional, national, sub-regional, corporate and professional institutions. Where we currently see the weakness in each exacerbating the weaknesses in others, we need governance reform in all areas supporting sustainable, low carbon versions of the good life. Avoiding the impending Global Carbon Crisis requires reform at all those levels:

- global (the post-Copenhagen process, international treaties on carbon and relevant cash transfers, the utilization of the unique legitimacy of the UN, Global Compact, Principles of Responsible Investment, the Earth Charter – and above all through the promotion of cross-cultural dialogue on sustainable versions of the good life)
- regional (regional organizations need to collaborate on environmental issues and in assisting each other to cope with climate change and environmental refugees)
- national (through the establishment of carbon taxes – or carbon trading if they must)
- professional (existing professions such as law, accounting and engineering should seek to build and apply their specialist knowledge bases so that they can assist their clients not only to comply with environmental laws but to become ethical entrepreneurs, seeking new ways to reduce the carbon emissions associated with their activity)
- corporate (corporations need to serve their communities and themselves by finding goods and services that support a sustainable good life for their customers, employees and shareholders – applying their entrepreneurship to developing new low carbon, zero carbon and negative carbon industries)
- these reforms will includes institutions at various levels that measure carbon emissions for taxing and/or trading and carbon abatements for trading schemes.

Conclusion

I have argued that we should not base our response to environmental crises on eighteenth- century-based ideas on environmental trust. We must focus on the ideas that followed – about the ultimate holders of power and the ultimate owners of capital thinking through their values about the kind of good life they want to secure in their 'pursuit of happiness' that has been widely asserted. But this is not just an exercise for the collective thought of a male householder with substantial property but the extended group of six billion people that subsequent political theory has identified as the true owners, not merely the supposed beneficiaries, of government.[51] An answer is unlikely to be found without them. An answer certainly cannot be institutionalized without them.

Values-based governance will require a lot more than the best ideas of Anglo-American lawyers over the last half millennium and the institutional governance measures developed in response to late 1980s crises. They will be respected and used but also transcended. Newton once said: 'if I have seen farther, it is by standing on the shoulders of giants'.[52] Equity lawyers are among those giants. But like Newton, we stand on their shoulders to look forward, not back. If not, the GFC will be followed by a GCC which may not only destroy our ability to live a good life but to live any kind of life at all.

Finally, I wish to emphasize that this is not a left wing agenda. Indeed, like championing of the rule of law, environmentalism is an issue that:

- should be bi-partisan;
- has traditionally been more aligned to conservative values and resonated more with the right than the left; but
- has been taken up by the left and rejected by much of the right who call themselves conservative.

When conservation began, it was conservative movement. It sought to conserve the essence of the countryside from industrial intrusion and pollution just as conservatives sought to preserve public institutions from radical entryism and transformation. It is no coincidence that conservative and conservation share the same root.

I do not see this, however, as 'anti-progress' or 'anti-development' – such as some 'back to nature' conservationists were in the nineteenth century and similar to some communards seeking a simpler rural autarchy in the twentieth century. The goal should be low-carbon industries that provide sustainable goods and services to further a sustainable version of the good life for their customers and decent fulfilling sustainable jobs for their employees.

---

51 The environment will not get a vote. But it should be the beneficiary of an informed one.

52 Newton (1676).

References

J. Bentham (1962 [1843]) 'Pannamonial Fragments', in J. Bowring (ed.), *The Works of Jeremy Bentham*, Russell & Russell Inc.

L. Feuerbach (2008 [1841]) *The Essence of Christianity (Das Wesen des Christentums)*, trans. George Eliot, Dover Publications.

J. Jastrow (1899) 'The Mind's Eye', 54 *Popular Sci. Monthly* 299.

J. Locke (1689) *The Two Treatises of Government*, 1924 edn, J. M. Dent.

J. Locke (1695) 'The Reasonableness of Christianity, as delivered in the Scriptures', in V. Nuovo (ed.), *John Locke: Writings on Religion*, Oxford University Press.

H. L. Mencken (1949) *A Mencken chrestomathy*, Knopf.

I. Newton (1676) *Letter to Robert Hooke*.

M. Nussbaum (1997) 'Capabilities and Human Rights', 66 *Fordham Law Review* 273.

OECD (1999) *Public Sector Corruption: An International Survey of Prevention Measures*.

J. Pope (2000) *Confronting Corruption: The elements of a National Integrity System (The TI Source Book)*, Transparency International.

J. Pope (2008) 'National Integrity Systems: the key to building sustainable, just and honest government', in B. Head, A. J. Brown and C. Connors (eds), *Promoting Integrity*, Ashgate.

W. R. Ransome and C. Sampford (2011) *Ethical and Socially Responsible Investment*, Ashgate.

C. Sampford (1990) *Law, Institutions and the Public Private Divide (Keynote address)*, Australasian Law Teachers Association Conference, Brisbane.

C. Sampford (1994b) 'Institutionalising public sector ethics', in N. Preston (ed.), *Ethics for the public sector: education and training*, Federation Press.

C. Sampford (1992) 'Law, Institutions and the Public Private Divide', 185 *Federal Law Review*.

C. Sampford (2000) *Ethical Standard Setting for Global Incentives: Towards an effective regulatory philosophy of global greenhouse response, Opening Plenary*, World Council of Churches Consultation on the Global Atmospheric Commons.

C. Sampford (2000) *Reconceiving the Rule of Law for a Globalizing World*.

C. Sampford (2008) *Global transparency: Fighting corruption for a sustainable future: From National Integrity Systems to Global Integrity Systems, commissioned conference discussion paper*, 13 IACC.

C. Sampford (2010a) 'Adam Smith's Dinner', in I. MacNeil and J. O'Brien (eds), *The Future of Financial Regulation*, Hart.

C. Sampford (2010b) 'Re-conceiving the Good Life: the Key to Sustainable Globalisation', 45 *Australian Journal of Social Issues* 13.

C. Sampford (2010c) *Towards a Global Carbon Integrity System: Learning from the GFC and avoiding a GCC*, 14th International Anti-Corruption Conference.

C. Sampford and S. Parker (1995) 'Legal Regulation, Ethical Standard-Setting, and Institutional Design', in S. Parker and C. Sampford (eds), *Legal Practice: Contemporary Issues*, Clarendon Press and Oxford University Press.

C. Sampford and D. Wood (1993) 'The future of business ethics: legal regulation, ethical standards setting and institutional design', in C. Sampford and C. Coady (eds), *Business Ethics and the Law*, Federation Press.

P. Singer (1993) *How are we to live? Ethics in an age of self-interest*, Text Publishing.

J. Stiglitz (2008) 'The Future of Global Governance', in N. Serra and J. Stiglitz (eds), *The Future of Global Governance*, Oxford University Press.

TI (2000) *TI Source Book 2000: Confronting Corruption: The Elements of a National Integrity System*, Transparency International.

S. Zifcak (ed.) (2005) *Globalisation and the Rule of Law*, Routledge.

Cases

Mabo and Others v Queensland (No. 2) (1992) 175 CLR 1

Legislation

The Declaration of Independence of the United States of America http://www.archives.gov/exhibits/charters/declaration_transcript.html

Chapter 5

Public Officials, Public Trusts and Fiduciary Duties

John Glover*

Public law concepts and private law remedies make uneasy bedfellows. Practical and conceptual problems arise when abuses of power by public officials are characterized as breaches of fiduciary obligation. However, the excesses of governments and their agents must be curbed. Many jurists have advocated public law adoption of the fiduciary canon.[1] Objects of this chapter are threefold. First, the traditional position in Anglo/Australian law will be stated, whereby public officials are rarely subject to fiduciary duties in the execution of public tasks. Secondly, the 'public trust' cases will be evaluated, together with developments in the United Kingdom, Canada and Australia. Finally, two conclusions will be drawn. Public law is properly *inclusivist* in the way that it controls the exercise of public power. Fiduciary law, by contrast, is *exclusivist*. Interests of some persons are preferred over the interests of other persons for indefensibly non-democratic reasons.

Fiduciary Relationships

Fiduciary Obligations

Fiduciary obligations are inferred from relationships found to be of a fiduciary nature.[2] Standards of acceptable conduct are imposed on one party for the benefit of another. 'Fiduciaries' are persons who are recognized as having a responsibility to preserve another's interests.[3] Fiduciary relationships can arise from statutes, as well as contracts and tort law, or the relation can be inferred remedially.[4]

---

*   I am indebted to Emeritus Professor Enid Campbell of Monash University for her advice on and criticism of an earlier draft of this chapter.

1   Criddle (2006); Fox-Decent (2005); Finn (1995b); Finn (1995a); Finn (1994); Finn (1992); Phegan (1980); Finn (1977).

2   See *Tito v Waddell (No 2)* [1977] 1 Ch 106, 230, Megarry VC.

3   Finn (1989), pp. 1, 2 and text below at n. 105.

4   See *Northern Land Council v Commonwealth (No 2)* (1987) 75 ALR 210, p. 215 (curiam); *Wik Peoples v Queensland* (1996) 187 CLR 1, at 96, Brennan J – statutes; *Kelly v Cooper* [1993] AC 205, *Hospital Products Ltd v United States Surgical Corporation* (1984)

Only two fiduciary duties, in strictness, exist in Anglo-Australian law. These are known as the conflicts and profits rules.[5] Both are of a negative nature.[6] Fiduciaries cannot enter transactions where they have a personal interest or duty which conflicts, or might conflict, with duties owed to persons whose interests they serve. Fiduciaries cannot take profits, advantages or gains from fiduciary office without the informed consent of those whom they serve.

Fiduciary rules are often referred to by their rationale. 'The distinguishing obligation of the fiduciary', according to Millett LJ in *Bristol and West Building Society v Mothew*, is that 'the principal is entitled to the single-minded loyalty of his fiduciary'.[7]

> ... the various obligations of the fiduciary merely reflect different aspects of his core duty of loyalty and fidelity. Breach of fiduciary obligations, therefore, connotes disloyalty or infidelity.

Commentators, particularly in North America, have developed the concept of 'fiduciary loyalty' into an independent doctrinal category.[8] A fiduciary duty of loyalty is said to exist. 'Loyalty', the word, does not permit of much deconstruction. The *New Shorter Oxford Dictionary* (1993) defines 'loyalty' as 'the fact or condition of being loyal; faithfulness to duty'. The duty of loyalty is a pleonasm for present purposes, which begs the question of which duty it is, to which the fiduciary must be loyal.

The loyalty idea, however, has taken hold with equity jurists. Equitable rules often require faithfulness to some purpose – the purpose of a power, the testator's purpose in a will, or the purpose for which a confidence is imparted. Back-formations from 'loyalty' have brought many of equity's rules within the fiduciary canon. *Company Directors*, by Austin, Ford and Ramsay, sets out the 'loyalty' approach to definition of fiduciary duties.[9] 'Because the relationship between a director and a company is a fiduciary relationship', the authors say, 'a high standard of loyalty is set for directors by principles of equity.'

---

156 CLR 41, at 97, Mason J – contracts; *English v Dedham Vale Properties Ltd* [1978] 1 WLR 93, *Duke Group Ltd (in liq) v Pilmer* (1999) 73 SASR 64 (SASC FC), revd. on appeal – remedial.

   5   See *Breen v Williams* (1996) 186 CLR 71, at 113 per Gaudron and McHugh JJ; and at 137–138 per Gummow J. *Pilmer v Duke Group Ltd (in liq)* (2001) 207 CLR 165 at 197–8 per McHugh, Gummow, Hayne and Callinan JJ, at 214 per Kirby J.

   6   See *Attorney-General v Blake* [1998] Ch 432 at 455 per Woolfe MR (CA); Finn (1989), n. 3, 25. Trustee-fiduciaries are excepted from this principle.

   7   [1998] Ch. 1 at 18 (CA), followed in *Arklow Investments Ltd v Maclean* [2000] 2 NZLR 1 at 4–6 (PC).

   8   Scott (1936); Scott (1949); Scott and Fratcher (1989), §495; Sealy (1963), p. 69; Maclean (1969), p. 218.

   9   Austin et al. (2005), at [5.3].

The standard of loyalty is reflected in a number of positive obligations, as well as in some negative ones.

The positive duties of loyalty of a company director include the duties:

- to act in good faith in the best interests of the company
- to act for proper corporate purposes
- to give adequate consideration to matters for decision and to keep discretions unfettered.

The negative aspects of the duty of loyalty are those which require a director to avoid conflicts of interest of various kinds.

Equitable rules which require 'loyalty', in other words, are within the fiduciary canon by reason of their rationale. Fiduciary status is signalled when an actor must be 'loyal'.

Fiduciary Status

Identification of fiduciaries is notoriously difficult to predict, even if fiduciary duties based in 'loyalty' are easily found. Orthodox reasoning can be 'status' or 'fact' based.[10] Priests in relation to penitents, partners in relation to other partners, solicitors to clients, trustees to beneficiaries, company directors to companies and persons connected in a few other ways comprise the set of status-based relations. After proof that an established, status-based relation existed, fiduciary characterization follows as a matter of law. However, in most cases, there is no established status. Fiduciaries are identified by an uncertain list of factors.[11] Ascendancy, power and control of property on the part of the stronger party, have been held to identify fiduciary relations – as have trust, vulnerability and reliance on the part of parties in need of the law's protection.[12] No definition appears. Argument can be plaintiff- or defendant-sided. Much has now been written about this.[13] At a high level of generality, one could say that persons in trusting relationships are identified as fiduciaries on account of what they agreed to, undertook, or are taken to have assumed. Beyond that there is no consensus. What a 'trusting relationship' is and what 'taken to have assumed' may include are matters which have eluded explanation.

---

10   Law Commission (1992), at [2.4.3]; Flannigan (1989); Sealy (1962).

11   See Law Commission (1992), n. 10, at [2.4.5]–[2.4.6].

12   For example, *Allcard v Skinner* (1887) 36 Ch D 145, at 134–5 (CA) (ascendancy); *Lac Minerals Ltd v International Corona Resources Ltd* [1989] 2 SCR 574 (power); *Reading v R* [1949] 2 KB 232 (CA) (entrusted property); *Bristol and West Building Society v Mothew* [1996] 4 All ER 698 (CA) (trust); *Guerin v R* [1984] 2 SCR 335 (vulnerability) and *United Dominions Corporation Ltd v Brian Pty Ltd* (1985) 157 CLR 1 (reliance).

13   Macpherson (1998); Glover (1995), p. 443.

Statutes, contracts and tort law are primary sources of obligation. The fiduciary canon is a 'gap-filler' in the private law which operates in the interstices of more primary ordering. Duties drawn from the fiduciary code mould themselves to, and must be consistent with, pre-existing allocations of rights, powers and duties. Discretions and duties arise, first, from contracts, statutes and property law and then are regulated with the assistance of the fiduciary canon.[14] Statutes, like contracts, take precedence over fiduciary obligations in establishing an order of rights, powers and obligations in the private law.[15]

Public Law and Public Officials

'Public officials' are persons who perform public functions and derive their authority and obligation to act from the Crown prerogative or a statutory source. Officials are 'public' if they act for, on behalf of, or for the benefit of society, or some part of it and, usually, public officials are paid out of funds publicly provided.[16] Public officials cannot be persons who owe purely private loyalties, or whose competence is too insignificant for public recognition.[17] Some public officials are elected and may be members of parliament or of subordinate law-making bodies, such as municipal corporations.[18]

Public officials and the bodies through which they operate are subject to a separate general law and statutory jurisdiction in the nature of judicial review – which is a much more stringent code than fiduciary law.[19] Judicial review is not limited to negative rules and proscribing things that public officers cannot be

---

14   See *Cubillo v Commonwealth* (2001) 112 FCR 455, [465] curiam (FC) – statute and *Cadbury Schweppes Inc v FBI Foods Ltd* (1999) 167 DLR (4th) 577, at 594–5, curiam Can SC – contracts; Finn (1989), p. 76; Easterbrook and Fischel (1993), p. 427.

15   See *Tito v Waddell (No 2)* [1977] 1 Ch 106, 139, Megarry VC and *Daly v Sydney Stock Exchange Ltd* (1986) 160 CLR 371, at 389–90, Brennan J – inconsistent contracts take priority to fiduciary law unless rescinded.

16   *R v Whitaker* [1914] 3 KB 1283, at 1296, Lawrence J (for the CCA), holding that an army colonel who accepted a bribe from army caterers was a 'public officer' in the sense of the *Prevention of Corruption Act* 1906: see Woolf, Jowett, Sueur (2007), at 3.041–3.045.

17   For example, *Ex parte Kearney* (1917) 17 SR (NSW) 578: a 'fettler and ganger' employed by the NSW railways performed tasks which were too minor.

18   See *R v Boston* (1923) 33 CLR 386, at 393–4, Knox CJ – member of NSW Legislative Assembly who received bribes liable as a public officer; also *R v White* (1875) 13 SCR (NSW) L 322.

19   See *De Smith* (2009), n. 16, Chap 3; Wade and Forsyth (2009), Part IV; and, regarding Aust federal decisions, the *Administrative Decisions (Judicial Review) Act 1977* (Cth) ss 3(2) and 10(1).

influenced by and cannot do.[20] The perspective is wider. Reasonableness and other prescriptive considerations are applicable when public power is regulated.[21]

Abuses of power by public officials in the United Kingdom and Australia are now sanctioned by judicial review of administrative action,[22] the criminal law,[23] the ballot box and the working of the political process. If fiduciary law principles are introduced into judicial review, decisions taken by public officials will be subjected to a further species of collateral challenge.[24]

Fiduciary law cannot regulate situations where a public official has no power to perform acts to which objection is taken. Rather, fiduciary law might regulate exercises of power flawed by actors' improper purposes, or actors' failure to have regard to relevant considerations, or having regard to considerations which could not legitimately be taken into account, unreasonableness (in the *Wednesbury* sense) and abuse of power. Procedural accommodation of fiduciary law claims within the United Kingdom[25] and Australian[26] schemes for administrative law will be assumed.

Two criticisms of existing judicial review must be borne in mind, applicable to public officials.[27] First, corrupt motives may not be probed sufficiently under the present law. Private and self-serving purposes actuating public officials do not always emerge in proceedings where applicants bear the burden of establishing what they allege. Secondly, the deterrence-value of administrative sanctions is small. Corrupt exercises of power are defended at public expense. Even if a given decision is set aside, 'the obloquy on the official is seldom great'.[28] Fiduciary law, by contrast, is focused on deterrence.[29] Economic analysts of fiduciaries law have

---

20 See Sealy (1963), pp. 119, 125–132 and MacLean (1996), n. 8.

21 As explained in *Associated Provincial Picture Houses Ltd v Wednesbury Corporation* [1948] 1 KB 223 (CA), at 229–30, Lord Greene MR; and see *De Smith*, n. 16, 11.036–11.039.

22 For example, employment discipline: see the *Public Service Act 1999* (Cth) and regulations, discussed in the text at n. 58.

23 For example, *Criminal Code Act 1995* (Cth), Pt 7.6 'Bribery and related offences'; civil, non-criminal employment discipline is also applicable: for example, pursuant to the *Public Service Act 1999* (Cth) and *Public Service Regulations*, discussed in text at n. 59.

24 See Campbell (1998), p. 272; Aronson (1998), p. 237; Jones (1993), p. 236.

25 Viz, the 'public law' procedure in the *Supreme Court Act 1981* (UK) c 54 s 31 and Order 53 of the Rules of the Supreme Court, as set out by Gordon (1996), at 1.002–1.023.

26 For example, federally, under the *Administrative Decisions (Judicial Review) Act 1977* (Cth), fiduciary law may base an application for review on the grounds that a decision is an 'improper exercise of a power' under para. 5(1)(e), or is 'otherwise contrary to law' under para. 5(1)(j).

27 Discussed by Mahoney (1996), pp. 17–23.

28 Mahoney (1996), pp. 17–23 p. 22.

29 See rhetoric used in, for example, *Bray v Ford* [1896] AC 44, at 51, Lord Herschell and *Meinhard v Salmon* 164 NE 545 (1928), at 546, Cardozo CJ (NY CA).

described the primary role of its sanctions as the protection of institutions and 'facilitative' relationships in private law.[30]

Criminal Law

Criminal law is the second public law level of restraint on public officers. States of mind and other criteria of intention are the foundations of liability for the criminal misuse of public powers. Crimes Acts and Criminal Codes have codified offences relating to bribery in some Anglo-Australian jurisdictions.[31] The general law otherwise applies, as set out in Sir James Stephen's *Digest of the Criminal Law.*[32] Article 142 of the *Digest* is headed 'Extortion and Oppression by Public Officers' and concerns misuse of office or discretionary power. Article 144 criminalizes frauds and breaches of trust committed by public officers in the course of their duties – whether or not the same would be criminal if committed against private persons. General law on public officer offences is vague and little used. The most important article in the 1787 impeachment of Warren Hastings fell under Article 142. After a seven-year trial, Hastings was acquitted by the House of Lords on all charges.[33]

Stephen cites *R v Bembridge*[34] as the principal authority for the second article. It is a significant decision and will repay closer examination. Bembridge was an accountant employed in the army paymaster's office. Embezzlement of public funds by an outgoing paymaster was concealed with his assistance. It was not proved, but the Court assumed, that Bembridge was 'privy to the cheat' and derived some private advantage. Lord Mansfield observed that Bembridge's was *prima facie* a civil wrong, comparable to a breach of trust:

> [A] trustee that embezzles money, or suffers it to be lost, is accountable for a civil injury and not for a public offence.[35]

---

30   Cooter and Freedman (1991), pp. 1048–56.

31   See the *Prevention of Corruption Acts* 1889 to 1916 and the *Honours (Prevention of Abuses) Act 1925* (c 72), s 1; in Aust, see *Criminal Code Act 1995* (Cth), Parts 7.5 and 7.6 'Unwarranted demands' and 'Bribery and related offences'; *Criminal Code Act 1899* (Qld) Chap. 13 'Corruption and abuse of office'; *Criminal Code Act 1924* (Tas), Chapter IX–'Corruption and abuse of office' and the *Criminal Code 1913* (WA), Part III–'Offences against the administration of law and justice and against public authority'.

32   Sturge (1950).

33   Massey (1860), v. 3, pp. 336–40 and *Abstract of the Articles of Charge, Answer, and Evidence upon the Impeachment of Warren Hastings, Esq* (1788), at 14–21, specif. the dealing with Cheyte Sing. See Macaulay (1913 edn), pp. 66–73, 121–3.

34   (1783) 22 State Tr 1 (Ct of Kings Bench).

35   *R v Bembridge* (1783) 22 State Tr 1, 155, at 155, Lord Justice Willes agreeing.

These were pre-Napoleonic times. Public affairs in 1781 were still, to a large degree, administered privately.[36] Though Bembridge was paid by the state, he was, at the same time, specifically employed by and answerable to whomever was the serving paymaster. Bembridge thus owed some loyalty to the person, in whose wrong he was complicit. Lord Mansfield continued:

> ... if a man accepts an office of trust and confidence, concerning the public, especially when it is attended with profit, he is answerable to the King for his execution of that office; and he can only answer to the King in a criminal prosecution; for the King cannot otherwise punish his misbehaviour, in acting contrary to the duty of his office, by whomsoever the office is given.

The dignity of states (and kings) is presumably inconsistent with proceedings brought as litigants against their own employees.

Where there is a breach of trust, a fraud or an imposition in a subject concerning the public, which, as between subject and subject, would only be actionable by a civil action, yet as that concerns the King and the public (I use them as synonymous terms), it is indictable.[37]

Use of criminal, rather than civil sanctions against corrupt public servants is still a normal, if uncommon, legal practice.[38] In *R v Jones*,[39] for example, the Victorian Supreme Court confirmed that an Assistant Controller of Supplies of Allied Works employed pursuant to the *National Security (Allied Works) Regulations* was properly convicted on seventeen counts of misconduct as a public officer. This was a common law offence. 'Fraudulent and corrupt' receipt of secret profits was proved. Receipt of secret profits is also a classic civil wrong of the disloyal fiduciary. O'Bryan J, however, overruled the accused's application for a demurrer based on the jurisdictional overlap and held that the Crown was right to proceed criminally.

Prosecutions have been uncommon in Australia and comparable jurisdictions.[40] Elected public officials, politicians, have more often been successfully made liable for various forms of the 'public official' wrong. United States authorities provide the most numerous collection of such authorities.[41] For example, *State of New*

---

36 Before the public service reforms of the nineteenth century.

37 *R v Bembridge* (1783) 22 State Tr 1, 155.

38 See *R v Baxter* [1851] 5 Cox CC 302; *R v Wright* [1891] 1 QB 747; *R v Carroll* [1913] VLR 380; public officer offences are hard to prove and little used: see Mahoney (1996), n. 27, p. 22.

39 [1946] VLR 300 at 303: the total received was only £156.

40 For example, counsel in *R v Jones* n. 39 referred to nine cases, only one of which was decided in the twentieth century.

41 A Lexis search on 27.01.2010 disclosed hundreds of 'misconduct in office' convictions, though criminal prosecution of public officials in the United States more commonly invokes the *Mail Fraud Act*, the *Wire Fraud Act* and the *Racketeer Influenced and Corrupt Organisation Act*.

Jersey v Furey[42] concerned the prosecution of a city commissioner and the Mayor of the City of Wildewood in New Jersey. These individuals were alleged to have misconducted themselves in elected office, by voting for and granting amusement licences to corporations in which they held direct and indirect interests. It was a wrong comparable to a fiduciary's conflict of interest. Confirming the convictions, the Court affirmed that:

> [A] public official acts corruptly and is guilty of misconduct in office when he has an undisclosed interest in a venture which comes before the body of which he is a member, and he acts in favour of that interest through the office which he holds.[43]

Criminal conspiracies to commit the public official offences are sometimes alleged. Evidential deficiencies may be overcome and the 'quid pro quo' aspect of bribery offences avoided in this proceeding.[44] Law enforcement officers must establish one of two things. Either there has been positive misconduct – a bribe is received in relation to particular acts or omissions – or there has been a conspiracy – whereby parties have agreed that the integrity of officials' discretionary judgements will later be suborned. Fewer things need to be proved in order to sustain conspiracy. In *R v Boston*[45] a member of the NSW Legislative Assembly was charged with a criminal conspiracy to use his official capacity for a corrupt purpose. The offence was established upon proof that the member received money to put pressure on the Minister for Lands. It was not necessary for conviction that the member be shown to have acted corruptly in an official capacity. The fact that the same pressure might have been exerted (lawfully) outside Parliament was of no account.

Government Enforcement of the Fiduciary Duties of Public Officers

Public officers are agents or employees of the state. Proscriptive fiduciary duties of employees apply and may be enforced by the appropriate representative of the state as principal or employer. Public officers are prohibited from receiving secret profits or entering conflicts between their personal interests and their official duties. Civil proceedings may be undertaken separately from or in addition to criminal prosecution. Fiduciary obligations enforced by the state are remedial

---

42   318 A 2d 783 (Sup Ct of NJ, App Div, 1974).

43   See n 42, at 787, Michels J (for the Court).

44   See *US v Espy* 989 F Supp 17, at 2–5 (Dist Ct DC 1997) – analogies from US doctrine; explained in Coffee (1998), pp. 428–31 and on conspiracy generally, see Brown et al. (2001), pp. 1279–1323.

45   (1923) 33 CLR 386, 343–4, Knox CJ, Isaacs and Rich JJ agreeing.

and do not pertain to the ordinary exercise of public or governmental functions.[46] Property rights asserted by governments arise independently of legislative or executive action.[47] Alternatively, contracts between the state and third parties entered through corrupt public officers may be set aside.[48]

Reading v R[49] was an instance of a public officer being required to disgorge gains that he wrongfully acquired. Reading was an army sergeant on overseas duty. Wearing his army uniform, he rode through Cairo on several occasions, seated prominently on a truck which carried contraband liquor. Local police inspection was thereby avoided. Reading received substantial amounts of money from the smugglers for these services. Later, he was duly court-martialled, imprisoned and reduced to the ranks. Proceedings were then commenced by the UK Crown to recover what remained of the bribes that Reading had received.

Denning J, at first instance, held that the bribes were money had and received to the Crown's use. Reading, he said, 'must not be allowed to enrich himself in this way'.[50] The Court of Appeal affirmed that the Crown should recover the bribes. Reading was found to be in a fiduciary relationship with the Crown as his employer 'in a very loose, or at all events, a very comprehensive sense'. The bribes had been obtained through use of the relationship and 'the uniform and the opportunities and facilities attached to it'.[51] In equity the bribes were property of the Crown.

Can Governments or Public Officers Owe Fiduciary Duties to the Public?

There is a risk that the law's integrity may be endangered if public officers are held to owe fiduciary duties to the public. First, parliamentary sovereignty is violated – to the extent that the fiduciary duties are enforced contrary to the terms of the public officers' empowering legislation.[52] Some things are not 'constitutionally cognizable' by the courts. '[T]he grundnorm with which the courts must work … is the sovereignty of parliament.'[53] Rights to enforce fiduciary obligations are

---

46   See Dawn Oliver, *Common Values and the Public-Private Divide* (1999), p. 81.

47   For the significance of this, see discussion of fiduciary duties owed to indigenous peoples in text at n. 103.

48   For example, see *Driscoll v Burlington-Bristol Bridge Co* (1952) 86 A 2d 201.

49   [1949] 2 KB 232 (CA), affirming [1948] 2 KB 268, Denning J, CA decision affirmed with little discussion, *Reading v Attorney General* [1951] AC 507.

50   Citing no authority: [1948] 2 KB 268, at 277.

51   [1949] 2 KB 232, at 236, curiam (Asquith LJ). See comparable reasoning in *Attorney General (Hong Kong) v Reid* [1994] 1 AC 324 (PC) – Crown has property in New Zealand land as the exchange-product of bribes received by a corrupt Crown prosecutor.

52   See Dicey (1920), pp. 58–68.

53   As Dickson CJ wrote, giving judgement for the Canadian Supreme Court in *Canada (Auditor General) v Canada (Minister of Energy, Mines and Resources)* [1989] 2 SCR 49, at [69] (re an auditor's access powers).

almost certainly not so 'deeply rooted in our democratic system of government and the common law' as to be a basis for overturning statute law.[54]

Secondly, governments must act in the public and not some sectional or personal interest. Laws must be applied equally and without unjustifiable differentiation.[55] Particular loyalty or consideration to a race, or class, or a religiously or otherwise identifiable group must be warranted by some standard inherent in democracy.[56] Procedural propriety in administration requires that fair decision-making procedures are observed. Individuals cannot be excessively advantaged or burdened consistently with the rule of law.[57]

A corresponding employment discipline, or a 'public sector morality', is applied to public officials.[58] Favoured treatment cannot be extended to one or more of the groups or classes of persons within the body of persons who are liable to the exercise of their powers. The Australian *Public Service Regulations* (1935) require Commonwealth officers to perform their duties 'with skill, care, diligence and *impartiality* … to the best of their ability' and 'treat members of the public … with courtesy and sensitivity to their rights, duties and aspirations'.[59]

Fiduciary obligations attributed to governments and public officials are radically inconsistent with these principles. The opposite of impartiality is mandated. Undivided loyalty amounting to positive discrimination must be shown in favour of the persons to whom fiduciary duties are owed.[60] Interests of other persons and of fiduciaries themselves must be subordinated to the trust invested in fiduciaries, or with which they have assumed to act.

Owing conflicting duties to two or more persons is prohibited by first principles of fiduciary law.[61] This is not mere rhetoric. Sometimes described as 'inveterate'[62] and 'inflexible',[63] absolute loyalty and the avoidance of conflicts are defining aspects of the fiduciary regime. Persons whose duties continually conflict must be regulated by some other and presumably a public law code.

---

54   See *Union Steamship Co of Australia Pty Ltd v King* (1988) 166 CLR 1, at 10 (curiam).

55   See *De Smith*, n. 16, at 11.069.

56   For example, pursuant to the terms of the *Racial Discrimination Act 1975* (Cth) and for the protection of Aboriginal rights: discussed Finn (1995), n. 1, p. 138.

57   See *De Smith*, n. 16, at 1–026; Mason (1995), pp. 127–8; Allen (1993), pp. 177–82; Jowett (1994), p. 3.

58   Finn (1992), p. 243.

59   'Duties of officers': reg. 8A (a) and (e) [emphasis added];

60   Scott (1989), n. 8, p. 540; Sealy (1962), p. 69; Maclean (1996), n. 8; Finn (1992), n. 3, pp. 27–8.

61   *Bolkiah v KPMG* [1999] 2 AC 222, at 234–5, Lord Millett, Lords Browne-Wilkinson, Hope, Cyde and Hutton agreeing; *Chan v Zacharia* (1984) 154 CLR 178, at 198, Deane J.

62   *Meinhart v Salmon* 164 NE 545, at 546 (1928), Cardozo J.

63   *Bray v Ford* [1896] AC 44, at 57, Lord Herschell.

Traditional Responses: The 'Public Trust' Cases

Trusts do not ordinarily arise under the terms of public legislation. Periodically, however, the courts have been called upon to distinguish between public and private fiduciary duties. The ambiguous category of 'public' trust is then involved. Public trusts are those where no private person or purpose benefits and the Attorney-General attends to enforcement on behalf of the state.[64] Trusts which are 'charitable' in the legal sense are the most common type.[65] For example, in *Director of Aboriginal and Islander Advancement v Peinkinna*,[66] the Privy Council advised that, whilst legislation did not grant income from bauxite mining on an Aboriginal reserve to the reserve's residents, the reserve's director could distribute the income through a charitable trust to the state's Aboriginal inhabitants.

'Non-charitable public trusts' is a smaller category. There is a presumption that legislation does not intend to impose trust obligations in respect of governmental or administrative functions on the Crown or its servants and agents.[67] This rule was not applied by the majority in *Registrar of the Accidents Compensation Tribunal v FCT*.[68] Compensation payments under the control of the Registrar were said to be held by him as 'a trustee in the strict sense' and liable for taxation accordingly. The function of providing accident compensation was neither 'governmental in nature', nor such as to involve 'the interests of government'.

Public trusts are sometimes inferred where public officials receive property not only in the performance of their administrative duties, but partly as a result of personal undertakings the officials gave in respect of that property. In this way, solicitors in *FCT v Harmer* were liable to be taxed as 'trustees' in respect of escrowed funds that they held for disputants as officers of the Court.[69]

Overtly 'political' public trusts were claimed in *The Skinners' Company v Irish Society*.[70] Claimants attempted to extract compensation payments and other benefits from the executive arm of government. The Court put the malleable trusts device to an uncharacteristic use: defence rather than subversion of the public fisc. *The Skinners' Company* involved breach of trust claims arising out of the

---

64 Ford et al. (2001), pp. 5230–5260.

65 Ford et al. (2001), n. 64, [19.060]–[19.690].

66 (1978) 17 ALR 129, 138, Lord Scarman, interpreting the *Aborigines Act 1971–1975* (Qld) and the *Aurukun Associates Agreement Act 1975* (Qld).

67 See *Kinlock v Secretary of State for India* (1882) 7 App Cas 616, *Tito v Waddell (No 2)* [1977] 1 Ch 106, at 211–216 and Ford et al. (2001), n. 64, p. 1020.

68 (1993) 178 CLR 145, at 164, Mason CJ, Deane, Toohey and Gaudron JJ, but cf. 181, Brennan, Dawson and McHugh JJ, (dis): no indication on these facts that the government intended to create a trust.

69 (1990) 94 ALR 541, at 545, Northrop J (FCA FC). The point was conceded on appeal to HC: (1991) 173 CLR 264.

70 (1845) 12 Cl & F 425, at 487–8, 8 ER 1474, at 1498–9 (HL); see cases below and *Civilian War Claimants Association Ltd v R* [1932] AC 14 and *Hereford Railway Co v the Queen* (1894) 24 SCR 1.

maladministration of a corporation founded by Royal Charter. Public purposes specified were the promotion in Ulster of 'the Protestant religion and the Protestant establishment'. Lands in Ireland which had been seized during the Protectorate were managed by the Society on behalf of twelve City of London corporations. Lord Lyndhurst in the House of Lords said that the Charter's specification of 'public objects of the greatest possible importance' meant that the trustees were 'public officers invested with a public trust'. As such, they were accountable only to the Crown. The corporations' maladministration claims brought on the theory of a breach of trust therefore failed.

Kinlock v Secretary of State for India in Council[71] followed *The Skinnners' Company* in the construction of another Royal Warrant. The Secretary of State for India was empowered to distribute a treasure which had been 'captured' during the Indian Mutiny.[72] All members of the House of Lords held that no private law rights or trust arose.[73] Lord Selborne went further and denied any applicability of a trust in the 'lower sense' – referring to the sort of relation administered by courts of equity. Rather, he said, any 'trust' implied in the Warrant could relate only to:

> higher matters, such as might take place between the Crown and Public Officers discharging ... duties ... belonging to the prerogative of the Crown.

Again, the claimants were unsuccessful.

Vice-Chancellor Megarry in *Tito v Waddell (No 2)*[74] held that the 'higher sense' of trust, regulated the British Crown's control of a phosphate-rich island colony. No private rights were found to be conferred by a 1928 'Mining Ordinance of the Gilbert and Ellice Islands'. The Ordinance provided that a resident Commissioner would establish formula for the payment of mining royalties to be held 'in trust' for the Islanders. Irregularities otherwise amounting to breach of fiduciary duty were shown. However, Megarry VC thought that it was 'far too wide and indefinite' to say that statutes imposing duties on the Crown relating to property carry with them fiduciary obligations 'as a general rule'. Other, absent, indicia were needed. The jurisdiction of equity was not attracted and the Islanders' claims to fiduciary compensation failed.[75]

---

71   (1882) 7 App Cas 619 (HL).

72   Viz, – 'specie, jewels and diamonds of priceless value ... the prize of Bandah and Kirwi': see Kayes and Malleson (1907), v. 5, pp. 140–41. In this account, the prize was seized without hostilities from a nine-year-old raj who was 'virtually a ward of the British government'. It was eventually distributed to persons including 'the Commander-in-Chief in India and his staff, who were hundreds of miles from the spot and whose action did not influence the capture'.

73   Lord Selborne LC, at 626, Lords O'Hagan and Blackburn separately agreeing.

74   [1977] 1 Ch 106, 211–16, 227–30.

75   See n. 74, at 230. Megarry VC observed: 'If there is a fiduciary duty, the equitable rules about self-dealing apply. But self-dealing does not impose the duty'.

Fiduciary consequences for the acts of public officers were also denied in an authority dealing with facts far removed from the Crown prerogative. *Swain v The Law Society*[76] concerned the Law Society's authority to take out and maintain a master policy of insurance, in respect of the professional liabilities of solicitors practising in England and Wales. Annually, the Society received a substantial rebate of the insurance brokers' commission and applied the same to its own purposes. Two solicitors argued that the Society received the rebate as fiduciary for the solicitors who paid insurance premiums and was required to account to them as constructive trustee. The House of Lords unanimously disallowed the solicitors' appeal against dismissal of the claim. Lord Brightman confirmed that in exercising the power conferred on it, the Society was performing a public duty and not a private duty to the premium-paying solicitors. This made a significant difference:

> [T]he nature of a public duty and the remedies of those who seek to challenge the manner in which it is performed differ markedly from the nature of a private duty and the remedies of those who say that a private duty is breached. If a public duty is breached, there is the remedy of judicial review. There is no remedy in breach of trust or equitable account ... .[77]

To assume otherwise was category-mistake, Lord Diplock said.[78]

Challenges to Fiduciary Orthodoxy on the Liability of Public Officials (and Governments)

United Kingdom Developments

Local authorities in the United Kingdom have been held to owe fiduciary or trust-like duties to their ratepayers. Discretionary expenditures must be undertaken and undertakings managed with due regard to the ratepayer interest.[79] The idea was first expressed in the 1920s. Councils were denied the right to make benevolent or philanthropic gestures with their ratepayers' funds. Redistributive policies of local authorities could not exploit disjunctures in the 'paying-voting-benefiting relationship'.[80] The House of Lords in *Roberts v Hopwood*[81] held that a Labour-controlled council acted contrary to law when it paid wages to its employees at rates well in excess of those which prevailed in the community. It was not 'businesslike', Lord Atkinson said, and in breach of the duty owed by local authorities to their

---

76 [1983] AC 598, referring to the *Solicitors' Act 1974* (UK), s 37.
77 See n. 76, at 618.
78 See n. 76, at 609.
79 See *Bromley London Borough Council v Greater London Council* [1983] AC 768.
80 See Loughlin (1996), p. 226.
81 [1925] AC 578, at 595–6, Lord Atkinson; and see Loughlin (1996), n. 80, pp. 209–12.

ratepayers – which was 'somewhat' like the duty owed by 'trustees or managers of property of others'. Local authorities are equally in breach of the duty if, for charitable or other reasons, they give the ratepayers' funds away.

Some thirty years later, the Court of Appeal in *Prescott v Birmingham Corporation*[82] used the 'fiduciary' word when it struck down a scheme to provide free omnibus transport to old persons within a municipal area. The Corporation's generosity with other peoples' money was illegally in breach of a specifically 'fiduciary' duty which the corporation was held to owe its ratepayers.

In 1981, the House of Lords confirmed this public law application of the fiduciary relationship in *Bromley London Borough Council v Greater London Council*.[83] The Labour-controlled GLC directed Bromley LBC to raise an additional rate, in order to pay for a 25 per cent reduction in bus and tube fares charged by the London Transport Executive. The GLC was aware that by raising this extra levy it would suffer a corresponding loss of grant from central government funds.

Prior to this new policy, the *Transport (London) Act 1969* (UK) had devolved control over transport fares and pricing levels to the GLC. Reduction in fare levels was also 'the main plank of the Labour party's manifesto in the GLC elections of May 1981'.[84]

Bromley was a Conservative-controlled outer London borough which was not served by any underground stations.[85] Disproportion and failure to consider the ratepayer interest were the bases of Bromley's application for judicial review of the GLC's decision to fund lower transport fares through a rate increase. The amount of the extra levy and loss of grant was alleged not to be justified by the value to the ratepayers of the fare-reduction benefit. In the circumstances, the proposal was said to be invalid as an unlawful breach of fiduciary duty owed by a local authority to its ratepayers.[86] All members of the House of Lords upheld the argument.

Lords Wilberforce and Scarman emphasized the GLC's statutory duty under section 1 of the *Transport (London) Act 1969* to provide 'integrated, efficient and *economic* transport facilities'.[87] Interpreted in the light of the 'accepted' fiduciary duty, both speeches concluded that the disproportion between the levy and the benefit invalidated the proposal. Lords Keith and Brandon found that parliament had evinced an 'overriding' intention in the 1969 Act that the GLC should have regard to 'ordinary business principles' in carrying on the undertaking. Business principles had clearly not been followed where, as here, the interests of the travelling public and the cost to ratepayers were so badly misaligned.[88]

---

82   [1955] 1 Ch 210, at 235, Jenkins LJ (for the Court).

83   [1983] AC 768.

84   See Dignan (1983), p. 630.

85   See Loughlin (1996), n. 80, p. 231.

86   See 'Bromley Council v Greater London Council: counsel's opinion' [1991] *PL* 499.

87   Emphasis added: [1983] AC 768, at 837, Lord Scarman, at 814–5, Lord Wilberforce.

88   [1983] AC 768, at 853, Lord Brandon, at 834–5, Lord Keith.

Lord Diplock's speech in *Bromley* was less concerned with statutory interpretation and was said to raise the fiduciary concept to the status of a 'quasi-constitutional principle'.[89] 'Categories' of person to whom the GLC owed duties in relation to public transport were identified to include several classes of person. Many ratepayers were either corporations or did not reside in the GLC area.

> [Ratepayers] constitute only 40 per cent of residents and that 40 per cent bears only 38 per cent of the total burden borne by all ratepayers.[90]

Lack of correspondence between ratepayer and passenger categories and loss of central funds meant that the GLC policy contained unacceptable 'disadvantages' for ratepayers and was therefore 'erroneous in law'.

> [A] local authority owes a fiduciary duty to the ratepayers from whom it obtains moneys needed to carry out its statutory functions and this includes a duty not to expend those moneys thriftlessly but to deploy the full financial resources available to it to the best advantage ... being ... the rate fund ... and the grants from central government respectively.[91]

Bromley's case represents a high-water mark of the fiduciary characterization of local authorities' relations with their ratepayers. It is followed sometimes a little reluctantly and the fiduciary notion is attributed with a kind of metaphor-status in the public law.[92] Recently, the House of Lords disapproved a comparable borrowing of a private law concept.[93] New Zealand courts followed,[94] and even extended the notion that local authorities are fiduciaries in relation to ratepayers,[95]

---

89   Loughlin (1996), n. 80, p. 232.

90   [1983] AC 768, at 829.

91   See n. 90.

92   For example, see *Equitable Life Assurance Society v Hyman* [2000] 2 All ER 331 (CA), at [18], Lord Woolfe MR: 'I would not accept that today any group such as the ratepayers can be singled out as the beneficiary of local government powers ... local authorities ... are very much in the same position they would be in if they had fiduciary powers conferred upon them'. See also *Hazell v Hammersmith & Fulham London Borough Council* [1990] 2 QB 697 (CA), at 768, Sir S Browne P.

93   In *R v East Sussex County Council ex p Reprotech (Pebsham) Ltd* [2002] 4 All ER 58 (HL), at [33]–[34], Lord Hoffman.

94   For example, *Waitakere City Council v Lovelock* [1997] 2 NZLR 385 (CA), at 397, Richardson P, Blanchard J agreeing.

95   In *Mackenzie District Council v Electricity Corporation of New Zealand* [1992] 3 NZLR 41 (CA) and *Barton v Masterton District Council* [1992] 1 NZLR 232 (HC) local authorities were held to be fiduciaries in regard to the levying of rates, in addition to rate expenditure.

though a critical note has now been sounded in that jurisdiction also.[96] Australia never accepted the idea.[97] Academic commentators have been, alternatively, unenthusiastic[98] and critical[99] of putting the fiduciary concept to this use.

A problem with using the fiduciary relationship to resolve local government problems may be that too one-sided a range of interests is consulted. Parties protected by the fiduciary relationships are privileged inconsistently with the proper exercise of discretions conferred by statute on public bodies, or on bodies exercising public functions. Taking *Bromley*[100] as an example, fiduciary categories overrode a discretion conferred on the GLC by the *Transport (London) Act 1969* and also the democratic mandate conferred in the preceding year's GLC elections.[101] Interests of passengers and ratepayers arguably needed to be balanced rather than prioritized. Equitable doctrines are unequal to the task. For fiduciary law to achieve fairness, it could proceed only by (unhelpfully) assuming that fiduciary obligations were owed all interested parties.[102]

Canadian Developments

A different public law role for the fiduciary relationship has been suggested by Canadian authority concerned with dealings between the Crown and representatives of indigenous peoples. In particular, the judgement delivered by Dickson J in *Guerin v The Queen*[103] has pointed the way to a governmental fiduciary liability based in the claimants' pre-existing property rights. Delbert Guerin was the Chief of an Indian Band which possessed valuable reserve land within the City of Vancouver. In the mid-1950s, officials of the Indian Affairs Branch of the Federal Department of Citizenship and Immigration had the real estate value of the land appraised and commenced negotiations for leasing it to a golf club. The appraisal was not put before the Band, nor were the lease terms offered by the club. Instead, a meeting was convened of Indian Affairs officials, together with Indian Band and

---

96　　See *Wellington City Council v Woolworths New Zealand Ltd (No 2)* [1996] 2 NZLR 537, at 546, curiam: 'fiduciary duty concept ... does not interfere with every exercise by local authorities of their discretionary powers ... [and is subject to] constitutional and democratic constraints'.

97　　Noted by Gummow J in *IW v City of Perth* (1996) 191 CLR 1, at 49.

98　　See Bailey (2004), at [4.92].

99　　See Dignan (1983), n. 84, p. 642; Griffith (1982), pp 218–9; Ratcliff (1996), pp. 52–3; Loughlin (1996), n. 80, pp. 234–7; Prosser [2000] *PL* 337, at 339 and Alder [2001] *PL* 717, at 726–7.

100　　[1983] AC 768.

101　　See Dignan (1983), n. 84, pp. 626–41.

102　　Finn (1977), pp. 56–63.

103　　[1984] 2 SCR 335, at 364: judgement of Dickson, Beetz, Chouinard and Lamer JJ; also, to similar effect, see the judgement of Ritchie, Macintyre and Wilson JJ at 339; facts taken from the report of the BC CA decision at (1982) 143 DLR (3d) 416.

golf club representatives. An annual rent was agreed and various other things were 'understood'. The Band then voted to surrender the land to the Crown, trusting that the government would lease it on terms 'most conducive to our welfare and that of our people'. Indian Affairs officials later accepted the golf club's agreement to terms significantly less favourable than those which had been understood. Band members were either not informed, or told that they had no choice. Not until 1970 were they given a copy of the lease.

Proceedings in *Guerin* for breach of trust were brought by the Chief and councillors of the Band. Justice Dickson in the Supreme Court of Canada held that the Band's unqualified surrender of the land left no basis for the trust alleged. In the alternative, however, Dickson J found that the Crown assumed a duty to the Indians comparable to that of a trustee when it accepted the surrendered land. Section s 18(1) of the *Indian Act 1952* (Can) gave the Crown a broad and unregulated discretion to deal with surrendered land. If and when an Indian Band surrenders its interest to the Crown, Dickson J said,

> a fiduciary obligation takes hold to regulate the manner in which the Crown exercises its discretion in dealing with the land on the Indians' behalf.

Then came a significant passage in the judgement. It has been followed on many occasions and is now the main Canadian articulation of fiduciary law in native title jurisprudence.[104] Justice Dickson began by noting that fiduciary obligations 'generally arise only with regard to obligations originating in a private law context'. Then he stated:

> [P]ublic law duties, the performance of which requires the exercise of a discretion, do not typically give rise to a fiduciary relationship As the 'political trust' cases indicate, the Crown is not normally viewed as a fiduciary in the exercise of its legislative or judicial function. The mere fact, however, that it is the Crown which is obligated to act on the Indians' behalf does not of itself remove the Crown's obligation from the scope of the fiduciary principle ... the Indians' interest in land is an independent legal interest. It is not a creation of either the legislative or executive branches of government. The Crown's obligation to the Indians with respect to that interest is therefore not a public law duty. Whilst it is not a private law duty in a strict sense either, it is nonetheless in the nature of a private law duty. Therefore, in this sui generis relationship, it is not improper to regard the Crown as a fiduciary.

---

104   See *Fairford First Nation v Canada (Attorney General)* [1999] 2 FC 48; *Opetchesant Indian Band v Canada* [1997] 2 SCR 119; *Delgamuuku v British Columbia* [1997] 3 SCR 1010; *Blueberry River Band v Canada (Dept of Indian Affairs)* [1995] 4 SCR 344; *R v Van der Peet* [1996] 2 SCR 507; *Roberts v Canada* [1989] 1 SCR 322; *Semiahmoo Indian Band v Canada* [1998] 1 FC 3; and *Chippewas of Nawash First Nation v Canada (Minister for Indian and Northern Affairs)* (1999) Ont Sup Ct (Lexis 53036).

An obligation 'in the nature of a private law duty' was triggered when the Crown assumed a duty to protect the claimant's 'independent legal interest'. Governments may become liable as fiduciaries if they are obliged to have regard to the rights of indigenous claimants which arise independently of the legislative or executive process. Property rights are the pre-eminent exemplar. Private economic interests are inevitably involved. More ambulatory indicia of 'power', 'dependency or 'vulnerability' express states or conditions, rather than interests of any kind, and were not employed in Justice Dickson's formulation.

Inferentially, the 'political trust' line of cases was distinguished. Because the relevant legal interest in *Guerin* was land which had never belonged to the Crown, Dickson was able to conclude that the Crown's obligation to the Indians with respect to that interest was 'not a public law duty'.

Justice Wilson reached a similar conclusion. She observed:

> I do not think that we are dealing with a purely public law context here. [A judge below] agrees that a Band has a beneficial interest in its reserve. I believe that it is clear from s 18 that that interest is to be respected and this is enough to make the political trust cases inapplicable.[105]

Placing emphasis on the words of s 18(1) of the *Indian Act*, she reasoned that the Crown owed the Indian Band a fiduciary duty which 'existed at large' before the time of the claim. The duty 'crystallized upon the surrender into an express trust of specific land for a specific purpose'.[106] In this way, the Indian claim was founded on a pre-existing right, a species of the 'independent legal interest' which Dickson J said was amenable to enforcement through private fiduciaries law.

A way was thus opened to impose private-law fiduciary obligations on the government without doing violence to trust law principles. For the purpose, however, Justice Wilson distinguished the UK 'political trust' cases in a rather narrow way. Whilst 'a Band has a beneficial interest in its reserve', the 'political trusts', or trusts 'in the higher sense',[107] she said, were all ultimately based in Crown property. Certainly this applied to the Irish lands in *The Skinners' Company v Irish Society*[108] and to the 'booty' in *Kinlock's* case.[109] Both land and 'booty' were forfeited to the

---

105　*Guerin v R* [1984] 2 SCR 335 at 351, judgement also of Ritchie and McIntyre JJ. Estey J wrote a separate judgement, concurring with Wilson J.

106　*Guerin v R* [1984] 2 SCR 335 at 355; s 18(1), (extracted at 348), provides that 'reserves shall be held by Her Majesty for the use and benefit of the respective bands for which they were set apart; and ... subject to ... the terms of any treaty or surrender, the Governor in Council may determine whether any purpose for which lands in a reserve are used or are to be used is for the use and benefit of the band'.

107　As referred to by Megarry VC in *Tito v Waddell (No 2)* [1977] 1 Ch 106, at 227.

108　(1845) 12 Cl & F 425, 8 ER 1474.

109　*Kinlock v Secretary of State for India in Council* (1882) 7 App Cas 619.

Crown as a consequence of successful military action. But *Tito v Waddell (No 2)*[110] and *Swain v The Law Society*[111] contained no equivalent interest. Royalties from use of the Islanders' land and solicitors' insurance premiums were the respective bases of the claim. Trust-like duties were denied in these cases because of the administrative inappropriateness of subjecting the Crown to fiduciary obligations. None of the judgements in *Guerin v The Queen*[112] responded to this point.

Writing in 1999, Justice Rothstein in *Fairford First Nation v Canada (Attorney General) (TD)*[113] made an extensive review of the authorities on fiduciary duties owed to indigenous peoples. He concluded that there is 'no indication that the courts have deviated from Dickson's approach in *Guerin v the Queen*'. Dickson's approach was consistent, Rothstein said, with the Supreme Court's subsequent treatment of indigenous fiduciary relationships in *R v Sparrow*.[114] Some commentators have maintained that *Sparrow* stands for the wider proposition that fiduciary duties to Indian Bands can be based on a 'power/dependency' relationship – independently of whether indigenous land rights are threatened with extinguishment through government action.[115] Justice Rothstein observes that nothing in *Sparrow* adds to or derogates from the view of Dickson J in *Guerin*. A requirement was simply added to the 'independent legal right' requirement, to the effect that governments must deal 'honourably' with Aboriginals and fully justify the use of extinguishment powers.[116]

Squamish Indian Band v R[117] involved an unsuccessful fiduciary duty claim by an Indian Band of the type made in *Guerin's* case. However, Simpson J, hearing the case at first instance, made a significant observation. 'It cannot be the case', he said, 'that each time legislation gives the Crown discretion to act, a private fiduciary duty or even a *sui generis* fiduciary duty applies.' Matters of public law, in fiduciary terms, generally do not import the 'reasonable expectation' that the Crown is acting for the sole benefit of the party affected. The Crown acts for everyone. Public fiduciary duties are exceptional. So, he thought, it followed that

> in matters of public law, discretion and vulnerability can exist without triggering
> a fiduciary standard. There would have to be special circumstances, other than

---

110   [1977] 1 Ch 106; cf. the unlikely contention that royalties paid for use of the Islanders' land 'were exclusively Crown property': Wilson J in *Guerin* [1984] 2 SCR 335 at 352.

111   [1983] AC 598.

112   [1984] 2 SCR 335.

113   [1999] 2 FC 48, <www.canlii.org/ca ... 23481>

114   [1990] 1 SCR 1075; cf. Petz and L Alcuitas-Imperial (1993), pp. 39–43; cf. *R v Marshall* [1999] 3 SCR 456 and see Edwards (2001), p. 107.

115   Mackelm (1997), p. 112.

116   Specifically, the possible negative effects of s 34 of the *Fisheries Act 1970* (Can) on the land rights of Indian Bands had to be justified in terms of the recognition given to indigenous land rights under s 35(1) of the *Constitution Act 1982* (Can).

117   [2000] FCJ No 1568, <www.canlii.org/ca ...29200>

those created by the legislation, to justify the imposition of a fiduciary duty on the Crown.

A similar view was expressed in a case where a claim was made that a fiduciary duty was owed by the Minister of National Revenue towards taxpayers generally. Denying this, Justice Dawson in *Harris v R*[118] said that the Minister had no discretion in the application of the *Income Tax Act*. More significantly, for present purposes, he quoted the view of Lord Diplock in *Swain v the Law Society*,[119] that public law duties were only susceptible of enforcement by public law remedies. There was, he said, no 'reasonable expectation' that the Crown could have been acting for the sole benefit of the party affected by a public Act.

Australian Authority

Australian acceptance of the Crown having fiduciary duties to Aborigines and Torres Strait Islanders is associated with the decision of the High Court in *Mabo v Queensland (No 2)*.[120] In the alternative to a native title claim, the Crown was alleged to be liable to the Meriam people as a fiduciary, or trustee, under a duty 'to recognise and protect their rights and interests in the Murray Islands'. The native title claim succeeded. In consequence, all judgments, except that of Toohey J, dealt only cursorily with the fiduciary point. Brennan J stayed within the case *Guerin* decided,[121] when he noted the possibility that a property-based fiduciary responsibility could arise

> if native title were surrendered to the Crown in the expectation of the grant of tenure to native title holders.[122]

Dawson J, in his dissenting judgement, likewise thought that a 'fiduciary or trust obligation' could not exist without an 'aboriginal interest existing in or over the land'.[123]

Justice Toohey, after some consideration, reached the conclusion that the Crown came under a 'fiduciary obligation ... in the nature of the obligation of a constructive trustee'.[124] He put this on two bases. First, assuming that the Crown

---

118   [2002] 1 CTC 243, at [187]–[188] (Fed Ct).

119   [1983] AC 598, at 609.

120   (1992) 175 CLR 1; issue discussed previously in *Northern Land Council v The Commonwealth* (1987) 75 ALR 616 (HC) and *Director of Aboriginal and Islander Advancement v Peinkinna* (1978) 17 ALR 17 (PC), discussed in text at n. 66.

121   [1984] 2 SCR 335.

122   *Mabo v Queensland (No 2)* (1992) 175 CLR 1, at 60; see also at 113, where Deane and Gaudron JJ briefly referred to 'equitable remedies' for interference with native title.

123   *Mabo v Queensland (No 2)* (1992) 175 CLR 1, at 166–7.

124   See n. 123, at 204.

had not previously extinguished traditional title, a fiduciary relationship could, he said, arise out of 'the *power* of the Crown to extinguish traditional title by alienating the land or otherwise'. Secondly, and regardless of whether traditional title continued, the Crown's 'course of dealings' with the Islands and the Islanders created the fiduciary obligation.[125] The second ground takes the matter a step further than Canadian orthodoxy. Existence of a power-dependency relationship was held to be sufficient to generate a fiduciary relationship between the government and an aboriginal people without any property qualification, or 'independent legal interest'. This is a large step. There are, perhaps, many groups within society who live in comparable conditions to the Meriam people.

The Crown's power to extinguish native title was also argued to have a fiduciary consequence in *Wik Peoples v State of Queensland*.[126] Chief Justice Brennan was the only judge to consider the point. Brennan found himself 'unable to accept' that

> a fiduciary duty can be owed by the Crown to the holders of native title in the exercise of a statutory power to alienate land whereby their native title in or over that land is liable to be extinguished without their consent and contrary to their interests.

The fact that the government had the power to alienate Aboriginal land and extinguish native title of itself was insufficient. *Wik* involved the construction of an enabling provision conferred on the Governor in Council by s 6 of the *Land Act 1910* (Qld). No indication was given of the interests to be considered or proper discretionary principles to be applied when the power was exercised. The Aboriginal people and their welfare were not referred to. Chief Justice Brennan observed that the words of the provision were 'inconsistent' with a requirement that the power should be exercised 'as agent for or on behalf of the indigenous inhabitants of the land to be alienated'. This, again, is within the ratio in *Guerin's* case.[127] At the High Court level, the dictum isolates the view of Justice Toohey in the *Mabo* case.[128]

A fiduciary obligation binding public officers and the Commonwealth government was argued in *Cubillo v Commonwealth*.[129] Two individuals asserted that they had been forcibly removed from Aboriginal families and wrongly detained in state institutions for several years. In addition to claims of false imprisonment and negligence, the Commonwealth was alleged to be liable for breach of fiduciary duties that were owed to each of them. Alternatively, the Commonwealth was alleged to be vicariously liable for breaches of fiduciary duty by two 'Directors of Aboriginal Affairs', empowered respectively pursuant to

---

125   *Mabo v Queensland (No 2)* (1992) 175 CLR 1 at 203 [emphasis in original].
126   (1996) 141 ALR 129, at 160–61.
127   *Guerin v R* [1984] 2 SCR 335.
128   (1992) 175 CLR 1.
129   (2001) 112 FCR 455 (FC), aff. (2000) 103 FCR 1, O'Loughlin J.

the *Aboriginals Ordinance 1918* (NT) and the *Welfare Ordinance 1956* (NT).[130] Fiduciary obligations were said to be owed directly by the Commonwealth on account of its powers over, and assumption of responsibility for, Aboriginal people in the Northern Territory. Claimants argued that they were vulnerable to the exercise of the Commonwealth's powers and to its relevant discretion. O'Loughlin J set out this part of the claim made as follows:[131]

> The fiduciary relationship between the Commonwealth and each Applicant was said to arise because of the role and functions of the Commonwealth's servant and agents in the removal and detention of the applicants and because of the Commonwealth's powers over, and its assumption of responsibility for, the Aboriginal people in the Northern Territory. It was also said to arise because of the powers, obligations and discretions of the Directors and the vulnerability of each applicant to the exercise of those powers and discretions. Next, it was said to arise because of the powers, obligations and discretions of the Administrator to administer the Northern Territory on behalf of the Commonwealth and in accordance with the instructions of the relevant Commonwealth Minister.

Indicia of power and dependency based the claim for a fiduciary relationship in *Cubillo*. No 'independent legal interest' was argued for in relation to the existence of a fiduciary relationship. Presumably, there was none relevant.[132]

It was in the breach, rather than the existence, of a fiduciary relationship, that a property interest was claimed in *Cubillo*. The Commonwealth was claimed to have breached its fiduciary duty *inter alia* in relation to the 'recognition' of the claimant's 'rights and interests ... arising under custom and tradition in relation to her mother's and her family's traditional lands'. O'Loughlin J set out the pleading whereby breach of fiduciary duty was alleged by reason of the Commonwealth acting where a conflict existed between its interests and those of the claimants.[133] Interests of the Commonwealth were said to include:

a. destroying the associations and connections of the Applicant with her Aboriginal mother, family and culture;
b. assimilating 'half caste' children, including the Applicant, into non-Aboriginal society;

---

130   Appointed and empowered pursuant to the *Aboriginals Ordinance 1918* (NT), ss. 5, 6, 13, 16 and *Welfare Ordinance 1956* (NT), s 17.

131   *Cubillo v Commonwealth* (2000) 103 FCR 1, at [1276].

132   Jones (2002), p. 73: 'Brennan CJ's formulation is easily applied in the Stolen Generation context. The existence of circumstances of power and vulnerability in *Wik* was the power of the Crown to extinguish native title. In the Stolen Generation context, it was the power of the state to remove children from their families'.

133   *Cubillo v Commonwealth* (2000) 103 FCR 1, at [1301].

c. providing domestic and manual labour for the European community of the Northern Territory;

d. breeding out 'half caste' Aboriginal people and protecting the primacy of the Anglo-Saxon community;

e. not being exposed to the risk of legal action by the Applicant against it or its employees, or any liability to pay damages or equitable compensation to the Applicant.

These conflicted with the applicant's interests in:

a. maintaining her associations and connections with her mother, family and culture;

b. achieving and maintaining recognition of her rights, interests and obligations arising under custom and tradition in relation to her mother's and her family's traditional lands;

c. not suffering psychological harm; and

d. being in a position to pursue her legal and equitable remedies against the [Commonwealth] and the Director of Native Affairs.

O'Loughlin J upheld the claimants' alleged fiduciary relationship, on the analogy of guardian and ward.[134] However, in succeeding paragraphs he found that the *scope* of that duty was limited to protecting the claimants' economic interests and did not extend to avoidance of the conflicts above itemized.[135] The effect of this was confirmed by the Full Court.[136] The decision thus conformed with the authority of *Breen v Williams*,[137] which provided that fiduciary duties were proscriptive and mainly protected economic interests. Reference was also made to the High Court decision in *Pilmer v Duke Group Ltd*.[138] Fiduciary law in *Pilmer* was described as complementary to, rather than a competitor with, contract law and tort. Accounting for profits, making good losses and avoiding conflicts of interest described a fiduciary's duties. The Commonwealth in *Cubillo* was not liable for breach of any of the direct fiduciary duties which were alleged.

Vicarious liability for the Commonwealth was also negatived factually. Neither of the Directors charged with the implementation of the regulations was shown to have failed to comply with the terms of the Ordinances. Relevantly, s 6 of the

---

134 The judgement is not clear: see *Cubillo v Commonwealth* (2000) 103 FCR 1, at [1287]–[1288] and [1300], following *Bennett v Minister for Community Welfare* (1992) 176 CLR 408; O'Loughlin J's decision on the point was examined by the FC *Cubillo v Commonwealth* (2001) 112 FCR 455, at [461].

135 *Cubillo v Commonwealth* (2000) 103 FCR 1, at [1291]–[1299], following *Breen v Williams* (1996) 186 CLR 71.

136 *Cubillo v Commonwealth* (2001) 112 FCR 455, at [463]–[469].

137 *Breen v Williams* (1996) 186 CLR 71, at 92, Dawson and Toohey JJ.

138 (2001) 180 ALR 249, at [77].

Aboriginals Ordinance provided that the Director of Native affairs was entitled at any time:

> to undertake the care, custody or control of any aboriginal or half caste, if, in
> his opinion it is necessary or desirable in the interests of the aboriginal or half
> caste for him to do so, and for that purpose may enter any premises where the
> aboriginal or half caste is or is supposed to be, and may take him into his custody.

The purpose of the Directors forcibly removing the claimants was not shown to be other than what the Directors were empowered to do. Ordinances, like statutes, are beyond challenge in equity proceedings.[139]

Conclusion

Private fiduciary sanctions for abuse of power by public officials may not yet be appropriate. Public law and private law are still too separate for this doctrinal coalescence to occur. In Anglo-Australian and comparable jurisdictions, there is too much inconsistency between defining aspects of fiduciary law and the community's 'interests of good administration'.

Many commentators have disapproved of the United Kingdom idea that there is a fiduciary relationship between local authorities and ratepayers. Final courts of appeal in Canada and Australia have been cautious in dealing with the fiduciary claims of indigenous peoples against public officials (and governments). The decision of the Canadian Supreme Court in *Guerin v The Queen*[140] remains the main authority on this breach of fiduciary relationships. Both of the judgements in *Guerin* that dealt with fiduciaries law placed a significant brake on the idea. Fiduciary liability for the Crown in Canada was limited to its vicarious or participatory dealings with land that the Indian Band surrendered. Authority still requires that a fiduciaries claim against the government be based in land, or an equivalent interest, 'not the creation of the legislative or executive branches of government'. The Australian High Court has yet to determine the conditions on which fiduciary relief for indigenous peoples will be forthcoming. Litigation in *Mabo v Queensland (No 2)*[141] and *Wik Peoples v State of Queensland*[142] was inconclusive.

Analytically, as has been seen, fiduciary obligations in a public law context are *exclusivist* and narrow. Fiduciaries must act only in the interests of the persons to whom fiduciary duties are owed.[143] Conflicting personal or third party interests

---

139   See text at n. 15.
140   [1984] 2 SCR 335.
141   (1992) 175 CLR 1.
142   (1996) 141 ALR 129.
143   See *Warman International Ltd v Dwyer* (1995) 128 ALR 201, at 208–9, curiam.

must be avoided.[144] Sharing is irrelevant. Efficiency is irrelevant. Loyalty is the animating idea.[145] Fiduciaries must act for those who trust them *to the exclusion of all others*. Little room is left for communitarian imperatives[146] and 'the interests of good administration'.[147]

Public law is *inclusivist*, by contrast. Service of the 'public interest' subordinates the private allegiances of public functionaries. Public power, whether exercised over cattle graziers,[148] indigenous peoples[149] or parliamentary interest groups,[150] originates differently from private power and responds to different policy imperatives.[151] The morality is different. As Lord Woolf says. There are 'higher standards to which public, not private, bodies should be required to conform'.[152] The modern code has been expressed by Professor Oliver.[153] There is no echo in equity jurisprudence. Public bodies, she says, [and public officers] are liable to 'special duties of legality, fairness and rationality'. These are duties of 'considerate decision making ... which would not rest on that body if it were private or exercising purely private functions'.

References

J. Alder (2001) 'Incommensurable values and judicial review: the case of local government', *PL* 717.

T. R. S. Allan (1993) *Law, Liberty, and Justice: The Legal Foundations of British Constitutionalism*, Oxford University Press.

M. Aronson (1998) 'Criteria for restricting collateral challenge', 9 *PL* 237.

R. Austin, H. Ford and I. Ramsay (2005) *Company Directors: Principles of Law and Corporate Governance*.

S. Bailey (2004) 'The structure, power and accountability of local government', in D. Feldman (ed.), *English Public Law*, OUP.

A. W. B. Bradley (1991) 'Bromley Council v Greater London Council: counsel's opinion', Win *PL* 499.

D. Brown, D. Farrier, D. Neal and D. Weisbrot (2001) *Criminal Laws*, 3rd edn, Federation Press.

E. Campbell (1998) 'Collateral challenge of the validity of governmental action', 24 *Mon L R* 272.

---

144 See *Chan v Zacharia* (1984) 154 CLR 178, at 205, Deane J.
145 *Hodgkinson v Simms* (1995) 117 DLR (4th) 161, 174, at 209, La Forest J.
146 See D. Oliver, n. 151, at 5–7.
147 Discussed by Lord Diplock in *O'Reilly v Mackman* [1983] 2 AC 237, at 282.
148 See *Northern Territory of Australia v Mengel* (1995) 185 CLR 307, discussed at n.
149 See *Mabo v State of Queensland (No 2)* (1992) 175 CLR 1, discussed at n.
150 See *R v Boston* (1923) 33 CLR 386.
151 See D. Oliver, n. 46, 4–7.
152 See H. Woolf (1995), 61.
153 D. Oliver, n. 46, 81.

J. Coffee (1998) 'Modern Mail Fraud: the restoration of the public/private distinction', 35 *Am Crim L Rev* 427, pp 428–31.

R. Cooter and B. Freedman (1991) 'Fiduciary relationship: its economic character', 66 *NYULR* 1045.

E. Criddle (2006) 'Fiduciary foundations of administrative law', 54 *UCLA L Rev* 117.

A. V. Dicey (1920) *Introduction to the Study of the Law of the Constitution*, Macmillan.

J. Dignan (1983) 'Policy making, local authorities and the courts: the 'GLC fares' case', 99 LQR 605.

F. Easterbrook and D. Fischel (1993) 'Contract and fiduciary duty', 36 *Journal of Law and Economics* 425.

B. Edwards (2001) 'Towards a bilateral fiduciary relationship: recognising mutual vulnerability in R v Marshall', 59 *U Toronto Fac L Rev* 107.

P. Finn (1977) 'Public officers: some personal liabilities', 51 *ALJ* 313.

P. Finn (1989) 'Contract and the fiduciary principle', 12 *UNSWLJ* 76.

P. Finn (1989) 'The fiduciary principle', in T. Youdan (ed.), *Equity, Fiduciaries and Trust*, Carswell.

P. Finn (1992) 'Integrity in government', 3 *PLR* 243.

P. Finn (1994) 'The abuse of public power in Australia: making our governors our servants', 5 *PLR* 43.

P. Finn (1995) 'A sovereign people, a public trust', in P. Finn (ed.), *Essays on Law and Government*, Law Book Company.

P. Finn (1995) 'The forgotten trust: the people and the state', in M. Cope (ed.), *Equity Issues and Trends*, Federation Press.

R. Flannigan (1989) 'The fiduciary obligation', 9 *Ox Jo of Leg Studies* 285.

H. Ford, W. A. Lee, M. Bryan, I. G. Fullerton and J. Glover (2010) *The Law of Trusts*, 4th edn, Lawbook Co.

E. Fox-Decent (2005) 'The fiduciary nature of state legal authority', 31 *Queen's LJ* 259.

J. Glover (1995) 'Wittgenstein and the existence of fiduciary relationships', 18 *UNSWLJ* 443.

R. Gordon (1996) *Judicial Review: Law and Procedure*, Polity.

J. Griffith (1982) 'Fare fair or fiduciary foul?', *CLJ* 216.

A. Jones (2002) A 'The state and the stolen generation: recognising a fiduciary duty', 28 *Mon L R* 59.

T. Jones (1993) 'The reform of judicial review in Queensland', *Civil Justice Quarterly* 256.

J. Jowett (1994) 'Is equality a constitutional principle?', 47 *CLP* 3.

J. Kaye and Colonel Malleson (1907) *History of the Indian Mutiny of 1857–8*, v. 5, Longmans.

Law Reform Commission (1992) *Fiduciary Duties and Regulatory Bodies*, Law Commission Consultation Paper No. 124.

Martin Loughlin (1996) *Legality and Locality*, Clarendon Press.

T. B. Macaulay (1913 edn) *Essays on Warren Hastings*, London.

P. Mackelm (1997) 'Aboriginal rights and state obligations', 36 *Alta L R* 97.

A. Maclean (1996) 'The theoretical basis of the trustee's duty of loyalty', 7 *Alta L Rev* 218.

B. Macpherson (1998) 'Fiduciaries: who are they?', 72 *ALR* 288.

J. Mahoney (1996) 'The criminal liability of public officers for the exercise of public power', 3 *TJR* 17.

K. Mason (1995) 'The rule of law', in P. Finn (ed.), *Essays on Law and Government, Vol 1 Principles and Values*, The Law Book Company.

W. Massey (1860) *A History of England during the Reign of George the Third*, v. 3, John J. Parker and Sons.

D. Oliver (1999) *Common Values and the Public-Private Divide*, Butterworth.

A. Petz and L. Alcuitas-Imperial (1993) 'Fiduciary obligations as a source of remedies against public officials: the aboriginal context and beyond', *Fiduciary Duties/Conflicts of Interest: the 1993 Isaac Pitblado Lectures*.

C. Phegan (1980) 'Damages for improper exercise of statutory powers', 9 *Syd L Rev* 93.

I. Ratcliff (1996) 'Controlling local authorities: the utility of English mechanisms of accountability for NSW', 2 *LGLJ* 43.

A. Scott, (1936) 'The trustee's duty of loyalty', 49 *Harv LR* 521.

A. Scott (1949) 'The fiduciary principle', 37 *Calif LR* 539.

A. Scott and W. Fratcher (1989) *The Law of Trusts*, 4th edn, American Law Institute.

L. Sealy (1962) 'Fiduciary relationships', *Camb L J* 69.

L. Sealy (1963) 'Some principles of fiduciary obligation', *CLJ* 119, 125–32.

L. F. Sturge (1950) Stephen, Digest. A Digest of the Criminal Law by Sir James *Fitzjames* Stephen, 9th edn.

W. Wade and C. Forsyth (2009) *Adminstrative Law*, 10th edn, OUP.

H. Woolf (1995) 'Droit public – English style', Spr *PL* 57.

H. Woolf, J. Jowett and A. Sueur (2007) *De Smith's Judicial Review*, 6th edn, Lord Woolf Publishers.

Cases

Allcard v Skinner (1887) 36 Ch D 145

Arklow Investments Ltd v Maclean [2000] 2 NZLR 1

Associated Provincial Picture Houses Ltd v Wednesbury Corporation [1948] 1 KB 223

Attorney General (Hong Kong) v Reid [1994] 1 AC 324

Attorney-General v Blake [1998] Ch 432. *Barton v Masterton District Council* [1992] 1 NZLR 232

Bennett v Minister for Community Welfare (1992) 176 CLR 408

Blueberry River Band v Canada (Dept of Indian Affairs) [1995] 4 SCR 344

Legislation

Aboriginals Ordinance 1918 (NT)
Aborigines Act 1971–1975 (Qld)
Administrative Decisions (Judicial Review) Act 1977 (Cth)
Administrative Decisions (Judicial Review) Act 1977 (Cth)
Aurukun Associates Agreement Act 1975 (Qld)
Constitution Act 1982 (Can)
Criminal Code 1913 (WA)
Criminal Code Act 1924 (Tas)
Criminal Code Act 1995 (Cth)
Fisheries Act 1970 (Can)
Honours (Prevention of Abuses) Act 1925
Prevention of Corruption Act 1906
Prevention of Corruption Acts 1989 to 1916
Public Service Act 1999 and regulations
Racial Discrimination Act 1975 (Cth)
Solicitors' Act 1974 (UK)
Supreme Court Act 1981
Welfare Ordinance 1956 (NT)

Chapter 6
Atmospheric Trust Litigation Across the World[1]

Mary Christina Wood[2]

Leading climate scientists warn that Earth is in 'imminent peril', on the verge of runaway climate heating that will impose catastrophic conditions on generations to come.[3] In their words, continued carbon pollution will cause a 'transformed planet'[4] – an Earth obliterated of some major fixtures, including the polar ice sheets, Greenland, the coral reefs, and the Amazon forest. The trajectory of civilization over the past century threatens to trigger the planet's sixth mass extinction – the kind that has not occurred on Earth for 65 million years.[5] Should business as usual continue even for a few more years, future humanity for untold generations will be pummelled by floods, hurricanes, heat waves, fires, disease, crop losses, food shortages, and droughts as part of a hellish struggle to survive in deadly greenhouse conditions.[6] In a world of runaway climate heating, these unrelenting disasters would force massive human migrations and cause staggering numbers of deaths – culminating in, as more and more analysts predict, humanity's own 'self-destruction'.[7] As author Fred Pearce states: 'Humanity faces a genuinely new situation … a crisis for the entire life-support system of our civilization and our species'.[8]

---

1   This article has adapted portions of the following works: Wood (2011 forthcoming); Wood (2009); Wood (2010).

2   Research assistance was provided by John Mellgren, Bowerman Fellow in the Environment and Natural Resources Law Program, University of Oregon School of Law. The author wishes to thank Tim Hicks, David Takacs, Joe Fox, Tim Duane, and Tom Athanasiou for helpful comments on an earlier draft.

3   Hansen et al. (2007a); Connor (2007).

4   Hansen (2006b).

5   Boitnott (2008).

6   See Ban Ki-moon (2009): 'The evidence is all around us. And unless we act, we will see catastrophic consequences including rising sea-levels, droughts and famine, and the loss of up to a third of the world's plant and animal species'; The University of New South Wales Climate Change Research Centre (2009); United States Global Change Research Program, Global Climate Change Impacts in the United States (2009); Lean (2007).

7   See Romm (2008b).

8   Pearce (2007), p. 239; see also Gore (2007): 'This is a moral issue, one that affects the survival of human civilization … . Put simply, it is wrong to destroy the habitability of our planet and ruin the prospects of every generation that follows ours.'

In order to stem global warming, the law must recognize and calibrate to the physical, chemical, and biological requirements for achieving climate equilibrium. Such requirements are set by nature, not politicians. Stated another way, averting climate disaster is a matter of carbon math, not carbon politics. Scientists warn that the world has only a short time to begin reversing global emissions of carbon before the planet passes a 'tipping point'[9] – a point at which dangerous feedback loops will unravel the planet's climate system despite any subsequent carbon reductions achieved by humanity.[10] The Ninth Circuit Court of Appeals has recognized the danger of the tipping point, stating in one climate case: 'Several studies also show that climate change may be non-linear, meaning that there are positive feedback mechanisms that may push global warming past a dangerous threshold (the 'tipping point')'.[11] In another climate case, the federal district court of Vermont found reliable the 'tipping point' theory of non-linear climate change advanced by NASA scientist James Hansen under the *Daubert* standard, stating: '[The] 'tipping point' theory posits that at a certain point the changes associated with global warming will become dramatically more rapid and out of control. ... [D]rastic consequences, including rapid sea level rise, extinctions, and other regional effects, would be inevitable with a two to three degrees Celsius warming expected if no limits are imposed and emissions continue at their current rate. *Such changes could happen quickly once a tipping point is passed.*'[12]

While just recently scientists believed the 'tipping point' would be triggered at 450 parts per million (ppm) of atmospheric carbon dioxide, the dangerous threshold is now thought to be at, or even well under, 350 ppm.[13] Present levels

---

9   Pearce (2007), pp. xxiv, 63, 75, 77–8; Ban Ki-moon (2009). 'T]his report shows that climate change is accelerating at a much faster pace than was previously thought by scientists. New scientific evidence suggests important tipping points, leading to irreversible changes in major Earth systems and ecosystems, may already have been reached or even overtaken.'

10   See Hansen (2007b) 'In the past few years it has become clear that the Earth is close to dangerous climate change, to tipping points of the system with the potential for irreversible deleterious effects.'; Hansen (2006), part 3: '[B]ecause of the global warming already bound to take place as a result of the continuing long-term effects of greenhouse gases and the energy systems now in use, ... it will soon be impossible to avoid climate change with far-ranging undesirable consequences. We have reached a critical tipping point.'; Hansen et al. (2007a), note 3, pp. 1925, 1949 (discussing positive feedback loops); Hansen et al. (2007b), p. 2303: '[W]e must be close to such a point, but we may not have passed it yet'.

11   See *Ctr. for Biological Diversity v. Nat'l Highway Traffic Safety Admin.*, 508 F3d 508, 523 (9th Cir 2008).

12   *Green Mountain Chrysler v Crombie*, 508 F Supp 2d 295, 313–17 (D Vermont 2007) (emphasis added) (stating also, 'Hansen's testimony is based on sufficient facts and data and reliable methods, applied reliably to the facts.'). See also *Daubert v Merrill Dow Pharmaceuticals, Inc.*, 509 US 579 (1993).

13   Hansen et al. (2008), p. 2; see Bill McKibben (2007); (2009) 'UN Scientists back "350" Target for CO2 Reduction'.

are at 390 ppm and climbing by 2 ppm a year.[14] Leading scientists warn that, if humanity follows business as usual for even another few years, it will 'lock in' future catastrophic global heating.[15]

These circumstances have hurled the Earth into a state of planetary emergency.[16] In 2007, the head of the United Nation's climate panel told the world: 'What we do in the next two to three years will determine our future. This is the defining moment'.[17] Immediate and decisive action to slash carbon pollution is imperative. Yet, despite this planetary crisis, there has been little action at either the international or national level. This may well be due to the fossil fuel industry's influence over political leaders, economies, and governmental systems worldwide.[18] While such political dynamics are too complex to explore here, exclusive reliance on the political branches for climate response now seems ill-advised.

---

14 Adam (2008); Hansen et al. (2011). While the 350 target has been exceeded, climate scientists still offer hope of atmospheric stability if the 'overshoot' is brief. Hansen et al. (2008): 'If the present overshoot of this target CO2 is not brief, there is a possibility of seeding irreversible catastrophic effects'.

15 See Hansen (2007b), '[I]gnoring the climate problem at this time, for even another decade, would serve to lock in future catastrophic climatic change and impacts that will unfold during the remainder of this century and beyond ... .'; Hansen (2007a): 'If we do follow that [Business as Usual] path, even for another ten years, it guarantees that we will have dramatic climate changes that produce what I would call a different planet...'; Hansen (2006a): 'How long have we got? We have to stabilize emissions of carbon dioxide within a decade, or temperatures will warm by more than one degree. That will be warmer than it has been for half a million years, and many things could become unstoppable'. A disturbing United Nations IPCC report indicates that the planet has already reached the danger point of atmospheric carbon dioxide equivalent concentrations, indicating that a decade is far too long to achieve significant greenhouse gas reduction. See Lamb (2007), discussing report and quoting climate scientist Tim Flannery, '[A]lso we have really seen an unexpected acceleration in the rate of accumulation of CO2 itself, and that's been beyond the limits of projection ... beyond the worst-case scenario. We are already at great risk of dangerous climate change – that's what the new figures say ... [I]t's not next year, or next decade; it's now'.

16 Spratt and Sutton (2008), pp. 222–40.

17 Rosenthal (2007), quoting Rajendra Pachauri. Of course, it bears noting that, while major scientific associations and the United Nations express urgency, some scientific views may land on the opposite side of the spectrum. In that vein, it is important to recognize that many scientific hypotheses relating to climate change, particularly those involving the tipping point, can not or may not be validated until it is too late; it is thus necessary to apply the precautionary principle in order to leave options open for humanity. Notably, in finding the testimony of Dr James Hansen reliable as to the tipping point, the federal District Court of Vermont stated, 'Hansen's predictions need not be certainties to be admissible under Rule 702, nor need his estimates of the timing and amount of sea level rise be exact to be admissible. ... There is widespread acceptance of the basic premises that underlie Hansen's testimony'. *Green Mountain Chrysler v Crombie*, 508 F. Supp 2d 295, 317–19 (D Vermont 2007).

18 See generally, Gelbspan (2004) p. 61; Goodell (2006).

This chapter explains a legal strategy called Atmospheric Trust Litigation (ATL)[19] that calls upon the judicial branches of governments worldwide to force carbon reduction on the basis of their fiduciary responsibility to protect the public trust. ATL seeks to accomplish, though decentralized domestic litigation in countries across the globe, what has thus far eluded the centralized, international diplomatic treaty-making process. The strategy draws upon fundamental principles of sovereign trust obligation to provide a framework that holds governments accountable for forcing carbon reduction within their own countries. The ATL approach is consistent with, and gives meaning to, the principles declared in the United Nations Framework Convention on Climate Change (UNFCCC), agreed to in 1992 by 192 nations, representing 'near universal' international membership.[20] Notably, in the United States, the UNFCCC still exists as a ratified treaty – which gives it Constitutional rank as the 'supreme law of the land'.[21] The ATL approach neither hinders nor forecloses any possibility for future international frameworks to address climate crisis, but rather, if successful, would infuse a strong fiduciary obligation into what has so far been a wholly discretionary diplomatic process.

This chapter begins first by describing government inaction worldwide and explaining the need for a swift legal strategy to hold sovereigns accountable for their pollution. Second, it explains the public trust doctrine, which provides the basis for ATL. It then describes the elements of an ATL claim and a remedy that would provide effective redress for government's recalcitrance. The chapter concludes with remarks on the judicial role in a time of planetary crisis. While ATL bears the risk of any untested strategy, it is perhaps the only macro approach that can empower courts to force emissions reductions within the limited time-frame that remains before the planet crosses critical climate thresholds.

Governmental Inaction

In December 2009, nations of the world gathered in Copenhagen, Denmark, for the United Nations Conference on Climate Change. Although the Conference resulted in a resolution that has been joined by many industrialized nations (including most of the major greenhouse gas emitters such as the United States, China, India, Brazil, Australia, and members of the European Union), the Copenhagen Accord is widely regarded as a failure.[22] It is not legally binding, and many pledges are

---

19   For additional materials on ATL, see *supra* n. 1.

20   United Nations Framework Convention on Climate Change (1992) .

21   United States Constitution, Art. VI.

22   United Nations Framework Convention on Climate Change, Copenhagen Accord, FCCC/CP/2009/L.7 (18 Dec. 2009), available at http://unfccc.int/resource/docs/2009/cop15/eng/l07.pdf; for commentary, see Vidal et al. (2009).

contingent on action taken by other nations.[23] A UN analysis showed that, even if the various national pledges were fulfilled, the total combined carbon reduction would still bring about a 3 °C temperature rise, capable of triggering catastrophic climate change.[24] The United States, for example, remains a recalcitrant global polluter, having offered only a meagre reduction proposal at the Copenhagen Conference. Though the United States emits nearly 16 per cent of the world's global greenhouse gases,[25] President Obama expressed a willingness to reduce United States greenhouse gas emissions by only 17 per cent (below 2005 levels) by 2020.[26] His choice of a 2005 baseline made the pledge appear far larger than it was –17 per cent equates to only a two per cent decline over 1990 levels.[27] Moreover, President Obama made the United States position contingent on the passage of energy legislation before the United States Congress – which, as of summer 2010, was a dead letter.[28]

Surely a robust, enforceable, international agreement would be the preferred vehicle for a solution to a global crisis such as planetary heating. It is understandable that so many climate advocacy groups poured resources and energy into influencing the world's governments to arrive at one. But the failure of the Copenhagen Conference calls into serious question the continued wisdom of relying on international negotiations as a mechanism to force pollution reduction. Even the Kyoto Protocol, which was negotiated in 1997, which included a broad range of signatories, and which contained numerical emissions reduction targets,

---

23   United Nations Framework Convention on Climate Change, Copenhagen Accord, FCCC/CP/2009/L.7 (18 Dec. 2009), available at http://unfccc.int/resource/docs/2009/cop15/eng/l07.pdf.

24   See Goldenberg et al. (2009). In-depth subsequent analysis of the Copenhagen pledges shows that they are something of a sham in that their total aggregate 'reduction' would amount to emissions above 1990 levels. As Tom Athanasiou, an analyst at EcoEquity, summarizes analysis of the Stockholm Environmental Institute in his blog post of 11 August 2010: 'In fact, even when conservative assumptions are used, *the Copenhagen pledges contain so many loopholes that, taken together, they sum to 21% of 1990 emissions, a number that entirely negates the pledges themselves!* So that the official, well-publicized global 2020 emissions reductions target of 12–18% actually means that emissions levels large enough to reach 3–9% above 1990 would be allowed. Which is ... actually more than the business-as-usual projection!). See *You Want Loopholes With That?*, available at http://www.ecoequity.org/2010/08/ (emphasis in original).

25   World Resources Institute, Climate Analysis Indicator Tool (2005) (includes land-use change and forestry).

26   Friedman (2010).

27   Duane (2010).

28   Letter from the United States Special Envoy for Climate Change to the Executive Secretary of the United Nations Framework Convention on Climate Change, 28 Jan. 2010, available at http://www.usclimatenetwork.org/resource-database/us-inscription-to-the-unfccc-on-the-copenhagen-accord.

largely failed.[29] The United States, one of the largest global polluters, never ratified its commitment of the Protocol. Moreover, few of the signatory countries to the Protocol ended up meeting their reduction commitments.[30] The fact remains that, due to the autonomy of nations and the lack of any world 'super-jurisdiction', there is no way to directly force sovereigns to reduce carbon emissions. Because the bottom line for international law is, unfortunately, voluntary compliance, exclusive reliance on international treaty negotiations to achieve global carbon reduction is perilous.

On the domestic level within various nations, one could hope for national legislation. For example, there have been enormous efforts to pass such legislation within the United States, a country that produces a lion's share of the globe's pollution and that has the most extensive set of environmental laws in the world. But the reality is that the US Congress remains beholden to the fossil fuel industry, which spent a whopping $514 million over eighteen months lobbying against a climate bill, until prospects for legislation came to a 'crashing demise' in summer, 2010.[31] Leading climate advocates admit that 'hope for any sweeping or comprehensive measure is probably gone'.[32] Even if a bill emerges, it is not likely to be adequate. The bills proposed thus far have fallen far short of providing sufficient reduction.[33]

The judicial branch should hold government to its legal responsibilities. So far, however, though many lawsuits have been filed, none have forced the carbon reduction needed to curb runaway atmospheric heating. In the US, for example, most lawsuits are structured around statutory mandates; plaintiffs have sued under the Clean Air Act, NEPA, the Endangered Species Act, and other statutes.[34] So far, these claims have not delivered any meaningful aggregate relief. In general, this may be because environmental statutory law (at least in the United States and perhaps in many other countries as well) has degenerated into an embrace of administrative political discretion.[35] The vast majority of agencies use their discretion to allow projects that cause significant environmental damage.[36] The statutes themselves are a major engine of environmental destruction: two-thirds of the greenhouse gas pollution in the United States is emitted pursuant to

---

29   Kyoto Protocol to the United Nations Framework Convention on Climate Change, 10 Dec., 1997, 37 I.L.M. 32.

30   Zarembo (2007).

31   (2010) 'The Changing Climate For Environmental Legislation'.

32   (2010) 'The Changing Climate For Environmental Legislation'.

33   See Coplan (2010) (surveying cap-and-trade proposals and concluding that they would allow emissions 'far in excess' of scientific recommendations).

34   For an overview of litigation, see generally Hildreth et al.; Hunter et al. (2009).

35   See Wood (2010a); Wood (2009b); Kennedy Jr (2005). (Federal agencies in the Bush II administration 'have given quick permit approvals and doled out waivers that exempt campaign contributors and polluters from rules or regulations'.).

36   See generally Collins (2010); see also Speth (2008).

government-issued permits.[37] Even where a statutory lawsuit is successful, it often fails to deliver meaningful relief. Remedies usually take the form of procedural remands to the agency, returning the matter to the same highly political process that produced the case in the first place. Moreover, statutes, which are typically narrow in scope, fracture government's overall climate responsibility into isolated, disjointed parts. Statute-based strategies – while nevertheless important in many respects – tend to diffuse the climate litigation effort and drain it of practical force in addressing the magnitude of climate crisis. Nuisance lawsuits are also micro in nature, targeted against only specific polluting parties. They arise in absence of adequate regulation and do not get at the underlying problem of government recalcitrance in addressing a mounting ecological calamity.[38]

Climate crisis demands broad, system-changing solutions and doctrines. The judiciary is potentially a crucial player in forcing carbon reduction because it tends to be a less politicized branch of government (in most, though certainly not all, countries) with power to order swift and decisive relief. But, for litigation to have any meaningful effect before the planet slips over irrevocable climate thresholds, litigators must present courts with macro-level claims that address government's full obligation to protect the atmosphere. Moreover, such claims must find their premise in government obligation, not discretion, which is readily hijacked by politically powerful interests. Finally, such claims must be linked to a premise that has global reach and transcends different legal systems and cultures.

The legal foothold for Atmospheric Trust Litigation (ATL) is the ancient public trust doctrine, which imposes a strict fiduciary obligation on government to protect natural resources in trust for the citizens.[39] The ATL strategy presents a macro-level approach to climate crisis by focusing on the atmosphere as a single asset in its entirety. It characterizes all nations on Earth as co-tenant sovereign trustees of that asset, bound together in a property-based framework of corollary and mutual

---

37   Kosloff and Trexler (2007).

38   In the United States, states and private parties have brought climate nuisance actions against major carbon polluters. In *Connecticut v American Electric Power*, 406 F. Supp 2d 265 (SD NY 2005), rev'd, 582 F3d 309 (2nd Cir 2009), cert granted, 131 S Ct 813 (2010), several states sought an injunction against major coal burning utilities. At the time of this writing, the suit was pending before the US Supreme Court. For additional nuisance actions, see *California v General Motors Corp.*, 2007 WL 2726871 (ND Cal 2007) (settled on appeal); *Comer v Murphy Oil Co.*, No. 05-CV-436L Q (SD Miss, 20 Aug., 2007), rev'd in part, 585 F3d 855 (5th Cir 2009), judgement vacated, rehearing en banc granted, subsequently dismissed (for lack of quorum), 607 F3d 1049 (5th Cir 2010); *Native Village of Kivalina v Exxon Mobil Corp.*, 663 F Supp 2d 863 (ND Cal, 2009), appeal pending before Ninth Circuit Court of Appeals. These cases are all based on the tort of public nuisance, which the Restatement defines as an 'unreasonable interference with a right common to the general public'. Restatement (Second) of Torts section 821B.

39   For sources and materials on the public trust doctrine, see Laitos et al. (2006), Chapter 8.II. For discussion of the public trust doctrine, see Sax (1970); Dunning (1989); Wood (2009b), part 1.

responsibilities. The seeds of the public trust doctrine are evident in legal systems worldwide and accessible to lawyers across the globe.

The Public Trust Doctrine

The public trust doctrine has flowed through countless forms of government through the ages of humanity. At its core, the doctrine is a declaration of public property rights as originally and inherently reserved through the peoples' social contract with their sovereign governments. Under this principle, the public holds a perpetual common property interest in crucial natural resources. Government, as trustee, must act in a fiduciary capacity to protect such natural assets for the beneficiaries of the trust, which include both present and future generations of citizens.[40] As the United States Supreme Court said in *Geer v Connecticut*, 'The ownership of the sovereign authority is in trust for all the people of the state; and hence, by implication, it is the duty of the legislature to enact such laws as will best preserve the subject of the trust, and secure its beneficial use in the future to the people of the state'.[41] The legislature is primary trustee; the executive branch, acting as agent of the trustee, is vested with the same public trust obligation.[42] As the Supreme Court of India once summarized the public trust doctrine in a landmark case:

> The State is the trustee of all natural resources which are by nature meant for public use and enjoyment. [The] public at large is the beneficiary of the seashore, running waters, airs, forests and ecologically fragile lands. The State as a trustee is under a legal duty to protect the natural resources. These resources meant for public use cannot be converted into private ownership.[43]

Having origins in indigenous systems, the principle finds expression in such venerable codes as the Institutes of Justinian[44] and the Magna Carta. It manifests in

---

40   See *Illinois Cent. R. Co. v Illinois*, 146 US 387, 455 (1892); *Arizona Ctr. for Law in the Pub. Interest v Hassell*, 837 P2d 158, 169 (Ariz Ct App 1991) ('The beneficiaries of the public trust are not just present generations but those to come').

41   161 US 519, 533–34 (1896).

42   See *Geer v Connecticut*, 161 US 519, 533–34 (1896) (legislature as trustee); *Center for Biological Diversity v FPL Group*, 2008 WL 4255789, slip op. at 6, 8 (Cal App 1 Dist, 18 Sept., 2008) (discussing public trust obligations of 'public agencies'); In *Re Water Use Permit Applications, Waihole Ditch Combined Contested Case Hearing* (hereinafter, *Waihole Ditch*), 94 Haw 97, 9 P3d 409 (Haw 2000) (applying public trust obligations to state agency).

43   *M. C. Mehta v Kamal Nath*, 1997 1 SCC (1997), at Par 34, cited in Takacs (2008), at n. 135–7 and accompanying text.

44   See Justinian, Institutes, 1.2.1, 2.1.1 (T. Sandars trans. 1st Am. edn 1876); discussed in *Arizona Center For Law In Public Interest v Hassell*, 837 P2d 158, 166, n. 12

a multitude of court decisions, constitutions, and statutes from around the world.[45] The endurance and prevalence of this doctrine is not at all surprising since it speaks to the most fundamental and intuitive rationale of government itself. Ranking among the most essential purposes of government is the necessity of protecting natural assets for the common benefit of the people and their society. The doctrine recognizes that, left in altogether private hands, increasingly scarce assets would be consumed with selfish intent to the detriment of all, ultimately leading to chaos and societal collapse. Indeed, this ancient and enduring principle has roots and reasoning that put it on par with the highest liberties of citizens living in a free society. As Professor Joseph Sax said decades ago in a landmark article, certain environmental interests protected by the public trust doctrine 'are so intrinsically important to every citizen that their free availability tends to mark the society as one of citizens rather than of serfs'.[46]

The Trust as an Attribute of Sovereignty

The public trust is best understood as a principle organic to government itself – an inherent constitutional restraint on legislative power, and a fundamental expression of legislative duty.[47] In a leading modern trust case, the Hawaii Supreme Court noted, '[H]istory and precedent have established the public trust as an inherent attribute of sovereignty …'.[48] As one commentator summarizes the field: 'The idea that public trust limits and powers inhere in the very nature of sovereignty is one consistent thread in public trust cases.'[49] Characterized as an attribute of sovereignty, the principle has force in the United States context, for example, at both the federal and state levels, though nearly all cases in the United States have involved the states (not surprisingly, because the states were

---

(Ariz App Div 1 1991) (discussing the 'ancient doctrine of common law' that 'restricts the sovereign's ability to dispose of resources held in public trust'); *1.58 Acres of Land*, 523 F Supp at 122 (D Mass 1981) ('Public trust theory has its roots in the Roman law'); Kennedy (2005), pp. 20–21 (stating that the Ancient Roman Code of Justinian 'guaranteed to all citizens the use of the "public trust", or commons – those shared resources that cannot be reduced to private property, including the air, flowing water, public lands, wandering animals, fisheries, wetlands, and aquifers').

45   See section in this chapter, *The Public Trust Doctrine in Legal Systems Around the World*.

46   Sax (1970) p. 484.

47   See Grant (2001), p. 851: explaining the public trust doctrine as part of the constitutional reserved powers doctrine, which prevents any one legislature from taking acts that would compromise a future legislature's ability to exercise sovereignty on behalf of the people.

48   *Waihole Ditch* 94 Haw 97, 130–31 (Haw 2000). See also Stevens (1980), p. 196: noting jurisprudence 'in the form of declarations that the public trust is inalienable as an attribute of sovereignty no more capable of conveyance than the police power itself'.

49   Coplan (2010), p. 311.

historically the primary managers of waters, wildlife, and other resources).[50] One federal district court that explored the dual federal and state roles in the context of tidelands concluded: 'Since the trust impressed upon this property is governmental and administered jointly by the state and federal governments by virtue of their sovereignty, neither sovereign may alienate this land free and clear of the public trust…'.[51] Increasingly, commentators urge application of the doctrine to federal resource managers.[52]

As a limitation on sovereignty and an expression of fundamental sovereign responsibility, the trust 'can only be destroyed by the destruction of the sovereign'.[53] In a landmark trust case, *Illinois Central Railroad v Illinois*, the United States Supreme Court declared that legislatures may not repudiate, abridge, or surrender their trust obligation:

> The state can no more abdicate its trust over property in which the whole people are interested … than it can abdicate its police powers in the administration of government and the preservation of the peace … . Every legislature must, at the time of its existence, exercise the power of the state in the execution of the trust devolved upon it. [54]

As a property-based counterweight to discretionary police power, the trust secures the people's rights to a sustained natural endowment. As one commentator describes, 'Its overarching principle … is that certain gifts of nature – pure air, clean water, a stable climate, and healthy ecosystems – belong to everyone and cannot be appropriated for exclusively private use'.[55] Under the public trust doctrine, government trustees may not allow private interests to cause irrevocable harm to critical public trust resources. Stated another way, government trustees, who serve only at the will of the public, may not allocate private rights to destroy what the people legitimately own for themselves and for their posterity. As the United States Supreme Court said in *Geer v Connecticut*:

---

50   See discussion at Turnipseed et al. (2010), p. 10: '[N]o one has forced the issue at the national level in the way that it has been pushed at the state level' (remarks of Patrick Parenteau).

51   *US v 1.58 Acres of Land*, 523 F Supp 120, 124 (D Mass 1981).

52   See, for example, Coplan (2010), pp. 313–15 (summarizing other scholarship and concluding that the doctrine applies to the federal government, stating, 'If … the public trust is essential to the nature of sovereignty and encompasses rights reserved to the people generally, then the doctrine applies equally to the sovereign federal government as it does to the sovereign state governments'); see also Glicksman (2008).

53   *1.58 Acres of Land*, 523 F Supp at 124.

54   *Illinois Central*, 146 US at 453, 460.

55   Turnipseed et al. (2010), p. 8 (remarks of Patrick Parenteau). As another commentator put it, the doctrine 'holds that some of Earth's riches should never be sequestered for private use, must be left for the public's enjoyment, and must be stewarded by those in power'. Takacs (2008), p. 711.

[T]he power or control lodged in the State, resulting from this common ownership, is to be exercised, like all other powers of government, as a trust for the benefit of the people, and not as a prerogative for the advantage of the government, as distinct from the people, or for the benefit of private individuals as distinguished from the public good ... . [T]he ownership is that of the people in their united sovereignty.[56]

The seminal public trust opinion in the United States is *Illinois Central Railroad Co. v Illinois*, where the Supreme Court announced that the shoreline of Lake Michigan was held in trust by the State of Illinois and could not be transferred out of public ownership to a private railroad corporation. In broad language expressing the public's fundamental right to natural resources, the Court stated:

[T]he decisions are numerous which declare that such property is held by the state, by virtue of its sovereignty, in trust for the public. The ownership of the navigable waters of the harbor, and of the lands under them, is a subject of public concern to the whole people of the state. The trust with which they are held, therefore, is governmental, and cannot be alienated ... .[57]

Unlike the permissive bent of administrative discretion, which accompanies most of statutory law, public trust law imposes a strict fiduciary obligation upon sovereign trustees to protect the people's trust assets from damage.[58] Under well-established principles of trust law, trustees may not sit idle and allow damage to occur to the trust. As one leading treatise explains, 'The trustee has a duty to take whatever steps are necessary ... to protect and preserve the trust property from loss or damage'.[59] Scores of cases emphasize this duty of protection.[60]

---

56   *Geer*, 161 US at 529. See also *Lake Michigan Federation v US Army Corps of Engineers*, 742 F Supp 441, 445 (D Ill 1990) ('[T]he public trust is violated when the primary purpose of a legislative grant is to benefit a private interest').

57   *Illinois Cent R Co*, 146 US at 455. The court also noted that parcels could be alienated 'when parcels can be disposed of without detriment to the public interest in the lands and waters remaining', at 453.

58   See, for example, *Geer*, 161 US at 534 ('[I]t is the duty of the legislature to enact such laws as will best preserve the subject of the trust, and secure its beneficial use in the future to the people of the state'); *State v City of Bowling Green*, 313 NE2d 409, 411 (Ohio 1974) ('[W]here the state is deemed to be the trustee of property for the benefit of the public it has the obligation to bring suit ... to protect the corpus of the trust property').

59   See Bogert (1987) § 99, p. 358; 76 Am Jur 2d Trusts § 404 ('One of the fundamental common-law duties of a trustee is to preserve and maintain trust assets. A trustee has the right and duty to safeguard, preserve, or protect the trust assets and the safety of the principal').

60   Wood (2009b), at notes 30–32.

The trustee's duty to protect the asset involves a corollary, active duty of vigilance to 'prevent decay or waste' to the asset.[61] The waste doctrine is a staple of property law and a jealous guardian of future interests. As one treatise describes the waste doctrine, it prohibits consumption of 'things belonging to the inheritance'.[62] Courts have readily granted injunctions against waste. As a leading old treatise on equity explains, courts enjoin waste to prevent 'great and irremediable mischief, which damages could not compensate, because the mischief reaches to the very substance and value of the estate, and goes to the destruction of it in the character in which it is enjoyed'.[63] A century and a half later, this description perfectly describes the effect of carbon pollution on the planet's atmosphere.

The duty against depleting the assets in a perpetual public trust forms a natural limit on the interests that any private parties can claim. Many have described the 'usufructuary' nature of private interests held in public trust assets. The landmark case, *Arnold v Mundy*, characterized public trust assets as 'things in which a sort of transient usufructuary possession, only, can be had'.[64] As Professor Karl Coplan explains:

> [T]he holder of usufructuary rights can only exploit the fruits of the property, and must not under any circumstances impair the productivity of the underlying asset ... . The interest is analogous to the interest of an income beneficiary of a conventional trust: the trustee may pay out the 'profits' of the trust, but must not invade the corpus.[65]

By prohibiting use of the asset in a manner that would invade the trust inheritance and thereby diminish the wealth available to future beneficiaries, the public trust doctrine, along with its companion waste prohibition, is well appointed to protect against generational theft.[66] Courts can readily apply these duties to government trustees of the public's enduring natural trust.[67] As the Hawaiian Supreme Court emphasized in a leading public trust case involving water resources: 'The check and balance of judicial review provides a level of protection against improvident

---

61   *Moore v Philips* 627 P.2d 831; Bogert, *supra* n. 59, § 99, at 358; 76 Am. Jur. 2d Trusts, *supra* n. 59, §§ 331, 404 (a trustee 'must not suffer the estate to waste or diminish, or fall out of repair').

62   See, for example, *Hill v Ground*, 114 Mo App 80, 343, 89 SW 343, 344 (Ct App Mo 1905).

63   Willard (1863), p. 382.

64   *Arnold v Mundy*, 6 NJ 1, 49 (1821); see discussion at Coplan (2010), p. 325.

65   Coplan (2010), p. 325; see also Coplan (2010), p. 324: '[T]he sovereign, as trustee, may distribute the income of public trust assets, but may not sell off the corpus'.

66   If a trustee of a term for years threatens to commit waste, the remainderman can maintain a suit to enjoin him. Restatement 2d of Trusts, § 200.

67   For an example enforcing the waste prohibition against the federal government in the context of Indian law, see *United States v White Mountain Apache Tribe*, 537 US 465, 475 (2003).

disposition of an irreplaceable res'.[68] The doctrine provides perhaps the only precise legal expression of an intergenerational equity principle. Because future generations do not vote, their interests are often trumped by the interests of the present generation, which holds the political clout. As Professor Coplan points out, in the case of perpetual trusts such as the public trust, '[t]rustees must routinely preserve trust assets for future beneficiaries even against the demands of current beneficiaries'.[69] In essence, the trust serves as a judicially imposed restraint to the powerful political inclination of government officials to over-indulge the living generation of citizens and at the expense of future citizens.

The Atmospheric Trust

While traditionally applied to water-based resources, the public trust doctrine has expanded its reach over time,[70] and commentators increasingly point out the logic of a trust approach to climate crisis.[71] In defining the scope of the trust endowment, courts have looked to the needs of the public as the primary guiding factor. As Professor Charles Wilkinson explains, 'The public trust doctrine is rooted in the precept that some resources are so central to the well-being of the community that they must be protected by distinctive, judge-made principles'.[72] At the time of the *Illinois Central* case, lakebeds served a vital function in supporting fishing, navigation and commerce. Describing the lakebed as property in which 'the whole people are interested', the Court reasoned: 'The trust with which they are held, therefore, is governmental ... follow[ing] necessarily from the *public character of the property*'.[73]

---

68 *Waihole Ditch*, 9 P3d at 455. '*Res*' refers to the assets in the trust.

69 Coplan (2010), p. 328; see also Bogert (1987), at § 404:'A trustee representing beneficiaries in succession is under a duty to successive beneficiaries to act with due regard to their respective interests and to preserve trust property for remainderpersons'.

70 See, generally, *Marks v Whitney*, 491 P2d 374, 380 (1971) ('In administering the trust the state is not burdened with an outmoded classification favoring one mode of utilization over another.'); Wilkinson (1989) (noting expansion).

71 See Torres (2002), p. 533: 'Properly understood ... the traditional rationale for the public trust doctrine provides a necessary legal cornerstone ... to protect the public interest in the sky'); Barnes (2006); Sax (1970), pp. 556–7 (urging application of doctrine to 'controversies involving air pollution'); Wood (2007), pp. 317–22: 'As technology, and the potential for cap-and-trade, makes aspects of the atmosphere subject to private ownership, the public trust doctrine should similarly evolve to include these interests in the public trust responsibilities of the sovereign'); Parenteau (2010).

72 Wilkinson (1980), p. 315.

73 *Illinois Central*, 146 US at 452–6 (emphasis added); see also *n.* 69 at 455 ('It would not be listened to that the control and management of the harbor of that great city – *a subject of concern to the whole people of the state* – should thus be placed elsewhere than in the state itself... ') (emphasis added).

As a legal doctrine, the public trust compels protection of those ecological assets necessary for public survival and community welfare. Courts have recognized an increasing variety of assets held in public trust on the rationale that such assets are necessary to meet society's changing needs. As the New Jersey Supreme Court said, '[W]e perceive the public trust doctrine not to be 'fixed or static', but one to be 'molded and extended to meet changing conditions and needs of the public it was created to benefit''.[74] Over time, the doctrine has reached new geographic areas, including water, ground water, wetlands, dry sand beaches, and non-navigable waterways.[75] In many states, it has pushed beyond the original societal interests of fishing, navigation and commerce to protect modern concerns such as biodiversity, wildlife habitat, aesthetics, and recreation.[76]

The essential doctrinal purpose expressed by courts in these public trust cases compels recognition of the atmosphere as one of the crucial assets of the public trust. The public interests at stake in climate crisis are incalculably more extensive and profound than the traditional fishing, navigation and commerce interests at the forefront of *Illinois Central*. Atmospheric health is essential to all civilizations and to human survival across the globe. As one climate analyst put it, carbon reduction is necessary for averting 'the end of life as we know it'.[77] Given the essential nature of air, it is unsurprising that numerous state constitutions and codes recognize air as part of the *res* of the public trust.[78] Moreover, federal statutory law includes air as a trust asset for which the federal government, states, and tribes can gain recovery of natural resource damages.[79]

---

74   See *Matthews v Bay Head Improvement Assoc*, 471 A2d 355, 365 (N J 1984) (citation omitted).

75   See, for example, *National Audubon Society v Superior Court of Alpine County*, 658 P2d 709, 719 (Cal 1983) (non-navigable tributaries); *Baxley v Alaska*, 958 P2d 422, 434 (Alaska 1998) (wildlife); *Matthews v Bay Head Improvement Ass'n*, 471 A2d 355, 358 (N J 1984) (dry sand area); *Robinson v Ariyoshi*, 658 P2d 287, 310 (Haw 1982) (groundwater); *Just v Marinette County*, 201 NW2d 761,769 (Wis 1972) (wetlands).

76   *Matthews*, 471 A.2d at 363; *National Audubon*, 658 P.2d at 719–22.

77   See Romm (2008a).

78   See, for example, *Her Majesty v City of Detroit*, 874 F2d 332, 337 (6th Cir 1989) (citing Michigan Act that codifies public trust to include 'air, water, and other natural resources'); Haw Const, art. XI, §1 (stating, 'All public natural resources are held in trust by the State for the benefit of the people', and 'the State and its political subdivisions shall conserve and protect Hawaii's ... natural resources, including land, water, air, minerals and energy resources ...'); LA Const, art. IX, §1 ('natural resources of the state, including air and water ... shall be protected...'); RI Const, art. I, §16 (duty of legislature to protect air), interpreted as codification of Rhode Island's public trust doctrine in *State ex. Rel. Town of Westerly v Bradley*, 877 A2d 601, 606 (RI 2005); *National Audubon Society v Superior Court of Alpine County*, 658 P2d 709, 720 (1983) ('purity of the air' protected by the public trust).

79   *CERCLA*, 42 U.S.C. § 9601 (2006) (defining air as among the natural resources subject to trust claims for damages).

The Roman origins of the public trust doctrine classified air – along with water, wildlife and the sea – as *'res communes'*.[80] In a public trust decision, *Geer v Connecticut*, the United States Supreme Court relied on this ancient Roman classification of *'res communes'* to find the public trust doctrine applicable to wildlife.[81] Just a few years later, the Court explicitly recognized the states' sovereign property interests in air and found such interests supreme to private title. In *Georgia v Tennessee*, the Court upheld an action brought by the state of Georgia against Tennessee copper companies for discharging noxious gases that drifted across state lines. The court declared: '[T]he state has an interest independent of and behind the titles of its citizens, in all the earth and air within its domain'.[82] Though the Court did not use the word 'trust', the decision essentially proclaimed air as the people's sovereign property.

In an article urging recognition of the atmosphere as a trust asset, Professor Coplan emphasizes the Roman roots of the public trust doctrine and Justinian's explicit coverage of air as *'res communes'*.[83] He points out that courts have extended the public trust doctrine to resources that previously seemed incapable of private ownership (like water and wildlife) as they became threatened with private exploitation[84] and notes that the same should be true of air: '[A]s governments seek to privatize rights to atmospheric assets through tradable emissions rights, the public trust doctrine should naturally extend to protect previously unpossessable interests in the atmospheric commons'.[85]

As yet, there is no precedent declaring these principles in the context of the atmosphere – certainly not surprising, as never prior to the modern industrialized era has humanity threatened the planet's entire climate system. These are new circumstances for courts and for society in general, and lawyers seeking exact precedent will be searching in vain. This is a time in human history when lawyers worldwide must draw upon timeless principles and extrapolate them in logical fashion to new circumstances. Throughout history, courts have found themselves in the position of declaring new law in response to unforeseen, often urgent, circumstances. The same principles that have informed all of the historic public trust cases apply with even greater force to the atmosphere. As the Supreme

---

80   See *Geer v Connecticut*, 161 US at 525 ('These things are those which the juris consults called 'res communes' – the air, the water which runs in the rivers, the sea and its shores ... [and] wild animals.'). See also Torres (2002), pp. 529–30 (discussing *res communes*).

81   See *Geer v Connecticut*, 161 US at 523.

82   *State of Georgia v Tennessee Copper Co*, 206 US 230, 237 (1907). The passage was cited in *Massachusetts v US Environmental Protection Agency*, 127 S Ct 1438, 1454 (2007).

83   See Coplan (2010).

84   Coplan (2010), p. 320: 'Once the *res communes* became susceptible to private ownership, but as yet unappropriated ..., the potential limitations on private ownership under the public trust doctrine became relevant...'.

85   Coplan (2010), p. 320.

Court said in applying the public trust to an unprecedented set of circumstances in *Illinois Central*, 'We cannot, it is true, cite any authority where a grant of this kind has been held invalid, for we believe that no instance exists where the harbor of a great city and its commerce have been allowed to pass into the control of any private corporation. But the decisions are numerous which declare that such property is held by the state, by virtue of its sovereignty, in trust for the public'.[86] Though conditions change with time, the basic task and the principles that guide courts remain constant. While air has not yet been the subject of trust litigation, modern courts have a solid legal rationale from which to draw in designating the atmosphere as a public trust asset.

The Public Trust Doctrine in Legal Systems Around the World

Government's obligation to protect natural resources for present and future generations is said to exist 'from the inception of humankind'.[87] This principle, declared forcefully by the Philippines Supreme Court in its landmark opinion, *Oposa v Factoran*, reflects a shared human understanding that ecological heritage, which is essential to human survival, is inviolate. Indeed, humankind's innate interest in survival and self-perpetuation suggests a doctrinal foundation of the trust redolent of natural law.[88] Notably, the petitioners in *Oposa* – children and their parents – characterized their right to self-preservation and perpetuation as 'the highest law of humankind – the natural law'. An early public trust case in the United States, *Arnold v Mundy*, also referred to the 'law of nature, which is the only true foundation of all the social rights' as a basis of the doctrine.[89] The United States Supreme Court in *Illinois Central* similarly declared: 'A state legislature cannot, consistently with the principles of the law of nature and the constitution of a well-ordered society, make a direct and absolute grant of the waters of the state, divesting all the citizens of their common right'.[90]

---

86   *Illinois Central RR v Illinois*, 146 US 387, 455 (1892).

87   *Juan Antonio Oposa v Fulgencio S Factoran, Jr*, GR 101083 (Sup Ct Phil 1993), as excerpted in Laitos et al. (2006), pp. 441–4.

88   For discussion of a natural law basis for the public trust, see generally Smith and Sweeney (2006); Yannacone (1975), pp. 615–53.

89   *Arnold v Mundy*, 6 NJL 1, 11 (NJ 1821).

90   *Illinois Central*, 146 US 387, 456 (1892). The same premise, deriving from natural law, found expression by the Supreme Court of Canada in a 2004 case, *British Columbia v Canadian Forest Products*, 2004 SCC 38. There, the court found the doctrine solidly a part of English common law as summarized in an influential treatise by H. de Bracton. As De Bracton described the doctrine, 'By natural law these things are common to all: running water, air, the sea and the shores of the sea … ', para. 75 (citing Bracton on the Laws and Customs of England 39–40 (1968)). As the Supreme Court summarized, the Crown is 'holder of inalienable "public rights" in the environment and certain common resources …', at para. 76.

The natural law underpinnings of the public trust trace back to its early articulation in Ancient Rome's *Institutes of Justinian*,[91] a document that informs many legal systems in the world. Compiled in 535 AD, the *Institutes of Justinian* ascribed the trust origins to natural law, 'the law which natural reason appoints for all mankind [that] obtains equally among all nations, because all nations make use of it'.[92] The *Institutes* declared: 'By the law of nature these things are common to mankind – the air, running water, the sea, and consequently the shores of the sea'.[93] This principle, both in its force and potential, manifests across varied legal systems. Professor Charles Wilkinson has observed the doctrine in the ancient societies of Europe, East Asia, Africa, as well as in Muslim Countries and Native American cultures.[94] He notes, '[t]he real headwaters of the public trust doctrine ... arise in rivulets from all reaches of the basin that holds the societies of the world'.[95]

The core principles of the public trust doctrine represent a crucial dimension of the sovereign politic. Fundamental to democracy, they are germane to any nation governed by the people. As Professor Coplan aptly describes, 'Public trust principles have been described as an essential attribute of sovereignty across cultures and across millennia'.[96] The public trust doctrine has developed extensively through common law in nations such as the United States and India, as well as in some combined civil/common law nations such as South Africa, and in some civil law nations as well.[97] Indeed, the public trust falls easily into a famous description offered by Justice Story in his leading treatise on equity. Explaining common underpinnings of far-flung and various legal systems, Justice Story commented:

> [T]here are in nature certain fountains of justice whence all civil laws are derived, but as streams; yet, that, like as waters do take tinctures and tastes from the very soils, through which they run; so do civil laws vary according to the

---

91 Kennedy (2005), pp. 20–21 (noting that the Ancient Roman Code of Justinian 'guaranteed to all citizens the use of the "public trust", or commons – those shared resources that cannot be reduced to private property, including the air, flowing water, public lands, wandering animals, fisheries, wetlands, and aquifers').

92 Justinian, Institutes, 1.2.1, 2.1.1 (T. Sandars trans. 1st Am. ed n. 1876).

93 Justinian, Institutes, 1.2.1, 2.1.1 (T. Sandars trans. 1st Am. ed n. 1876).

94 Wilkinson (1989), pp. 429–31.

95 Wilkinson (1989), p. 431.

96 Coplan (2010), p. 311.

97 See Takacs (2008), p. 713: 'In the peripatetic manner that has come to characterize it, the Public Trust Doctrine migrated with the Corpus Juris Civilis throughout Europe, to both civil law and common law regimes'; Takacs (2008), pp. 740–48 (describing South Africa regime); see also Ankersen (2003), p. 813, n. 27 (describing civil law analog to public trust doctrine in Latin America); Gleason and Johnson (1995), p. 76: 'The public trust doctrine, having roots in ancient Roman law, appears in many legal systems'.

regions or governments, where they are planted, though they proceed from the same fountains.[98]

The common thread in all public trust iterations is a public property right and corollary sovereign obligation. In all nations, a sovereign property interest emerges from government's control over a particular territory. Where the sovereign derives its power from the people (as distinguished from a totalitarian government or despotic monarchy), this governmental property interest is necessarily that of a trust, held in fiduciary capacity on behalf of the people. It may not explicitly be called a 'trust' in all countries, but the sovereign character of ownership is such that the trust construct serves as a useful analogue even in nations that lack the nomenclature developed in common law. A sovereign trust distinguishes a democracy serving at the will of the public from a government that effectuates the interests of an oligarchy at the expense of the citizens. As Justice Finn of Australia observes: 'Sovereignty and trust probably are best seen as expressions of intrinsic qualities of our democracy. In this, they properly can be described as 'constitutional principles''.[99] In the same vein, Professor Karl Coplan notes that the public trust of the United States is 'equally enforceable as part of the social contract underlying the constitutional bargains of federalism and popular sovereignty ... . Public trust limits inhere in sovereignty, and these limits are reserved to the people'.[100]

The *Oposa* opinion from the Philippines Supreme Court illuminates the natural law force of the trust doctrine in a country beset with environmental turmoil. In *Oposa,* the Court faced a lawsuit brought by children and their parents to prevent the federal government from allowing private logging corporations to cut down the last remaining old-growth forest in the country. Invoking the trust to enjoin any further logging, the Court rendered a sterling pronouncement that, indeed, finds resonance through nearly all political systems:

> Needless to say, every generation has a responsibility to the next to preserve that rhythm and harmony for the full enjoyment of a balanced and healthful ecology .... [T]he right to a balanced ecology ... belongs to a different category of rights [than civil and political rights] altogether for it concerns nothing less than self-preservation and self-perpetuation ... the advancement of which may even be said to predate all governments and constitutions.
>
> As a matter of fact, these basic rights need not even be written in the Constitution for they are assumed to exist from the inception of humankind. If they are now explicitly mentioned ... it is because of the well-founded fear of its framers that unless the right to a balanced and healthful ecology and to health are mandated as state policies by the Constitution itself ... the day would not be too far when

---

98   Story (1866), p. 732 (paraphrasing Lord Bacon).
99   Finn (1995).
100   Coplan (2010), p. 311.

all else would be lost not only for the present generation, but also for those to come – generations which stand to inherit nothing but parched earth incapable of sustaining life.[101]

The public trust doctrine is particularly vibrant in India, no doubt one of the most crucial players in the climate context because of its enormous current pollution and future development aspirations that are tied to fossil fuels.[102] In India, the Supreme Court extended the right to life found in the Constitution to include the right to a healthy environment,[103] and declared the public trust as part of the law of the land.[104] The Court first applied public trust principles with regard to the protection and preservation of natural resources in *M. C. Mehta v Kamal Nath and Others*, finding that the state government had violated the trust in granting a lease on riparian forestland.[105] The Court has invoked the public trust in several other cases as well, ranging across varied contexts.[106] Indian public trust jurisprudence relies heavily on cases and scholarship from the United States.[107] In a 2010 case, the Court noted that government's power is 'vested in trust by the people'[108] and warned of institutional degeneration in natural resources policy: '[T]he problems arise because exploitation of those resources occurs without appropriate supervision by the State as to the rates of exploitation, equitable distribution of the

---

101   *Juan Antonio Oposa v Fulgencio S Factoran, Jr*, GR No 101083 (Sup Ct Phil 1993), as excerpted in Laitos et al. (2006), pp. 443–4.

102   Watts (2010).

103   The Constitution of India was amended in 1976 to expressly address environmental quality. Part IVA, art. 51A, cl. (g) requires every citizen of India 'to protect and improve the natural environment including forests, lakes, rivers and wild life, and to have compassion for living creatures'; and Part IV, art. 48A, entitled *Protection and improvement of environment and safeguarding of forests and wild life*, states: 'The State shall endeavour to protect and improve the environment and to safeguard the forests and wild life of the country'. Article 21 of the Constitution of India states that '[n]o person shall be deprived of his life or personal liberty except according to procedure established by law', Const. of India Part III, art. 21.

104   *M. C. Mehta v Kamal Nath and Others*, ¶ 34 (1997 1 SCC 388), WP 182/1996 (2000.05.12).

105   See n. 104.

106   See *Th. Majra Singh v Indian Oil Corporation* AIR 1999 J&K 81; *M. Builders Pv. Ltd v Radhey Shyam Sahu and Others*, 1999 6 SC 464, AIR 1999 SC 2468; *Karnataka Industrial Areas Development Board v C Kenchappa*, AIRSCW 2546 (India 2006); see also Razzaque (2001), p. 231.

107   See, for example, *Fomento Resorts and Hotels Ltd v Minguel Martins*, Civil Appeal Nos 4154, para. 32, 35 (S Ct India 2000), available at *http://www.elaw.org/node/3731*; *Reliance Natural Resources Ltd v Reliance Industries Ltd*, Civil Appeal No. 4273 of 2010, slip op at 207, para. 97 (S Ct India 2010) (hereinafter, *Reliance Natural Resources*), available at: http://www.legallyindia.com/images/stories/docs/cases/RIL_v_RNRL_supreme_court.pdf.

108   *Reliance Natural Resources*, see n. 107, at 125, para. 5.

wealth it generates, collusions between the extractive industry and some agents of the State and the consequent evisceration of the moral authority of the institutions of the State'.[109] In 2000, in *Fomento Resorts v Minguel Martins*, the Court declared a sweeping reach of the public trust doctrine, clearly encompassing air:

> [The public trust doctrine] primarily rests on the principle that certain resources *like air,* sea, waters and the forests have such a great importance to the people as a whole that it would be wholly unjustified to make them a subject of private ownership. These resources are gifts of nature, therefore, they should be freely available to everyone irrespective of one's status in life. The public trust doctrine enjoins upon the Government to protect the resources for the enjoyment of the general public rather than to permit their use for private ownership or commercial purposes. This doctrine ... mandates affirmative State action for effective management of natural resources and empowers the citizens to question ineffective management thereof. The heart of the public trust doctrine is that it imposes limits and obligations upon government agencies and their administrators on behalf of all the people and especially future generations.[110]

Throughout all of its public trust jurisprudence, the India Supreme Court has expressed an abiding concern for the future generations. In its 2010 *Reliance Natural Resources* case, the Court declared:

> The concept of people as a nation does not include just the living; it includes those who are unborn and waiting to be instantiated. Conservation of resources, especially scarce ones, is both a matter of efficient use to alleviate the suffering of the living and also of ensuring that such use does not lead to diminishment of the prospects of their use by future generations.[111]

In the *Fomento Resorts* case, the Court drew upon the work of Professor Joseph Sax to conclude:

> [T]he Public Trust Doctrine, of all concepts known to law, constitutes the best practical and philosophical premise and legal tool for protecting public rights and for protecting and managing resources, ecological values or objects held in trust. The Public Trust Doctrine is a tool for exerting long-established public rights over short-term public rights and private gain. *Today, every person exercising his or her right to use the air, water, or land and associated natural*

---

109 *Reliance Natural Resources*, see n. 107, at 132 para. 12.
110 *Fomento Resorts*, see n. 107, at para. 32. In the *Reliance Natural Resources* case, the Court declared a public trust over natural gas reserves, noting: 'It is now a well established principle of jurisprudence that the true owners of "natural wealth and resources" are the people as a nation...'. *Reliance Natural Resources*, see n. 107, at 200, para. 88.
111 *Reliance Natural Resources*, see n. 107, at 205–06, para. 9.

> *ecosystems has the obligation to secure for the rest of us the right to live or*
> *otherwise use that same resource or property for the long term and enjoyment*
> *by future generations.*[112]

These public trust cases provide hope that, even as India's political leaders refuse to commit to carbon reduction, the courts will recognize an atmospheric trust responsibility on the part of government.

In Canada, the public trust doctrine gained explicit recognition in a 2004 case, *British Columbia v Canadian Forest Products Limited.*[113] There, the Canadian government sought damages against a private logging company for a fire that swept through public forests. Detailing the origins of the public trust doctrine, as well as its companion *parens patriae* doctrine (which allows the government to sue on behalf of the public), the Court recognized that they form an appropriate common law basis for recovering natural resource damages to a public resource.[114] Finding declarations of public rights and ownership in 'running water, air, the sea and the shores of the sea' in both the *Institutes of Justinian* and H. de Bracton's influential treatise on English law, the Court stated: 'By legal convention, ownership of such public right was vested in the Crown, as too did authority to enforce public rights of use ... ':

> Since the time of de Bracton it has been the case that public rights and jurisdiction over these cannot be separated from the Crown. This notion of the Crown as holder of inalienable 'public rights' in the environment and certain common resources was accompanied by the procedural right of the Attorney General to sue for their protection representing the Crown as *parens patriae. This is an important jurisdiction that should not be attenuated by a narrow judicial construction.*

Citing favourably American law declaring a public trust, and noting that the public trust and *parens patriae* doctrines have supported successful United States common law claims for monetary natural resource damages in absence of statutes, the Court acknowledged that the trust raises important policy questions, including 'the Crown's potential liability for *inactivity* in the face of threats to the environment, [and] the existence or non-existence of enforceable fiduciary duties owed to the public by the Crown... '.[115] Finding the case at hand 'not a proper appeal for the Court to embark on a consideration of these difficult issues', as the case in the lower court had been framed around the right of the Crown to make a claim as 'any other landowner' for loss of timber value, the Court deferred a detailed analysis of

---

112　*Fomento Resorts, supra* n 107, at par. 32 (emphasis added).

113　*British Columbia v Canadian Forest Products*, 2004 SCC 38, para. 64–83, available at http://scc.lexum.umontreal.ca/en/2004/2004scc38/2004scc38.html.

114　See n. 113 at para. 71–81.

115　See n. 113 at para. 79–81.

the public trust for another time. Clearly, however, the *Canadian Forest Products* case indicates receptivity to further trust litigation.

In Australia, a country that has common law roots, one would expect the public trust to be a pillar of environmental jurisprudence. Instead, however, the idea of government trust obligation is just beginning to crystallize. As Justice Finn writes in a book on the trust underpinnings of governance, Australia rapidly advanced from being a colony under the control of England to its modern administrative state, without a sufficient corresponding progression in legal doctrine to redefine the essential character of government.[116] The nation seemingly skipped a chapter of political thinking that infused the American democracy (as he points out, 'the judges of the 17th and 18th centuries were unable to draw the treasonable conclusion that public power came directly from the people').[117] He explains that 'the casualty in legal thought' from this progression was a failure of the legal system to drape government officers and agencies with fiduciary trust obligations as the servants of the people.[118] But, while until recent times, trust ideals have had 'little resonance' in Australian political and legal thought, Justice Finn suggests that the fiduciary conception is inevitably part of the political embrace of popular sovereignty:

> However we may wish to interpret [the] common law, we cannot now ignore the inexorable logic of popular sovereignty. If the powers of government belong to and are derived from the people, can the donees of those powers under our constitutional arrangements properly be characterized in terms other than that they are the trustees, the fiduciaries, of those powers for the people? Though separated by more than two centuries, our answer should be that of the American colonists after the Revolution. I would formulate it in this way: ...
> The institutions of government, the officers and agencies of government, exist for the people, to serve the interests of the people, and, as such, are accountable to the people ... . Sovereignty and trust probably are best seen as expressions of intrinsic qualities of our democracy.[119]

While the public trust concept is embryonic in some countries, and a mature doctrinal force in others, the foundation of the public trust applies to the majority of nations whose citizenry celebrates and honours fundamental assumptions of democracy. Indeed, a growing chorus of legal voices notes the commonality of public trust principles among nations and urges their application to vexing and

---

116 Finn (1995), pp. 10–11.

117 Finn (1995), p. 10.

118 Finn (1995), pp. 11–12: '[L]egal principle all but surrendered its place as a force defining the nature and end of government itself'.

119 Finn (1995), p. 15.

unprecedented problems of global ecology.[120] Yet there is a persistent perception that the trust doctrine, whose most detailed elucidation is through common law, does not apply to 'civil law' countries. The assumption derives from an overly rigid view of legal expression and the common tendency of lawyers to overlook doctrine's roots, which extend to the very bedrock of human civilization. It is true that civil law nations do not have any concept of private trusts.[121] Unfortunately, this has often led lawyers in civil law countries to brush aside the public trust doctrine, presuming it not applicable to their systems. The public trust doctrine, however, speaks to sovereignty and public ownership, not to private trust arrangements. As international public trust scholar Peter Sand points out, the 'functional equivalents of public trusteeship' are evident in many civil law systems.[122] For example, classic civil law countries such as Germany, France, Switzerland and the Scandinavian nations have laws declaring public ownership interests in waterways, shorelines, and/or wildlife.[123] Such laws reflect a quintessential understanding residing at the very core of the trust – a concept of public ownership that has maintained a steady pulse through time, still iterated in many civil law countries through discrete laws proclaiming public rights in natural resources.

While many scholars focus on the common law iterations of the public trust doctrine – understandably so, since courts have often been the pace-setters in both establishing environmental rights of citizens and announcing their fundamental trust basis – the trust concept is by no means confined to common law alone. A remarkably fluid precept, it manifests in countless and varied iterations, has manifold origins, and proliferates across the globe through multiple routes. The landmark United States case, *Arnold v Mundy*, ascribed the trust to an amalgamation of law, including the law of nature, civil law, and English common law.[124] By necessity, the doctrine has adapted and moulded to new sovereign circumstances as nations have changed their governing character. As Professors Goble and Freyfogle note in their extensive analysis of the doctrine's transformation from England to the United States, '[T]he doctrine of royal prerogative ownership of submerged lands

---

120   See Nanda and Ris (1976), p. 306: (inventorying trust concepts in other countries, and concluding, 'The principles of public trust are such that they can be understood and embraced by most countries of the world'); Sand (2004), pp. 57–8 (suggesting trust principles as framework for international law, stating, '[A] transfer of the public trust concept from the national to the global level is conceivable, feasible, and tolerable ... . The essence of transnational environmental trusteeship ... is the democratic *accountability* of states for their management of trust resources in the interest of the beneficiaries – the world's "peoples" ... .') (emphasis in original); Turnipseed et al. (2010).

121   Turnipseed et al. (2010), p. 12.

122   Turnipseed et al. (2010), p. 12.

123   Turnipseed et al. (2010), p. 12. For an analysis of the public trust concept in German law, see Kube (1997).

124   *Arnold v Mundy*, 6 NJL 1, 76–77 (1821). For an example of a modern case tracing the doctrine, see *1.58 Acres of Land,* 523 F Supp at 122 ('Public trust theory has its roots in the Roman law').

thus was *transformed in the transition from monarchy to republic* into the doctrine of state sovereign ownership in which the state held the lands as trustee for the real sovereign, the people'.[125] At times in various countries, the public trust doctrine has rested in dormancy, only to be resurrected to fit new circumstances. As David Takacs describes the doctrine's progression in South Africa,

> [I]n the same year that India's Supreme Court mandated its Public Trust Doctrine, the South African government disinterred its own moribund Public Trust Doctrine, which had been buried through decades of apartheid regimes whose leaders felt no need to act to preserve resources for the majority of the public.[126]

South Africa now has one of the leading iterations of the doctrine worldwide.

Through the course of human civilization, this remarkable doctrine has come to life through the efforts of lawyers, judges, citizens, and legislators – all of whom recognize its arresting potential and universal force in compelling governmental obedience to a timeless moral covenant with both present and future generations of citizens.[127] At many times in jurisprudential history, the doctrine has emerged from historic precedent to address, often urgently, a diminishing endowment of natural resources that are crucial for public welfare. Climate emergency portends unparalleled resource scarcity – it forces a leap beyond applicable experience. As award-winning journalist, Ross Gelbspan, has written, 'There is no body of expertise – no authoritative answers – for this one. We are crossing a threshold into uncharted territory'.[128] As explained earlier, lawyers and judges worldwide must respond boldly and creatively, drawing upon fundamental legal concepts such as the public trust doctrine to logically address circumstances that elude conventional legal approaches.

Today's lawyers in civil law countries certainly have the means to unearth the public trust doctrine from their own jurisprudential history and mould it to their modern legal architecture. In doing so, they may use existing environmental statutes as legal hooks for atmospheric trust lawsuits seeking to protect the Earth's climate. Nearly all countries have laws requiring protection of the environment, and 117 nations have constitutions that express environmental obligations and/or

---

125   Goble and Freyfogle (2002).

126   Takacs (2008), p. 743.

127   *Alliance to Protect Nantucket Sound Inc v Energy Facilities Siting Board*, 457 Mass. 663, 702 (Mass SCt 2010), available at http://www.massreports.com/SlipOps/Default.aspx (case involving tidelands) (Marshall, CJ, concurring and dissenting) ('The public trust doctrine stands as a covenant between the people of the Commonwealth and their government, a covenant to safeguard our tidelands for all generations for the use of the people... ').

128   Gelbspan (2007).

ecological rights held by the people.[129] In fact, many nations have constitutional provisions that codify the trust precept.[130] Such statutory or constitutional provisions can fasten an atmospheric trust lawsuit to the existing legal structure of the nation. Where existing laws affirm public rights in water resources or public beach access but are silent as to air, lawyers can describe these water and coastal resources as proxies for climate crisis, explaining that hotter planetary temperatures and rising atmospheric concentrations of carbon dioxide will cause droughts, water

---

129   Takacs (2008), at n 78. As the district court noted in *1.58 Acres of Land,* the public ownership roots of the doctrine are evident with respect to submerged lands in many countries. See *1.58 Acres of Land,* 523 F Supp at 123 ('Historically, no developed western civilization has recognized absolute rights of private ownership in such land as a means of allocating this scarce and precious resource among the competing public demands').

130   South Africa, for example, ratified a constitution in 1996 that declares, in Section 24: 'Everyone has the right: a) to an environment that is not harmful to their health or well-being; and b) to have the environment protected, *for the benefit of present and future generations,* through reasonable legislative and other measures that: i) prevent pollution and ecological degradation; ii) promote conservation; and iii) secure ecologically sustainable development … ' South African. Constitution 1996 section 24 (emphasis added), cited in Takacs (2008), n. 154 and accompanying text. Kenya's Constitution, revised in 2010, declares the right of every person 'to a clean and healthy environment, which includes the right (a) to have the environment protected for the benefit of present and future generations through legislative and other measures …'. Kenya Constitution. Art. 42, available at http://www.nation.co.ke/blob/view/-/913208/data/157983/-/l8do0kz/-/published+draft.pdf. Kenya's High Court has upheld public trust principles in environmental cases, comparing the right to a healthy environment to the right to life. Waweru v. Republic, Misc. Civil Application No. 118 of 2004, at 689 (High Court, at Nairobi, 2 March, 2006) ('Living … takes place in some environment and therefore the denial of wholesome environment is a deprivation of life.' And '[i]n the case of land resources, forests, wetlands and waterways to give some examples the Government and its agencies are under a public trust to manage them in a way that maintains a proper balance between the economic benefits of development with the needs of a clean environment.'), available at http://www.chr.up.ac.za/index.php/browse-by-subject/339-kenya-waweru-v-republic-2007-ahrlr-149-kehc-2006-.html. In Ecuador, a Constitutional referendum adopted in 2008 gave inalienable rights of Nature to exist and persist and regenerate. See Koons (2008). In April, 2011, Bolivia was poised to pass a similar set of laws, known as the Law of Mother Earth. See Vidal (2011). Section 13 of the Ukraine Constitution makes explicit reference to public ownership of atmosphere:

The land, its mineral wealth, *atmosphere,* water and other natural resources within the territory of Ukraine, the natural resources of its continental shelf, and the exclusive (maritime) economic zone, are objects of the right of property of the Ukrainian people. Ownership rights on behalf of the Ukrainian people are exercised by bodies of state power and bodies of local self-government within the limits determined by this Constitution. Every citizen has the right to utilize the natural objects of the people's right of property in accordance with the law. Property entails responsibility. Property shall not be used to the detriment of the person and society. The State ensures the protection of the rights of all subjects of the right of property and economic management, and the social orientation of the economy. Const. of Ukraine, art. 13.

scarcity, disruption of natural hydrological cycles, rising tides, coastal flooding, and ocean acidification. Through briefs and other writings, lawyers may infuse the statutory and constitutional provisions at their disposal with trust principles and contextualize them in the deep understandings of popular sovereignty that arouse both citizens and judges.[131] Lawyers can urge judges to graft trust principles onto existing statutory or constitutional provisions, or to anchor their interpretation of such provisions to underlying trust concepts.

In this manner, lawyers worldwide can unite in their capacity to declare sovereign ecological obligations and public ownership of crucial planetary assets. In countries across the globe, lawyers may launch hundreds of different legal actions – some based on statutory law, some on common law, some on Constitutional law, some on customary or religious law, some on natural law – to express the trust obligation on the part of all governments to protect conditions necessary to sustain life on Earth for both present and future generations. Only by characterizing the atmosphere in its planetary entirety will Humanity arrive at an adequate regime of carbon reduction. Without such a singular focus that encompasses all nations of the world in joint and collective sovereign responsibility, the climate movement is at risk of degenerating into fractured and diffused efforts set adrift from any core, unifying principle. The public trust doctrine, by presenting a fundamental basis transcending national and cultural differences, provides the most promising framework by which citizens of different nations can establish carbon reduction responsibility against their own governments as part of a unified global approach. Indeed, the UNFCCC, negotiated in 1992 and signed by most nations of the world, provides an umbrella legal framework for applying the public trust concept to climate change by calling upon nations to 'protect the climate system for the benefit of present and future generations of humankind'.[132]

The Public Trust and Shared Assets: A Sovereign Co-Tenancy

One of the great strengths of the trust doctrine in addressing climate crisis is that it draws upon a property framework that creates logical rights to shared assets of a trans-boundary nature. It is well established that, with respect to trans-boundary trust assets, all sovereigns with jurisdiction over the natural territory of the asset have legitimate property claims to the resource.[133] Property law arranges these

---

131   Indeed, in some nations like India, the Philippines, and South Africa, where the public trust is law of the land, it is inextricably woven into constitutional declarations. See generally Takacs (2008) (discussing India and South Africa).

132   UNFCCC, S. Treaty Doc. No. 102–38, Art. 3, p. 1 (1992), available at http://unfccc.int/resource/docs/convkp/conveng.pdf.

133   States that share a waterway, for example, have correlative rights to the water. *Arizona v California*, 373 US 546, 601 (1963). Similarly, states and tribes have co-existing property rights to share in the harvest of fish passing through their borders. *Washington v Washington State Commercial Passenger Fishing Vessel Ass'n*, 443 US.658, 676–79

interests into a co-tenancy. A co-tenancy is 'the ownership of property by two or more persons in such manner that they have an undivided … right to possession'.[134]

Courts have used the co-tenancy model on the sovereign level to describe shared interests to migrating salmon. In Indian treaty litigation in the United States, the Ninth Circuit declared that the tribes and the states have 'something analogous to a co-tenancy in the off-reservation fishery', finding that each sovereign class had an 'equality of right' under the treaties to the migrating fish: '[T]he state and the tribes stand in similar positions as holders of quasi-sovereign rights in the fishery, and … the federal courts are, when necessary, arbiters of those rights'.[135]

A bedrock principle in any co-tenancy is the correlative duty not to 'waste' the common asset. Acts that amount to permanent damage to the common property are held to constitute waste.[136] This parallels the waste prohibition that applies to trustees of property, as described earlier. As the Ninth Circuit described the duty in the treaty fishing cases:

> Co-tenants stand in a fiduciary relationship one to the other. Each has the right to full enjoyment of the property, but must use it as a reasonable property owner. A cotenant is liable for waste if he destroys the property or abuses it so as to permanently impair its value. A court will enjoin the commission of waste … .

(1979). See also *Idaho ex rel Evans v Oregon*, 462 US 1017, 1031 n1 (1983) (O'Connor, J., dissenting) (noting 'recognition by the international community that each sovereign whose territory temporarily shelters [migratory] wildlife has a legitimate and protectable interest in that wildlife').

134   Black's Law Dictionary 1477 (8th edn 2004); Singer (1994); see also 20 Am. Jur. 2d Co-tenancy and Joint Ownership § 1 (1995).

135   *Puget Sound Gillnetters Ass'n v U S Dist Court*, 573 F 2d 1123, 1128 (9th Cir 1978) (holding that the treaty established 'something analogous to a co-tenancy, with the tribes as one co-tenant and all citizens of the Territory (and later of the state) as the other'); *United States v Washington*, 520 F2d 685, 686, 690 (9th Cir 1975) (applying co-tenancy construct, by analogy, to Indian fishing rights). The court recognized that, as applied to a fishery on the sovereign level, not all of the 'rights and incidents of a common law co-tenancy necessarily follow' as they would in the case of a co-tenancy in land. *Puget Sound*, 573 F2d at 1128, n. 3. But the court nevertheless found the analogy helpful and used it to guide its allocation of the migratory fishery. While the Indian fishing cases drew upon treaty language (reserving tribal fishing rights 'in common' with the states) to find a co-tenancy, the concept is equally applicable to non-treaty situations in which different sovereigns share assets of transitory or migratory character. In the Indian fishing cases, the treaties were crucial for establishing that the tribes held *any* property interest in the fishery, as the states had become governing sovereigns with territorial jurisdiction. In the case of national sovereigns (as oppose to domestic native nations), this property interest springs automatically from the assertion of exclusive sovereignty over a particular territory.

136   Hopkins (1896), p. 342; Walsh (1947), § 131, at 72.

By analogy, neither the treaty Indians nor the state on behalf of its citizens may permit the subject matter of these treaties to be destroyed.[137]

These principles are readily extrapolated to the atmosphere, a natural asset that (like a migratory fishery) transcends sovereign borders. Within a sovereign property framework, all nations on Earth are co-tenant trustees of the global atmosphere.[138] This conception is reinforced by the UNFCC, which essentially declares a commonly held atmospheric trust obligation.[139] From this property framework, two separate duties arise. First is the sovereign duty that each government, as trustee, has towards its own citizens to protect the atmospheric asset and prohibit waste of their natural inheritance. Second is the duty owed by each nation towards all other nations, arising from the sovereign co-tenancy relationship, to prevent waste to their common asset, the atmosphere. The two duties merge into a uniform obligation, incumbent on all governments, to reduce atmospheric emissions.

Atmospheric Trust Litigation

By characterizing the atmosphere in its entirety as a defined trust asset, ATL is designed as a macro-level legal strategy to enforce scientifically based prescriptions for carbon reduction.[140] It seeks to impose concrete, quantitative carbon reduction requirements on governments worldwide. As co-trustees of the world's atmosphere, all sovereign nations are bound by the fiduciary obligation to ensure overall health of the asset. The various agencies and sub-jurisdictions of government, as agents of the trustees, are similarly bound. Fiduciary standards are defined by objective, not political, criteria. Scientific prescriptions for achieving climate equilibrium form the yardstick for the atmospheric fiduciary obligation. The judicial role is to compel the political branches to meet their fiduciary obligation through whatever measures and policies they choose, as long as such measures sufficiently reduce carbon emissions within the required time frame. The courts' role is not to supplant a judge's wisdom for a legislature's approach, but

---

137   See *Washington*, 520 F2d at 685.

138   For the concept of a 'planetary trust', see Weiss (1984); Finn (1995).

139   See UNFCCC, n. 132.

140   See Torres (2002), p. 532: 'The public trust doctrine supplies a broad framework that supports the establishment of a mechanism ... to supervise the government dealings in relationship to the carrying capacity of the atmosphere'. It should be noted, however, that a carbon prescription standing alone, even if faithful to the best science, will likely not solve the global warming crisis. It is evident that society must deploy multiple strategies to arrive at carbon reduction. Nevertheless, many strategies will likely fail absent a clear framework of legal responsibility that forces carbon reduction. A carbon prescription mandating regular cuts on a path to a near-zero-carbon endpoint forces a transition that would otherwise fail due to inertia and political impasse.

rather to police the other branches to ensure fulfilment of their trust responsibility in accordance with the climate imperatives of nature.[141] By linking to scientific prescriptions as the measure of fiduciary responsibility, the ATL approach is aimed at divesting the world's leaders of their assumed prerogative to take action only according to their political objectives.

Carbon Math and Orphan Shares

A core task in defining the sovereign fiduciary obligation is determining how much carbon reduction must occur, and within what time-frame, in order to achieve climate equilibrium over the long term. There are several steps to arriving at a carbon reduction trajectory that would meet the fiduciary obligation to protect the atmosphere. To begin with, climate goals are often expressed in terms of limits on planetary heating, and this is the logical starting point for determining the atmospheric trust obligation. Some amount of heating is already beyond humanity's control as a result of past carbon releases to the atmosphere.[142] This heating is 'in the pipeline' and cannot be called back by humanity unless technology is developed that will draw down carbon, a prospect that remains unlikely. As the National Oceanic and Atmospheric Administration explains: '[T]he climate change that takes place due to increases in carbon dioxide concentration is largely irreversible for 1,000 years after emissions stop'.[143]

Leading climate scientists warn that, if the Earth heats beyond 2 °C, the planet will pass irrevocable thresholds, rendering much of life on Earth impossible.[144] Even with 2 °C heating, scientists predict significant risk of intolerable impacts, such as the Greenland icecap melting and a consequential rise in sea level that would displace millions of people worldwide.[145] As Dr James Hansen explains: 'Global warming of 2 °C or more would make Earth as warm as it had been ... three million years ago [which] caused sea levels to be about twenty-five meters (eighty feet) higher than they are today'.[146] Many vulnerable island states and African nations reject any 2 °C target, as such an increase would obliterate or devastate their nations. A coalition of 112 nations now calls for a limit of 1.5 °C

---

141   See *Lake Michigan Federation v US Army Corps of Engineers*, 742 F Supp 441, 446 (N. D Ill 1990) ('The very purpose of the public trust doctrine is to police the legislature's disposition of public lands. If courts were to rubber stamp legislative decisions ... the doctrine would have no teeth. The legislature would have unfettered discretion to breach the public trust as long as it was able to articulate some gain to the public').

142   See Parenteau (2010), see n. 68 at (draft) 2–3 (citing reports and explaining unavoidable heating); see also Hansen (2008a).

143   Solomona et al (2009), p. 1704 cited in Parenteau (2010), p. 3.

144   McCarthy (2009).

145   Copenhagen Diagnosis (2009), p. 49.

146   Hansen (2009), p. 13.

heating.[147] By all normative principles of justice, the atmospheric restoration goal must be set at a level that maintains planetary fixtures and the climate system in support of humanity in all co-tenant nations, not just the industrialized nations. As a guiding principle that mirrors the atmospheric fiduciary obligation, the UNFCCC declared a universal responsibility to avoid 'dangerous anthropogenic heating' of the planet[148]– which now appears to demand a limit of a 1.5 °C increase over pre-industrial temperatures.[149] Moreover, to the extent that uncertainty surrounds climate policy, the UNFCCC called for a precautionary approach, which should lead courts to err on the side of a more aggressive reduction trajectory.[150]

The next step is to translate that heating limit (1.5 °C) into atmospheric concentrations of carbon dioxide. To limit heating at 1.5 °C, leading scientists maintain that the atmosphere's carbon dioxide concentration should reduce *below* 350 parts per million (ppm) over the long term.[151] As Dr Hansen and his colleagues conclude:

> If humanity wishes to preserve a planet similar to that on which civilization developed and to which life on Earth is adapted, paleoclimate evidence and ongoing climate change suggest that CO_2 will need to be reduced from its current 385 ppm to at most 350 ppm.[152]

---

147   See *Countries for 150 ppm / 1.5 C.*, 350.org., available at http://www.350.org/sites/all/files/Countries_Endorsing_350_ppm.pdf; see also *UN Scientist Backs '350' Target for CO2 Reduction*, Yahoo News (25 Aug. 2009). For the perspective of one island nation that risks total submersion from rising sea levels as a result of planetary heating, see President Nasheed, Maldives, Address by His Excellency President Nasheed at the High Level Conference on Climate Change: Technology Development and Transfer

New Delhi, India (October 22, 2009), available at http://www.newdelhicctechconference.com/InauguralSession/Speech-PresidentofMaldives.pdf.

148   UNFCCC, Art. 2.

149   As James Hansen (2008a) writes, ('[T]he safe level of atmospheric carbon dioxide is no more than 350 ppm (parts per million), and it may be less ... [T]he oft-stated goal to keep global warming less than two degrees Celsius (3.6 degrees Fahrenheit) is a recipe for global disaster, not salvation'. Summarizing his climate research, he concludes, 'the safe level of atmospheric carbon dioxide is no more than 350 ppm (parts per million), and it may be less'.

150   See UNFCCC, Art. 3.3 ('The Parties should take precautionary measures to anticipate, prevent or minimize the causes of climate change and mitigate its adverse effects. Where there are threats of serious or irreversible damage, lack of full scientific certainty should not be used as a reason for postponing such measures ... ').

151   Hansen et al. (2008); (2009) '*UN Scientist back '350'...*', p. 1; see also Rockström et al. (2009b), p. 32.

152   Hansen et al. (2008); (2009) '*UN Scientist back '350'...*', p. 1.

The global average CO_2 concentration – which now exceeds *390 parts per million* – is the highest in 650,000 years of geologic history.[153] Currently, humanity is on a lock-step 'Business As Usual' (BAU) track that, if continued for several decades, is projected to heat the planet a disastrous 4–7 °C by 2100.[154] In the words of climate scientists, BAU will 'loc[k] in climate change at a scale that would profoundly and adversely affect all of human civilization and all of the world's major ecosystems'.[155]

It is eminently clear that continued greenhouse gas pollution by any nation on Earth constitutes 'waste' to the common asset. Yet, the waste principle can only gain quantitative meaning when scientists translate the <350 ppm atmospheric restoration goal into a global prescription of carbon dioxide reduction. In 2007, the Union of Concerned Scientists produced a report, *How to Avoid Dangerous Climate Change: A Target for US Emissions Reductions*, that called for a four per cent annual reduction in greenhouse gas emissions in the industrialized world starting in 2010.[156] Though this target was structured as a scientific prescription that could be judicially adopted as a sovereign fiduciary obligation to restore the atmosphere, the report's trajectory was calibrated to a goal of 450 ppm of atmospheric carbon dioxide concentration (and, even so, presented only a 50 per cent chance of limiting the temperature rise to 2 °C, a risk that would not satisfy any fiduciary standard, which incorporates a measure of prudence and caution in managing assets).[157] Just as the report issued, new data poured in showing the disintegration of the Arctic sea ice and the accelerated melting of Greenland's massive ice sheet.[158] As Dr James Hansen told colleagues at the American Geophysical Union annual conference in December 2007, 'The evidence indicates we've aimed too high – that the safe upper limit for atmospheric CO_2 is no more than 350 ppm'.[159]

In May 2011, Dr Hansen and other leading scientists issued a path-breaking paper that set forth a trajectory of global carbon reduction that could return the atmosphere to equilibrium at 350 ppm. They presented projections showing that a global decline of six per cent in fossil fuel emissions, beginning in year 2012, would lower the atmospheric concentration of CO_2 to 350 ppm by the end of the century, assuming a corresponding major effort to extract roughly 100 Gigatons of CO_2 (GtC) from the atmosphere through reforestation and improved forestry and

---

153   US Department of Commerce (2008). This compares to a pre-industrial revolution (1880) average of 280 ppm. See also Parenteau (2010).

154   'The Copenhagen Diagnosis', Ki-moon (2009), p. 49.

155   'The Copenhagen Diagnosis', Ki-moon (2009), p. 49; See also Rockstrom et al. (2009b), pp. 472–5: (temperature increase of 6°C 'would severely challenge the viability of contemporary human societies').

156   Union of Concerned Scientists (2007).

157   Union of Concerned Scientists (2007), p. 3. A trustee must exercise 'reasonable prudence' in managing trust assets. See Bogert (1987), p. 366.

158   McKibben (2007).

159   See McKibben (2007).

agricultural practices (they deemed 100 GtC the 'largest practical extraction').[160] As some climate analysts have aptly described, a return to 350 ppm involves an 'Emergency Pathway' of carbon emissions reduction, requiring a scale of effort that 'can only correspond to a societal mobilization with few if any peacetime precedents ...'. [161]

Charting a pathway of reduction, of course, is not exact science. Any trajectory is tied to a probability of meeting the stated goal (or, on the flip side, a risk of not meeting it). A reduction path with a 50 per cent probability of limiting global heating to 1.5 °C will look different from a path carrying a 90 per cent probability. Moreover, there may be bumps in the trajectory as investments are made and technology is developed – it may not be a straight line headed downward. But most important to understand, trajectories might change as science develops. Just as the 450 ppm goal, now considered dangerously high, was accepted by many leading climate thinkers only half a decade ago, so might a 350 ppm limit be shown by future science as still dangerously high, or a certain trajectory of reduction not aggressive enough. As science pours in demonstrating harm from climate change, courts may have to modify their quantitative standard of trust protection.[162] This kind of mid-stream adjustment is not at all unusual, for courts face the prospect of natural change in nearly all environmental cases. In cases involving fisheries, for example, courts have recognized the impossibility of accurately predicting fish runs, and have thus emphasized over and over again that 'precise mathematical equality' in allocating fish between states and tribes is unnecessary for the judicial remedy.[163] The fact that some facts are 'not susceptible of rigid pre-determination' does not defeat a court's ability to craft a remedy. A court's broad power of equity includes generous latitude for estimation, approximation, and adjustment.[164]

---

160 Hansen et al. (2011) at 10–11. An earlier paper by other analysts had set forth a similar global trajectory, but without reliance on the assumptions of carbon drawdown through reforestation and improved soil practices. That analysis, without the carbon drawdown, presented a trajectory in which global emissions could peak in year 2011, decline an average of ten per cent a year, reach near-zero emissions by 2050, and lead to stabilization of atmospheric CO2 concentrations of 350 ppm by 2100. Baer et al. (2009), *A 350 ppm Emergency Pathway*, pp. 1–4.

161 Baer et al. (2009), p. 4.

162 Baer et al. (2009), p. 5, discussing uncertainties and emphasizing that the 350 ppm emergency trajectory should be reviewed and updated as new science emerges. The authors point out the possibility, for example, that time could reveal that 'Earth's climate system is even less tolerant of elevated CO_2 concentrations than we currently fear', or that 'the oceanic and terrestrial sinks that we're counting on to absorb our emissions are declining even faster than we currently fear', either of which could indicate the need for steeper reduction.

163 *US v Washington*, 384 F Supp 312, 343 (WD Wash 1974). See discussion in Mulier (2006), at n. 165 and accompanying text.

164 See *United States v Washington*, 384 F Supp 312, 346 (WD Wash 1974).

A key factor in determining how much annual reduction is necessary is the year in which the reduction starts. Leading climate scientists stress that humanity's delay in reducing emissions 'drastically increases' both the speed at which emissions must be cut and the amount of emissions reduction required.[165] If humanity waits too long to bend the rising curve of emissions (that is, reach the global 'peak' in emissions and head the trajectory downward), the slope of necessary emissions reduction becomes so steep a descent that it may be impossible to achieve.[166] The most feasible pathways remaining have a starting point of *now*.[167] Moreover, as many scientists warn, delay is dangerous, because it pushes the planet closer to the unknown tipping point.[168] Courts should adopt the recently developed 350 ppm scientific prescription – which calls for a global trajectory of six per cent carbon emissions reduction beginning in 2012 (along with 100 GtC extraction) – as a general atmospheric fiduciary obligation shared by all co-trustee sovereigns on Earth. *This global trajectory is the marker to which courts around the world may calibrate in assigning carbon reduction pathways to sovereign trustees in their own jurisdictions.*[169]

In the big picture, this planetary carbon reduction can only be met by every nation taking responsibility for the problem. Stated another way, the necessary global emissions reductions will be achieved only if reductions among all nations add up so as to satisfy the required 'carbon math'. Each industrialized nation must carry out its proportion of the overall planetary carbon reduction, or it will leave a major, deadly 'orphan share' on the doorstep of the world. An orphan share is a share of liability for which the liable party does not take responsibility. In the context of carbon reduction, any significant orphan share is likely to defeat global efforts to reduce emissions adequately in the short timeframe needed. No nation is equipped to adopt a significant orphan share left by another sovereign. Therefore,

---

165   Ban Ki-moon (2009) 'Copenhagen Diagnosis', See also Union of Concerned Scientists (2007), p. 2: (noting the 'costs of delay are high', requiring accelerated emissions reductions).

166   See Baer et al. (2009), p. 4 (delineating a global carbon reduction trajectory to achieve 350 ppm, contemplating a peak in emissions in 2011 and a ten per cent annual reduction, and noting, 'if the 350 pathway is defined to have a global peak that's a mere four years later – if emissions continue to rise until 2015 – then the subsequent decline would have to reach a nearly unimaginable rate of 20 per cent per year').

167   Ban Ki-moon (2009), 'Copenhagen Diagnosis', p. 51, fig. 22 (depicting different reduction trajectories based on start dates, but calibrating the 2C goal rather than 1.5).

168   See Ban Ki-moon (2009), p. 7: 'The risk of transgressing critical thresholds ("tipping points") increases strongly with ongoing climate change. Thus waiting for higher levels of scientific certainty could mean that some tipping points will be crossed before they are recognized'; Hansen (2009), p. 171: 'If the world does not make a dramatic shift in energy policies over the next few years, we may well pass the point of no return'; see also Hansen (2008a).

169   See *infra*: *The Anti-Waste Doctrine Applied Differentially to Carbon Reduction.*

a bedrock principle of atmospheric trust liability must be the inexcusability of orphan shares and partial orphan shares.

Causes of Action

The trust framework presents two causes of action, available to different classes of parties, to enforce the atmospheric fiduciary obligation. The first is an action by citizen beneficiaries against their governmental trustees for failing to protect their natural trust. It is well settled that beneficiaries may sue the trustee to protect their property.[170] In the US, for example, citizens are positioned to bring trust actions against their states or the federal government.[171] The second is an action brought by one sovereign trustee against another for committing waste to common property. Co-tenants have a right against other co-tenants for waste.[172] Nations, states, or tribal sovereigns may bring an action for waste against other nations, states or tribal sovereigns. This may be possible both in international courts, and in some cases through domestic jurisdiction. Waste and breach of trust claims find grounding within the same basic property framework; both link to the scientific prescription of carbon reduction as the expression of duty.

As with any claim, of course, a myriad of issues may bar recovery. Litigants must navigate potential barriers such as standing, sovereign immunity, pre-emption, political question doctrine, ripeness, jurisdiction, and intervention, among others. This chapter does not discuss such hurdles, as they vary considerably with the context in which the particular legal claim is brought. In charting the broad terrain of atmospheric trust litigation, however, it should be noted that courts recognizing the enormity of climate crisis and the crucial role of the judiciary may approach these barriers with a leniency that is not characteristic of past decisions. At its core, the unparalleled force of the public trust doctrine is its mandate to preserve resources for future generations. It appoints the court to police the legislature and agencies in their management of trust assets. The substantive underpinning of the doctrine thus creates powerful arguments in defence of many potential barriers.

---

170 See Bogert (1987) § 154 at 551: 'If the trustee is preparing to commit a breach of trust, the beneficiary need not sit idly by and wait until damage has been done. He may sue in a court of equity for an injunction against the wrongful act'.

171 *Marks v Whitney*, 491 P2d 374, 381 (Cal 1971) (private citizens have standing to sue under public trust, though a court may raise the issue on its own). Of course issues of sovereign immunity may arise in such suits, and general constitutional requirements of standing apply.

172 63C Am. Jur. 2d Property § 31; *Chosar Corp v Owens*, 370 SE2d 305, 307–08 (Va 1988) (co-tenants who allowed mining without consent of all other co-tenants were liable for waste); *Anders v Meredith*, 1839 WL 525 (NC 1839); see also *US v Washington*, 520 F2d at 685 (discussing waste in context of sovereign co-tenancy in migrating fishery).

The Anti-Waste Doctrine Applied Differentially to Carbon Reduction

To review, the essential legal starting point in atmospheric trust litigation is the fiduciary obligation to protect and prevent 'waste' to the asset. These principles bind all sovereign co-tenant trustees – that is, all nations on Earth. In the end, domestic courts must define the fiduciary obligation with a mind towards the planetary prescription for carbon reduction – it must be fully met. In an uncomplicated world that we can only imagine, the most straightforward conceptual way of accomplishing this planetary reduction would be an across-the-board mandate on all countries to reduce their own emissions according to the planetary reduction trajectory. Thus, for example, if the planetary prescription calls for a six per cent reduction in *global* emissions per year (tied to a specified uniform baseline year), then every country would have to reduce emissions by that amount. This method, indeed, would ensure that the carbon reduction 'adds up' to the required amount. Looking at it slightly differently, if each and every piece of the 'pollution pie' is reduced by a fixed amount, the pie as a whole will shrink by the same amount. If nothing else, this shows that the courts must be attentive to the planetary prescription as a marker trajectory.

The reality, however, is that countries stand on remarkably different footing, on a number of levels, regarding their carbon emissions. Some have minimal carbon emissions per capita; asking these nations to shoulder the same immediate proportionate burden as the countries with much larger amounts of per capita emissions would not only be patently unfair but impractical as well, because such reduction may compromise citizens' basic living needs. Moreover, some countries (like the United States and other industrialized nations) have contributed vastly greater amounts of historic pollution, while other countries have contributed only minimal amounts. Other key differences, described below, exist as well among nations. These challenges complicate the task of allocating liability for carbon reduction. Nevertheless, the task can be avoided only at peril to humanity's collective future. Recognizing the many disparities among countries of the world, the UNFCCC called for 'common but differentiated responsibilities' in reducing carbon emissions.[173] Unfortunately, though the UNFCCC declaration announced a useful umbrella concept, diplomatic negotiations have utterly failed to untangle the basic differences that speak to climate responsibility. Given the multitude of differentiating factors and the number of nations engaged in diplomatic processes, it is likely that the circumstantial complexity will continue to stymie efforts towards an international compact. Indeed, key differences have created an impasse between the North and the South (wealthy versus developing countries) in international negotiations.[174] As noted at the outset, it may well be that the task simply overwhelms international diplomacy.

---

173   UNFCCC, Art. 3, Par. 1.
174   Baer et al. (2008), pp. 83–90.

Despite conventional assumptions favouring international processes, judges are hardly novice to the task of formulating principled factors and arranging them into a coherent liability scheme. Their discipline, training, and processes are all geared towards applying basic principles of fairness to sets of complicated, and often disputed, facts. They regularly allocate liability among multiple players in complex natural settings and do so with the public interest in mind. For example, in the hazardous waste context, courts may impose responsibility on dozens or even hundreds of parties for cleaning up a contaminated site. They allocate scarce water supplies among hundreds or even thousands of competing claimants in river basins. In many different types of case, courts have determined the 'fair share' of responsibility that various sovereign parties must assume relative to other parties[175] – which, after all, is the basic task of an international carbon reduction scheme.

Of course, no liability scheme is perfect, but judicial tradition allows for arriving at rough approximations of justice rather than insisting on precise formulations that could hopelessly drag out the process. The very nature of equitable power allows judges to craft remedies to fit the circumstances. The remedies might not always carry out the various parties' rights in precise or perfect form. As one federal appellate court emphasized in approving a plan for carrying out sovereign rights to a shared treaty fishery, the remedy is an 'amalgam of delicate balancing, gross approximations and rough justice'.[176] And, as one famous treatise on equity observes, 'Courts of Justice aim at practical good and general convenience rather than at theoretical perfection'.[177] The point is important for atmospheric trust litigation remedies. Because so little atmospheric 'space' remains for further carbon pollution (that space having been largely consumed by industrialized countries), even the highest conceivable amounts of reduction by an industrialized country (most notably, the United States) will not fully compensate for its total contribution to the pollution.[178] The reality should not deter courts.

---

175   The 'fair share' was a standard for the allocating of fish between states and tribes in the Pacific treaty fishing cases, for example. See, generally, Blumm and Steadman (2009), p. 666. Courts also devised a 'fair share' standard to reflect municipal obligations to meet regional low-income housing needs. *See S. Burlington County, NAACP v Township of Mount Laurel* (*Mt Laurel I*), 336 A2d 713, 724 (NJ 1975) (municipality's fair share of the present and prospective regional need).

176   *US v Oregon*, 913 F 2d 576, 580–81 (9th Cir 1990).

177   Willard (1863), p. 151.

178   See Baer et al. (2008), Executive Summary. Such liability for past pollution would best be satisfied through large international payments flowing from the industrialized nations to the developing nations. Those nations need financial compensation and technology transfers in order to accomplish carbon transition. *Id.* Though such transfers are undoubtedly part of a global climate scheme, they fall outside the classic purview of a domestic court. Developing nations in need, however, might draw upon their status and authority as co-tenant sovereign trustees of the atmosphere to pursue natural resource damages against large private corporations, which are largely responsible for carbon pollution of the atmosphere. Needless to say, such nations would have to find ways of

Carbon reduction must commence immediately – further delay risks catastrophe. The fact that any one domestic court cannot solve the whole problem should not dissuade judges. As the United States Supreme Court emphasized in *Massachusetts v EPA*, '[A plaintiff] need not show that a favourable decision will relieve his *every* injury'. As another court reasoned in allowing citizens to challenge greenhouse gas pollution for its effect on climate change:

> Particularly in environmental and land use cases, the challenged harm often results from the cumulative effects of many separate actions that, taken together, threaten the plaintiff's interest. The relief sought ... need not promise to solve the entire problem, any more than a legislative body is forbidden to enact a law addressing a discrete part of a problem rather than the entire problem.[179]

To assess the carbon reduction responsibility of their own nations, domestic courts should refer to the planetary prescription for carbon reduction as a marker. Ultimately, any climate prescription will likely call for near-zero carbon emissions to be achieved by a future date, most likely 2050.[180] Courts of various countries, then, must impose a timeline for reducing emissions with an endpoint that achieves near-zero emissions. Courts that shorten the timeframe for achieving near-zero emissions will effectively impose a steeper curve of reduction.[181] While on one hand it may seem an inherently arbitrary task for a court to devise a timeframe, on the other hand, courts do this all the time. In nearly every enforcement case, judges have to impose timeframes for their remedies. They do so on the basis of reason and equity, both of which should govern climate responsibility. At least five factors inform a judicially-imposed carbon reduction path. All of these factors have surfaced in international discussions and literature, but they have never been arranged into a basic formula of responsibility.

The first factor is the country's global share of carbon emissions. The United Nations compiles data that shows every nation's slice of the carbon pollution pie.[182] The United States and China lead as the top two polluters in the world, each with about a 20 per cent share.[183] The top ten polluters account for roughly

---

enforcing such monetary judgements in the polluters' home countries where the assets are located. Such an initiative is not included in this description of ATL, but it complements the trust strategy contained herein for holding governments accountable for carbon reduction.

179　*NEDC v Owens Corning Corporation*, 434 F Supp 2d 957, 968 (D Or 2006).

180　See Baer et al. (2009), pp. 1–4.

181　By way of illustration, an 80 per cent reduction of pollution by 2020 will require far more aggressive annual reduction than achieving the same by 2040, assuming the same baseline year from which the reduction is calibrated.

182　See United Nations, Millennium Development Goals Indicator, Carbon Dioxide Emissions (CO2), Thousand Metric Tons of CO2 (CDIAC), available at http://mdgs.un.org/unsd/mdg/SeriesDetail.aspx?srid=749&crid=.

183　For a user-friendly presentation of global greenhouse pollution data, see *List of Countries by Carbon Dioxide Emissions*, available at http://en.wikipedia.org/wiki/

two-thirds of the world's total carbon pollution.[184] Judges in those countries should impose a particularly aggressive timeframe for carbon reduction, paying heed to the planetary baseline as a minimum reduction trajectory. They should search the outer bounds of feasibility – bearing constant mind to the reality that, if carbon reduction does not happen in their countries, major orphan shares of carbon will loom, shares potentially capable of plunging the planet into full catastrophe.

The second factor is historical emissions. In other legal contexts, historical practices alone are the basis of liability. If the atmosphere were a straightforward hazardous waste dump, for example, it could be cleaned up with money from an overall clean-up fund. The various liable parties would contribute money according to their equitable share.[185] The problem with climate crisis, however, is that there is no way to simply pay for 'cleaning up' the atmosphere, because there is no technology to artificially 'draw down' carbon. Unlike a relatively simple clean-up of a hazardous waste site, the historical emissions factor is not easily translatable into quantified responsibility for carbon reduction. It is, however, an equity factor that can reinforce an aggressive trajectory of reduction. (Certainly, on this level, the United States bears a colossal share of responsibility.)

A third factor is the country's 'per capita emissions'. This is the amount of carbon dioxide emitted on average by each person in the particular country.[186] The numbers vary widely, generally reflecting the dichotomies between the industrialized and emerging nations. The American lifestyle, for example, produces nearly 20 metric tons of carbon dioxide emissions per capita on average.[187] India, on the other hand, produces only about 1.16 metric tons of carbon dioxide pollution

---

List_of_countries_by_carbon_dioxide_emissions#List_of_countries_by_2007_emissions (hereinafter *List of Countries*). Data from 2007 showed that China had about a 22 per cent share, and the U.S. had nearly a 20 per cent share. The UN compiles detailed data on the greenhouse gas emissions of most countries of the world. See UNFCC, *GHG Data from UNFCC*, http://unfccc.int/ghg_data/ghg_data_unfccc/items/4146.php.

184   See *List of Countries, supra* n. 183.

185   A common law doctrine known as contribution assigns responsibility and liability among joint tortfeasors based on equitable factors. See generally, 18 Am Jur 2d Contribution § 1. Principles of joint liability allow courts to apportion liability among joint contributors. See Restatement 2d of Torts, § 433(a), Apportionment of Harm to Causes; see also *Burlington Northern & Santa Fe Ry v United States*, 129 S Ct 1870 (2009) (apportioning harm in hazardous waste cleanup context). In hazardous waste cleanups, courts use several factors to equitably allocate shares of liability. See, for example *Action Mfg Co v Simon Wrecking Co*, 428 F Supp 2d 28, 93–94 (ED Penn 2006).

186   See *List of Countries by Carbon Dioxide Emissions Per Capita*, available at http://en.wikipedia.org/wiki/List_of_countries_by_carbon_dioxide_emissions_per_capita; see also Union of Concerned Scientists, *Each Country's Share of CO2 Emissions*, available at http://www.ucsusa.org/global_warming/science_and_impacts/science/each-countrys-share-of-co2.html (using 2006 data from Energy Information Agency, Department of Energy).

187   See n. 186.

per person on average.[188] Both nations are among the top five polluters of the planet (reflecting a huge population disparity between the two).[189] The implication of this per capita emissions data is that some countries are using more than what could be considered their 'fair share' of carbon. Countries on the low-carbon-per-capita end of the spectrum include Brazil, Vietnam, the Philippines, Ghana, and Bangladesh, for example, all of which produce under two metric tons per capita.[190] On the far other end (along with the United States) are Luxemburg, Australia, Canada, Kuwait, and Qatar, for example. These countries have what could be called 'grossly unsustainable' per capita rates, ranging from 15 tons per capita up to Qatar's off-the-charts rate of 56.2 tons per capita.[191] These higher rates reflect both excessive consumption and capacity for dramatic reduction, both of which justify imposing a steep trajectory of carbon reduction.

A fourth factor is not easily obtainable from United Nations data, but is directly relevant to the waste doctrine. It is the purpose behind the activity causing the carbon emissions. The public trust doctrine examines the purpose behind using critical public resources,[192] and the traditional waste doctrine looks askew on extravagant uses, particularly in situations of scarcity. Water courts and allocation statutes, for example, prioritize domestic use of water over commercial use when water is in short supply.[193] Courts faced with atmospheric trust litigation can, and should, make similar judgements. As scientists emphasize, the planet has a limit to the amount of carbon pollution that it can absorb before crossing into a realm of catastrophic, and irrevocable, planetary heating. Some scientists have tried to quantify that amount in terms of a carbon budget for the world. Viewed in this way, the available 'space' for remaining carbon pollution is a scarce resource in itself. Like a river with too little water, the atmosphere has multiple demands, and courts must inevitably prioritize those calls on the resource. The trajectory of carbon reduction should be steeper for non-essential uses. To contend otherwise

---

188   See n. 186.

189   See n. 186

190   See n. 186.

191   See n. 186.

192   See, for example, *Robinson v Ariyoshi*, 65 Haw 641, 674, 658 P2d 287, 310 (Haw 1982) (public trust duty includes duty to 'assure that the waters of our land are put to reasonable and beneficial uses' and to 'assure the continued existence and beneficial application of the resource for the common good').

193   See, for example, OREGON. REV. STAT. ANN. § 540.140 (2010) ('When the waters of any natural stream are not sufficient for the service of all those desiring the use of the same, those using the water for domestic purposes shall, subject to such limitations as may be prescribed by law, ... have the preference over those claiming such water for any other purpose, ... and those using the water for agricultural purposes shall have the preference over those using the same for manufacturing purposes'); Amos (2008) ('When water rights with the same priority date are in mutually exclusive conflict, domestic uses have preference over all others, and agricultural uses have preference over manufacturing uses').

ignores the reality that there is simply not enough 'space' left for all of the carbon pollution that 'Business As Usual' produces.

In this vein, it is helpful to organize the carbon emissions into three categories that logically carry descending order of priority: 1) domestic; 2) transition; and 3) luxury. Much like in the water context, 'domestic carbon' connotes pollution made in service of basic human needs – which would describe the cook stoves of India or Tibet, or energy needed to heat homes for basic health and comfort. 'Transition carbon' signifies the pollution needed to create new infrastructure for a low-carbon society (pollution resulting from the manufacture of bicycles, windmills and solar panels, for example). 'Luxury carbon' denotes pollution caused by non-essential and frivolous needs. The English government recently tagged this type of carbon pollution by rejecting new airport runways near London that would serve 'binge flying' – described to include 'jetting off to weekend homes in Spain and bachelor parties in Prague'.[194] As the damaging activity becomes more excessive relative to the basic needs of the general population, it becomes less tolerable from an equity and waste standpoint. Just as water courts may curtail luxury uses of water in times of scarcity, so might courts impose a steep trajectory of reduction on luxury carbon emissions. By the same reasoning, countries with large populations that emit most carbon pollution in service of domestic purposes and basic human needs may formulate trajectories that provide for meeting those needs, while spending available transition carbon in furtherance of a new low-carbon infrastructure. The waste doctrine is sufficiently elastic to accommodate such considerations, if courts are perspicacious enough to recognize them.

The fifth factor is the recalcitrance of the sovereign in taking responsibility for its carbon pollution. Here again, countries stand in markedly different positions. While some have made great efforts to reduce their pollution, others (like the United States) have made little progress. In the past, courts overseeing environmental cleanup settlements have delivered what is called a 'penalty for recalcitrance', which amounts to an extra monetary amount imposed as punishment for not settling or co-operating earlier.[195] The same approach may be suited to the climate context. The gross abdication of responsibility on the part of huge polluters like the United States has pushed the planet precariously close to the tipping point. There is little doubt that the failure of the industrialized nations to cut their pollution sooner will result in untold damage to the Earth's resources. The penalty for recalcitrance should translate into a steeper, more urgent, emissions reduction trajectory for such countries.

By way of summary, in order to save the planet, judicial decisions worldwide must set the various nations on a course of *aggregate carbon reduction* that meets the planetary prescription set by climate scientists. The UNFCCC set the framework for 'common but differentiated responsibilities', but failed to quantify

---

194   See Rosenthal (2010).

195   See, for example, *Simon Wrecking Company*, 428 F Supp 2d at 99–100 (noting availability of recalcitrance penalty but finding it not suitable for that case).

the various nations' obligations in concrete terms. Through the waste doctrine, courts can give quantitative definition, and enforcement, to this international standard. Just as contemplated by the UNFCCC, the waste doctrine – deriving from longstanding equitable concepts – moulds itself to variable circumstances.[196] The factors explained above indicate parameters of flexibility.

Much of the work in extrapolating individualized sovereign trajectories for carbon reduction from these equitable factors has already been done in a leading analysis, the *Greenhouse Development Rights Framework* (GDRF), prepared by the Stockholm Environment Institute and EcoEquity.[197] The GDRF presents a framework of responsibility based on the UNFCCC's standard of common and differentiated responsibilities (which, again, also reflects an anti-waste standard).[198] The GDRF template is fixed by two parameters – 'responsibility and capacity' – that bring some definition to the first four factors explained above. 'Responsibility' is a proxy for the country's contribution to atmospheric contamination based on historic cumulative emissions since 1990.[199] 'Capacity', generally derived from national financial data, reflects the ability of the nation to carry out carbon reduction without threatening the basic survival capacity of its population. To this end, the capacity parameter excludes the income demanded by the necessities of daily life.[200] 'Capacity' is a more nuanced approach than mere per capita income average, because it 'takes explicit account of the unequal distribution of income within countries'.[201] Using these two factors, statistical data for individual countries are assembled into a straightforward 'Responsibility Capacity Index' or RCI, which represents each country's logical share of the global 'ecological debt'.[202] Then, based on each nation's RCI, the report delineates individualized carbon reduction trajectories for each nation.[203] Crucially, the pathways are correlative and cumulative, *designed to calibrate to, and collectively meet, a planetary trajectory for necessary carbon reduction.* The RCI approach thereby offers a global distribution scheme of carbon reduction – one tied to the logical factors that both quantify the duty against waste and also give concrete meaning to the UNFCCC's standard of 'common but differentiated responsibilities'. As the authors describe, the GDRF 'proceeds in the only possible way, by operationalizing

---

196   See, generally, Willard (1863), p. 370 (describing different standards of waste between England and the United States).

197   Baer et al. 2008.

198   Baer et al. (2008), Executive Summary, p. 16.

199   Baer et al. (2008), Executive Summary, p. 18 (but excluding emissions attributable to consumption below the development threshold).

200   Baer et al. (2008), Executive Summary, pp. 16–17 (defining 'capacity' as total national income excluding income below a 'development threshold').

201   Baer et al. (2008), Executive Summary, p. 18.

202   Baer et al. (2008), Executive Summary, pp. 17–19; see also Baer et al. (2008), pp. 93–8 (Table A1, presenting RCI for all countries).

203   Baer et al. (2008), pp. 76–9.

the official principles of the [UNFCCC]'.[204] Moreover, the formula is designed to move forward through time, incorporating change within a stable framework of equity. As a template for allocating carbon reduction, the GDRF can yield revised trajectories as data changes both in respect to the nations' individualized circumstances[205] and the globe's atmospheric needs (which may be re-defined as more scientific data emerges).[206]

In sum, the waste principle creates a domestic legal framework that positions nations of the world in an equitable relationship with each other while calibrating to planetary requirements of carbon reduction. Principles of trust responsibility can yield quantitative measures applicable to all governments of the world, enforceable through domestic processes in nations having independent judicial branches. The GDRF allows domestic courts to assign individual sovereign responsibility as part of a macro, uniform approach to global carbon reduction. Of course, ATL will not culminate in successful judgements in every nation. There will still be enforcement gaps leaving 'orphan shares' of liability for carbon emissions reduction. Nevertheless, use of the atmospheric trust framework described herein should impose a frame of fundamental sovereign obligation that exposes orphan shares to the citizens of the world and thereby advances efforts to hold governments accountable in both the domestic political realm and in the context of international negotiations. This approach gives domestic force, through the courts, to the UNFCCC's expressed duty to carry out carbon reduction according to 'common but differentiated responsibilities' – a principle agreed to by nearly every nation on Earth, yet persistently obviated by international diplomatic impasse.

Relying on a judicially created property law framework to give meaning to treaty obligations is not unprecedented. In the United States, there is a rich history of courts interpreting and enforcing broad treaty obligations as to natural resources shared between sovereigns. The Indian fishing cases, originating out of fish allocation disputes between states and tribes, gave rise to a vast and much celebrated body of case law that gave detailed interpretation to basic treaty language. The language was sparse, reserving Indian rights to take fish *'in common with'* the non-Indian settlers.[207] Without the willingness of courts to define, in practical terms, the rights and responsibilities arising from this basic treaty obligation, Indian fishing that had endured 10,000 years surely would have been extinguished by industrialization. No state legislature was willing to carry out the

---

204   Baer et al. (2008), p. 16.

205   For example, the authors show how China's burden would triple between 2010 and 2030 if it continues its rapid growth and wealth increase. See Baer et al. (2008), p. 19.

206   The trajectories in the GDRF were calibrated to the former goal of 2 °C., which, as previously noted, will not be sufficient to protect the planet's major fixtures. The authors are currently in the process of recalibrating the RCI, based on updated data, to the 350 trajectory associated with the 1.5 °C degree goal.

207   See Blumm and Steadman (2009), p. 654 (citing treaty language) (emphasis added).

treaty promise that the United States had solemnly agreed to. In recent times, these same federal courts have advanced their common law jurisprudence to confront the modern crisis of extinction. Still relying on the basic treaty promise and the property inferences they spawned, courts have defined duties of environmental protection of the imperilled fish and the waterways that sustain them.[208] Even as some courts today hesitate to confront climate change, it is worth remembering that, based on just three key words lodged in the Indian treaties, United States courts constructed a co-tenancy framework, declared a waste principle, and created detailed, practical remedial structures to give force and effect to sovereign property rights.[209] Today, courts confront basically the same task. In order to save the planet from catastrophic heating, they must give force to the public trust responsibility towards the atmosphere, which finds elucidation in the principled language of the UNFCCC.

While the suggested analysis may seem over-simplified in response to multifarious policy concerns and complex science, the urgency in launching planet-saving efforts requires a decisive and straightforward approach that the judicial branches of government can spearhead across the planet through atmospheric trust decrees. The fact that any one court cannot enforce a global reduction scheme – because its jurisdiction is domestic – should not dissuade courts of any nation. As the United States Supreme Court recognized in *Massachusetts v EPA*, the climate problem can be tackled on the domestic level despite the lack of jurisdiction over other nations: 'Nor is it dispositive that developing countries such as China and India are poised to increase greenhouse gas emissions substantially over the next century: A reduction in domestic emissions would slow the pace of global emissions increases, no matter what happens elsewhere'.[210]

In this manner, atmospheric trust litigation invokes a decentralized judicial strategy to achieve what has long eluded the centralized (and thus far ineffectual) diplomatic system of treaty negotiation. Of course, any decentralized approach may yield variable results across jurisdictions. But variability may be neither an outrageous nor unwelcome prospect. It is perhaps time to recognize that, while uniformity is well suited to times of stability, times of crisis might be best met by innovation and experimentation. If anything, the decades-long quest for a centralized international climate compact has caused dangerous stagnation in domestic response and threatens to draw standards down to the lowest common denominator. Atmospheric trust litigation breaks the mould by inviting judicial innovation in defining and enforcing carbon emissions reduction at the domestic level, worldwide. Moreover, the global marker trajectory provides a baseline that offers a consistent starting point to this process.

---

208   Blumm and Steadman (2009): (describing treaty litigation as it moved into habitat protection); see also Blumm and Swift (1998).

209   For background, see Blumm and Steadman (2009).

210   *Massachusetts v Environmental Protection Agency*, 549 US 497, 500 (2007).

Forms of Relief

It is important to design a remedy for atmospheric trust litigation with a view towards providing the macro relief necessary to fulfil the sovereign's share of global emissions reduction. Both declaratory and injunctive relief are appropriate in the atmospheric trust context.[211] As to the first, a declaratory judgement is a straightforward remedy that can greatly advance the task of clarifying responsibilities of governments worldwide in addressing climate crisis. A declaratory judgement should clearly iterate the following principles:

1. all governments have a fiduciary obligation, as trustees, to protect the atmosphere as a commonly shared asset;
2. all governments bear liability for reducing carbon emissions;
3. this fiduciary obligation is organic to government and permits no orphan shares or partial orphan shares of responsibility;
4. the fiduciary obligation is enforceable by the citizen beneficiaries of the trust representing present and future generations;
5. the fiduciary obligation and the concomitant duty to prevent waste is enforceable by co-tenant trustees;
6. the fiduciary obligation calibrates to a scientific prescription for carbon reduction designed to restore global carbon dioxide levels to below 350 parts per million.[212]

Judicial declarations setting forth the trust framework for atmospheric obligations in this manner may alone spur some carbon reduction, because the mere declaration of responsibility from a constitutional branch of government furnishes citizens with the conceptual tools they need to hold their governments accountable.

Declaratory relief should, if possible within the governing legal system, be accompanied by suitable injunctive relief that allows courts to provide a remedy on a macro level without invading the province of the political branches.[213] Courts

---

211 Sovereign trustees are also positioned to pursue natural resource damages against atmospheric polluters, but that claim is outside the scope of this chapter. The claim might be confused with, but is theoretically quite distinct from, nuisance cases brought by states against major polluters. For an overview of climate nuisance theory, see Merrill (2005). Natural resource damages are common in oil spill cases such as the BP oil spill (off the Gulf of Mexico), and the Exxon Valdeez spill (off the coast of Alaska). For a brief discussion of natural resource damages in the context of climate, see Wood (2008).

212 While declaring macro sovereign responsibility, this iteration of the atmospheric trust principle seemingly satisfies the rule that a declaratory judgement should not be a 'general admonition', but must be narrowly crafted to define a duty according to 'concrete facts presented in a particular dispute'. *United States v Washington*, 2007 US Dist LEXIS 61850 *23 (WD Wash. 2007).

213 *Weinberger v Romero-Barcelo*, 456 US 305, 312 (1982) (the basis for injunctive relief is a finding of irreparable injury and the absence of an adequate legal remedy).

have emphasized that the core purpose of the public trust doctrine is to police the other branches of government in their disposition of public assets.[214] In the United States, by drawing on traditional relief available against co-tenants and trustees for misuse of property, courts may require carbon accountings and enforceable carbon budgets as remedies for sovereign breach of the atmospheric fiduciary obligation without reaching into the law-making purview of the other branches. In other countries, these same tools may be extrapolated to fit the unique judicial function within those systems.

An accounting is a traditional remedy springing from the equitable powers of the court in both the co-tenancy and trust contexts.[215] It is a judicial process whereby co-tenants or trustees must account for expenses and/or profits in connection with the trust; it allows beneficiaries to ensure the proper management of their trust assets.[216] Accordingly, courts have held that 'any beneficiary, including one who holds only a present interest in the remainder of a trust, is entitled to petition the court for an accounting'.[217] The accounting also plays a role in the co-tenancy context, in that each co-tenant is responsible for his share of the expenses, and is due his share of the profit from the property.[218] The scope of an accounting

---

214   *See Ariz Ctr for Law in the Pub Interest v Hassell*, 837 P2d 158, 169 (Ariz Ct App 1991) ('The check and balance of judicial review provides a level of protection against improvident dissipation of an irreplaceable res'); see also *Lake Michigan Federation v United States Army Corps of Engineers*, 742 F Supp 441, 446 (N D Ill 1990) ('The very purpose of the public trust doctrine is to police the legislature's disposition of public lands').

215   See, for example, *Evans v Little*, 271 SE 2d 138, 141 (Ga 1980) (co-tenancy); *Willmon v Koyer*, 143 P 694, 695 (Cal 1914) ('As an incident to a co-tenancy relationship, either co-tenant has a right to demand of the other an accounting as to rents and profits of the co-tenancy, which of course, involves the right of one co-tenant to have refunded to him by the other his proportion of any expenditures made for the benefit of the common property'); *Zuch v Conn. Bank & Trust Co*, 500 A2d 565, 568 (Conn App 1985) ('As a general matter of equity, the existence of a trust relationship is accompanied as a matter of course by the right of the beneficiary to demand of the fiduciary a full and complete accounting at any proper time') (citations omitted); *Cobell v Norton*, 240 F3d 1081 (DC Cir 2001) (*Cobell VI*) (accounting against federal government for mismanagement of Indian trust funds).

216   *Evans*, 271 SE 2d at 141.

217   In re *Estate of Ehlers*, 911 P2d 1017, 1021 (Wash App 1996) (citing *Nelsen v Griffiths*, 585 P2d 840, 843 (Wash App. 1978)).

218   See, for example, *Garber v Whittaker*, 174 A 34, 37 (Del Super Ct 1934) ('Tenants in common of the legal title to land are ordinarily entitled to the use, benefit and possession of such land, including their just and proper shares of the rents and profits therefrom'); *Willmon v Koyer*, 143 P 694, 695-96 (Cal 1914) ('The rule is that when one tenant in common has paid a debt or obligation for the benefit of the joint property, or has discharged a lien or assessment imposed upon it as a common burden, he is entitled as a matter or right to have his co-tenant, who has received the benefit of it, refund to him his proportionate share of the amount paid'); see also *Stoebuck and Whitman* (2000) (where a co-tenant derives income from a use of land that permanently reduces its value, the co-tenant must account to the other co-tenants); *White v Smyth*, 214 S.W.2d 967, 978 (Tex. 1948) ('When it

must include 'all items of information in which the beneficiary has a legitimate concern'.[219] In the financial context, this means a statement 'in clear and concise terms of the nature and value of the corpus of the trust ... and the amount and location of any balance or remainder'.[220]

In the context of atmospheric trust litigation, an accounting would take the form of quantifying carbon emissions and tracking their reduction over time. This form of accounting is an extrapolation from the traditional remedy in two ways. First, it is applied against a sovereign trustee, not a private trustee. It is well established in the US, however, that a sovereign defendant may be subject to an accounting for mismanagement of a trust. In the federal Indian law context, recently, the US government was subject to a multi-billion dollar accounting action for its mismanagement of tribal trust funds.[221] Second, a carbon accounting draws upon a tool developed in the financial context and tailors it to the natural context. Commentators have advocated trust accountings in the natural resources area, and such a leap should be well within the imagination of judges.[222] Courts have essentially engaged in natural 'accountings' in the environmental context before, without using the label. For example, courts are quite familiar with assigning monetary value to resources and awarding natural resource damages to governmental trustees.[223] Moreover, in determining rights to fish runs or water resources shared between states and tribes, courts delve deeply into the quantitative aspects of beneficial resource use.[224]

The accounting required by atmospheric trust litigation consists of a judicially-supervised quantification of the amount of greenhouse gas pollution emitted by the sovereign defendant. The accounting establishes the current carbon pollution emitted on the particular jurisdictional level (local, state, or federal) so as to define a baseline and then tracks progressive reduction over time. Modern modelling is capable of quantifying a carbon footprint on virtually any scale, from individual to global. Much of the necessary data has been developed and is already accessible. In the United States, for example, the Department of Energy maintains data on the overall carbon emissions of all 50 states.[225] While inevitably there will be areas

---

is claimed that a co-tenant in possession of ... property has become liable to his co-tenants for profits accruing from his productive operations, the usual mode of settling the account is to charge him with all his receipts and credit him with all his expenses, thereby ascertaining the net profits available for distribution [among co-tenants]').

219   *Zuch*, 500 A2d at 568.

220   *Zuch*, 500 A2d at 568.

221   See *Cobell v Kempthorne*, 455 F3d 317, 319–21 (DC Cir 2006) (describing background of litigation).

222   See Torres (2002), p. 547 (calling for accounting); Babcock (2009), p. 410.

223   See, generally, *Coeur D'Alene Tribe v Asarco Inc.*, 280 F Supp 2d 1094 (D Id 2003).

224   See, generally, Wood (2000).

225   Raw data for state carbon dioxide emissions are available from the Energy Information Administration, US Department of Energy, *Energy Emissions Data and Environmental Analysis of Energy Data*, available at http://www.eia.doe.gov/environment.html.

of dispute regarding some emissions sources, particularly mobile sources, the methodology for measuring jurisdictional carbon footprints will continue to be refined as professional standards emerge in the field of carbon accounting.

Carbon accounting allows co-tenants and beneficiaries of the trust to evaluate government's measures to protect the atmospheric trust. Without such an accounting, the legislatures can easily mislead the public into thinking that carbon reduction measures are adequate – whereas in fact they may grossly exceed the planet's capacity to assimilate pollution. As Professor Coplan points out in his article on the atmospheric trust, legislative cap-and-trade bills in the US have not been tied to any sustainable capacity of the atmosphere; instead, they have offered a scheme to privatize pollution rights 'far in excess of the IPCC-determined maximum sustainable emissions'.[226] He notes: 'In essence, the sovereign 'trustee' would invade the public trust 'corpus' to make distributions in excess of the sustainable yield of the atmospheric 'trust''.[227] A judicially supervised accounting process would expose such legislative misappropriations of the people's atmospheric trust.

While an accounting remedy provides the means whereby a beneficiary or co-tenant can measure the performance of a governmental trustee, additional injunctive relief is necessary to enforce the sovereign duty to restore the natural trust where it has been damaged. At a very simple level, the fiduciary obligation to reduce carbon pollution (per a judicially set trajectory) can be carried out through a 'budget' for carbon reduction over time that sets forth quantifiable mileposts. The jurisdiction must develop a plan containing measures calibrated to bring about such reduction.[228] Enjoining the sovereign to do so does not invade the prerogatives of the other branches, because the court does not dictate to the trustee how to accomplish the carbon reduction. Cities, counties, and states have wide latitude in devising plans that are tailored to the unique circumstances of their jurisdiction. In this respect, the trust remedy may strike the ideal balance between necessarily potent, macro judicial enforcement and traditional deference to the political branches.

Because carbon reduction must occur steadily over the long term, a court must maintain on-going jurisdiction over an ATL case. Continuing jurisdiction is a regular feature of litigation requiring protracted remedies against government institutions.[229] Over time, the court must receive periodic progress reports, a common feature of accounting cases. These reports inform the court and the

---

226   Coplan (2010), p. 330.
227   Coplan (2010), p. 330.
228   Proposed British legislation provides an example of a 'carbon budget'. (2007) 'Britain Proposes Bold Environmental Legislation That Could Pave Way for Post-Kyoto Pact'.
229   In the Pacific Northwest treaty litigation, for example, federal courts have recognized the need for ongoing jurisdiction, which has lasted over three decades. See, for example, *United States v Washington*, 384 F Supp 312, 346 (WD Wash 1974) (acknowledging, in context of a treaty fishing dispute between a state and tribes, that the remedy in face of changing conditions is best carried out through 'retention of continuing

beneficiaries whether the trustee is making adequate progress in accordance with the budget and plan. Trust accountings are usually performed on a regular basis, such as quarterly, biannually or annually, and contain an updated inventory of the *corpus* or *res* (the assets of the trust).[230] In the case of ATL suits, the reports would show the jurisdiction's progress in reducing carbon during the reporting period. In view of the narrow window of time remaining before climate thresholds are crossed, courts are justified in requiring carbon accounting reports every quarter or, at the very least, every six months. Such reports should be posted on websites so as to be easily accessible to the sovereign's own citizen beneficiaries, world citizens, and to other ATL courts.

While some judges may be overwhelmed by the novel and all-encompassing context of carbon reduction, it is important to bear in mind that the envisioned judicial role is much the same as in other natural resource contexts where courts enforce management and/or recovery of diminished natural assets. Again, the treaty fishing cases of the late 1960s and 1970s provide a model. The recalcitrance of state fishery managers against allowing tribal harvest was so embedded in state governance that the district courts of Oregon and Washington tasked themselves with detailed supervision of tribal and state salmon harvests.[231] The federal district court of Oregon created a consent decree structure whereby the states and tribes developed a judicially supervised and enforceable plan for future harvest of the salmon.[232] More recently, in the ESA lawsuits over the imperilled Columbia River salmon, the same court has assumed a rigorous role overseeing the development of a fish recovery plan pursuant to a process of multi-sovereign consultation structured by the court.[233] Courts have supervised broad plans to address exclusionary zoning, racial desegregation, and prison reform.[234] While courts must be cognizant of appropriate judicial boundaries in structuring relief for trust violations, they arguably have wide latitude in requiring sovereigns to develop

---

jurisdiction, more appropriate than overly-detailed judicial predetermination') (citation omitted); see also Blumm and Steadman (2009) (summarizing litigation).

230   See, for example, *Fraser v Southeast First Bank of Jacksonville*, 417 So2d 707, 708 (Fla 5 Dist App 1982) (citing Florida statutes); *Cobell v Norton*, 240 F3d 1081, 1086 (DC Cir 2001) (*Cobell VI*) (quarterly reports in Indian trust litigation against the federal government).

231   See *Puget Sound Gillnetters Ass'n v US Dist. Court*, 573 F 2d 1123, 1133 (9th Cir 1978); *United States v Washington*, 520 F2d 685, 686, 693 (9th Cir 1975); see also discussion at Wood (2006), pp. 10176–7.

232   See discussion at Wood (1998).

233   See Wood (2006), pp. 10175–6.

234   See *Coleman v Schwarzenegger*, 2009 WL 2430820 (ED Cal 2009) (Three Judge Federal Court, August 4 2010) (prison reform case), available at http://www.caed.uscourts.gov/caed/Documents/90cv520o10804.pdf; *Southern Burlington County, NAACP v Township of Mount Laurel*, 336 A2d 713 (NJ 1975) (Mt Laurel I) (land use reform); *Southern Burlington County, NAACP v Township of Mount Laurel*, 456 A2d 390 (NJ 1983) (Mt Laurel II)(same); see also discussion in Singer (2006).

enforceable plans for proper trust management.[235] Judicial consent decrees are an ideal tool for holding sovereign defendants to court-imposed obligations, yet allowing for regulatory expertise in developing and carrying out a detailed remedy structure.

Although courts will not be able to enforce every minute detail of a carbon reduction plan, many courts have it well within their power to force carbon reduction through discrete injunctive measures tailored towards obvious carbon sources. An injunction may contain 'backstops' that consist of measures the court will mandate if the budget is not carried out. The broad realm of environmental and land use litigation provides precedent for many measures that may serve as effective backstops. Such measures might include, for example, injunctions prohibiting new coal-fired plants, large-scale logging, recreational vehicle use on public lands, airport expansions, sewer hook-ups, issuance of air pollution permits, and a myriad other activities.[236] It is within the traditional province of courts of equity to devise relief to remedy the harm.[237] Of course, the ultimate enforcement mechanism is to hold government officials personally in contempt of court for failure to carry out court-ordered fiduciary duties.[238]

---

235   See *Cobell v Kempthorne*, 455 F3d 317, 330–31 (DC Cir 2006) (reviewing reversals of district court remedies in an Indian trust accounting case); *Cobell v Norton*, 283 F Supp 2d 66 (D DC 2003), rev'd on other grounds, 392 F3d 469 (DC Cir 2004).

236   See, for example, *United States v Metropolitan Dist Comm'n*, 757 F Supp 121, 128–29 (D Mass 1991), *aff'd*, 930 F2d 132 (1st Cir 1991) (moratorium against sewer hook up); Jeffery J Matthews, (1999) (discussing injunctions imposing moratoria against sewer hookups); *American Motorcyclist Ass'n v Watt*, 543 F Supp 789, 798 (C D Cal, 1982) (enjoining off-road vehicle use because agency plan did not comply with the statute); *Pacific Rivers Council v Thomas*, 30 F3d 1050 (9th Cir 1994) (enjoining the US Forest Service from proceeding with projects under land resource management plans prior to ESA consultation); *Lane County Audubon Soc'y v Jamison*, 958 F2d 290, 294 (9th Cir 1992) (enjoining the BLM from new timber sales until ESA consultation was completed); *Thomas v Peterson*, 753 F2d 754 (9th Cir 1975) (enjoining construction of road until agency prepared biological assessment); *Oregon Natural Desert Assn v Singleton*, 75 F Supp 2d 1139 (D Or 1999) (permanently enjoining domestic livestock grazing in all 'areas of concern'). While most of the precedent for such injunctions is grounded in claims brought under statutory law, the relief awarded is typically not statutorily mandated, but rather devised by a court to afford a meaningful remedy.

237   See *Alaska Ctr for the Env't v Browner*, 20 F3d 981, 986 (9th Cir 1994) ('The district court has broad latitude in fashioning equitable relief when necessary to remedy an established wrong'); *Weinberger v Romero-Barcelo*, 456 US 305, 311 (1982) ('The essence of equity jurisdiction has been the power of the [court] to do equity and to mould each decree to the necessities of the particular case. Flexibility rather than rigidity has distinguished it').

238   See, for example, Gouras (2008): (district court threatened to hold US Agriculture Undersecretary Mark Rey with contempt of court and jail time for the agency's 'systematic disregard of the rule of law').

The above discussion provides mechanisms by which a court may enforce aggressive and sustained carbon reduction in a particular sovereign jurisdiction. As the prior discussion noted, this task is both urgent and feasible for the industrialized world. The ATL remedy might be immediately aimed at different objectives in the developing world. As noted earlier, while all counties of the world stand as co-trustees of the atmosphere, developing nations are situated differently from the industrialized nations. Trajectories for carbon reduction for the developing countries will look much different, in part because such nations presently contribute a smaller amount of the global pollution and because they lack the finances to afford rapid transition to a low-carbon economy. This, however, does not mean that atmospheric trust litigation is any less important in these countries. To the contrary, it is equally imperative for several reasons.

First, the atmospheric trust, as a macro approach to climate, is truly global. The framework requires judicial acknowledgement that every sovereign, not just willing sovereigns, bear responsibility for atmospheric health. The declaratory judgements issued by courts in developing countries can be instrumental in setting forth the principles of the atmospheric trust and creating the framework for global reduction. Such declarations of principle may be persuasive to courts of industrialized nations. Second, if the planet is to achieve near-zero carbon emissions, every country must ultimately participate in the effort, despite the differing timeframes for doing so. It will be important for courts of developing nations to require their governments to take the initial step of planning a transition to near-zero carbon emissions and setting a cost on funding that transition. Third, this cost quantification will send a signal to the industrialized world on how much funding assistance is necessary, and in this sense, court-ordered reduction plans may have a significant impact on the diplomatic realm. Fourth, such plans may provide a basis for natural resource damage recovery actions by developing nations against atmospheric polluters in their own domestic legal systems. By recognizing a trust framework, all sovereigns theoretically have grounds for recovering damages from third parties who destroy the trust. It is a settled principle of trust law that trustees have the affirmative duty to recoup monetary damages against third parties that destroy trust assets.[239] These funds, gained from the private fossil-fuel industry polluters, could substantially help finance the zero-carbon economic transition in the developing world.[240]

Fifth, ATL suits that establish a carbon reduction trajectory in developing countries will likely have an impact on the energy assumptions driving economic growth scenarios in those countries. The political branches in developing nations may be more inclined to invest in renewable energy than in the cheaper fossil fuel energy if faced with an ATL judgement forcing carbon reduction over the long term. In other words, an ATL suit is not just aimed at reducing existing pollution levels in the industrialized nations, but it is also aimed at thwarting investment

---

239   Restatement (Second) of Torts § 177 (1959).
240   Natural resource damage suits are beyond the scope of this chapter.

decisions that would send future pollution rates soaring in the developing world. Sixth, ATL suits in developing countries are crucial for addressing the reforestation and agricultural practices that make up the 'drawdown' component of the scientific prescription for returning the atmosphere to 350 ppm equilibrium. As noted earlier, such practices must be implemented worldwide in order to achieve the 100 GtC necessary for a return to 350 ppm. Developing countries are situated to protect their forests, reforest their logged lands, and implement soil practices in the short term even if their transition to a zero-carbon economy will occur over a much longer term. Also, for nations with significant deforestation pressures, the ATL suits are crucial to enjoin further logging. And finally, ATL suits in developing nations may seek orders requiring the governments of those countries to plan adaptation strategies for climate disruption. While adaptation plans do not directly achieve carbon reduction, they spotlight the irrevocable damage that climate change will bring. By necessarily detailing the global heating differences associated with different levels of carbon in the atmosphere, adaptation plans of developing countries may fortify aggressive mitigation efforts in the industrialized world, as wealthy countries see the consequences of a world hurtling towards climate chaos. Moreover, these adaptation plans will likely present a platform of equity that may advance diplomatic efforts to hold industrialized nations accountable for their fair share of climate recovery.

Conclusion

As the planet approaches climate points of no return, trial attorneys and community lawyers across the globe should unite in a worldwide legal movement to hold recalcitrant governments responsible for reducing greenhouse gas emissions. Atmospheric trust litigation is a high-stakes, high-yield strategy that offers three crucial features currently missing from any other approach. First, it galvanizes a sheer moral force that eludes most procedure-laden, highly technical, statutes. The atmosphere is an endowment to which today's children, as well as remote future generations, have a legitimate moral and legal claim as beneficiaries. Government's failure to safeguard this priceless asset amounts to gross breach of fiduciary duty and, ultimately, generational theft. Second, the trust framework overhauls government's posture from one based on political discretion (allowing officials to ignore the problem) to one based on strict fiduciary obligation (requiring government to protect crucial public property assets belonging in common to the citizens). Third, the trust framework brings a new conceptual macro-approach to a global problem that embroils virtually every nation on Earth. By defining the atmosphere as common property, the trust positions all nations of the world in a logical relationship towards each other and towards their common atmosphere. Specifically, this approach:

1. yields a carbon reduction pathway tied to planetary atmospheric requirements (nature's own laws);
2. characterizes an atmospheric fiduciary obligation incumbent on all sovereigns; and
3. packages the scientific prescription for carbon reduction in a way that lends itself to enforcement in the domestic courts of the world's sovereigns.

Such a macro approach is essential in order to solve the problem on the level that it presents itself. Statutory approaches, which are micro in nature, present a dangerous risk of failing to conceptualize, much less address, atmospheric pollution in its entirety. The trust approach appeals to domestic courts to apply the force of law in singular yet orchestrated and coherent fashion to protect global common property – the atmosphere – that belongs to the present and future generations of the planet.

Inevitably, atmospheric trust litigation will be criticized on the basis that it invites courts to overstep their function and intrude into a matter best left to the political branches. If the world could rewind several years of time, that criticism would carry far more weight. But after two futile international climate treaty negotiations in the past five years and the refusal of most polluting nations to pass meaningful domestic legislation, climate crisis screams out for a reality check.[241] Children born in 1992, the year nations of the world signed the UNFCCC, have arrived at adulthood with no meaningful action to protect their planet – and their future survival. Withering droughts in Australia and Africa; infernos in Greece and California; devastating floods in Pakistan, India, and the American Midwest; ferocious hurricanes on the Gulf Coast; searing heat waves in Russia and France; melting glaciers in Montana, Tibet, Alaska, Chile, and Argentina; thawing polar icecaps and permafrost; worldwide ocean acidification; and the tortuous impending drowning of island-nations across the planet – these all portend a future unfolding because governments worldwide have abdicated their most fundamental duty to safeguard the natural inheritance held in sovereign trust for all generations. As the Secretary General of the United Nations warned the world in 2009: 'Climate change is happening. The evidence is all around us. And unless we act, we will see catastrophic consequences ... . The time for hesitation is over'.[242] Given the stakes of planetary heating for the youth alive today, much less their future children, one would expect legislatures and agencies worldwide to respond with emergency haste to climate crisis, rendering litigation altogether unnecessary. Instead, political leaders and governing institutions around the world still push the world on a deadly Business as Usual course. Having squandered any further opportunity

---

241 Over two decades have passed since the United States Congress was notified by the nation's leading climate scientist that, with 99 per cent confidence, the planet was warming dangerously due to human-caused greenhouse gas pollution, yet Congress has still failed to act to control one of the largest national shares of global pollution. See Hansen (2009), p. xv.

242 See Ban Ki-moon (2009).

for slow, incremental policy, the political branches still demonstrate a shocking reticence to the threat of runaway heating that imperils all of humanity. Courts are a last resort – but a resort nonetheless.

Atmospheric trust litigation challenges lawyers and judges to take fundamental principles of public trust law and apply them in coherent fashion in a new and urgent context so as to arrive at a uniform, quantifiable measure of governmental responsibility to reduce carbon. At a time in history when thinkers worldwide are calling for new, innovative technologies and practices to address climate crisis, lawyers should pioneer promising, if untested, legal constructs to address carbon loading of the atmosphere. Ultimately, the public trust doctrine and the primordial rights that infuse it are part of a populist manifesto that surfaces through the generations of Humanity, no less revolutionary for our time and our crisis than was the forcing of the Magna Carta on the English monarchy in 1215,[243] or Mahatma Gandhi's great Salt March to the sea in 1930.[244] The difference between those great struggles and the present crisis is one of global scale, mind-blowing urgency, and the unthinkable consequences of failure to future generations. By bringing an orchestrated campaign of public trust lawsuits in countries throughout the world, lawyers may gather the most powerful elements of public property rights into one revolutionary global legal movement that not only forces governmental protection of the atmosphere but also catalyses citizen environmental democracy and advances universal human rights.

There should be no doubt that judges have it well within their ability to issue decisions forcing carbon reduction. In past eras, judges have called forth logic and principled reasoning to formulate law in response to unprecedented circumstances. As Justice Holmes wrote, the common law is '[t]he felt necessities of the times'.[245] The power of equity to provide relief is formidable, described by Justice Story in his famous treatise as that 'admirable intervention of judicial polity, which interposes preventative guards against impending dangers and mischiefs, and which does not [wait] until the destructive blow has been dealt'.[246] Judges in many civil law systems have analogues that can carry powerful trust principles forward

---

243 The Magna Carta is often cited as a source of the public trust doctrine, as it forced the monarchy to open access to resources such as navigable waterways. For discussion, see Cronin and Kennedy (1999), pp. 139–42.

244 The British had imposed a heavy tax on salt and exercised a monopoly over its production and sale. The common people were forbidden from collecting salt, which was vital for preservation and other needs. Cast in public trust terms, the British government fully alienated an element of the public trust corpus to corporate interests. Gandhi rejected the British position forbidding the people from harvesting a natural resource and consequently led a non-violent march to the sea for the purpose of collecting salt. So many people were arrested that the jails overflowed and the British had to change the law and accept their right to collect salt. For a summary, see Lal (1949).

245 Holmes (1881), p. 1.

246 Story (1836), p. 203.

as well. As John Willard wrote in his Treatise on Equity Jurisprudence, 'Equity must have a place in each system of jurisprudence, in substance if not in name'.[247]

Unfortunately, after decades of modern environmental law, even many judges in common law systems are now so accustomed to issuing rulings within detailed confines of legislation or regulations that they may have lost an inclination to construct meaningful remedies using their powerful traditional prerogatives of equity. Nevertheless, history tells us that conditions of impossibility often inspire imagination and courage. The irrevocable damage associated with the climate tipping points is unprecedented, far beyond the ability of any future legislature to mitigate or repair. Handed the right complaint, there will no doubt be path-breaking judges who, no matter what their nationality, recognize this epochal moment in the course of human civilization and will exert their judicial authority to protect the globe's atmosphere – for the sake of not only their nation's citizens, but also for the many billions of people dependent on Earth's life systems for all time to come.

Postscript

As this chapter was going to press, Atmospheric Trust Litigation (ATL) materialized in a 'hatch' of lawsuits and petitions filed on behalf of youth plaintiffs simultaneously across the United States in the first week of May, 2011. The actions, co-ordinated by Our Children's Trust, consist of a federal lawsuit, nine state law suits, 39 petitions for state rule-making, and one notice of intent to sue – a legal campaign covering all 50 states in the US.[248] Internationally, an ATL lawsuit was filed in Ukraine, with subsequent legal actions planned in other countries as well.[249] The suits and petitions declared a sovereign atmospheric trust duty and called upon government to produce carbon accountings and plans for annual emissions reductions of at least six per cent (as well as to embark on reforestation and improved soil practices to achieve carbon drawdown from the atmosphere). The orchestrated legal actions calibrated the government's fiduciary obligation to the scientific prescription produced by Dr James Hansen and other scientists in early May 2011, which set a path for returning the planet to atmospheric equilibrium at 350 ppm by the end of the century through aggressive carbon emissions reduction, reforestation, and improved soil practices.[250]

---

247   Willard (1863), p. 38.

248   The complaints and petitions are compiled on the website of Our Children's Trust. See http://www.ourchildrenstrust.org/legal-action/lawsuits (lawsuits) and http://www.ourchildrenstrust.org/legal-action/petitions (petitions). For media coverage, see http://www.ourchildrenstrust.org/media.

249   The Ukraine complaint is posted at: http://www.ourchildrenstrust.org/sites/default/files/Ukraine%20ATL%20English.pdf.

250   See Hansen, et al. (2011).

This initial 'hatch' of ATL actions corresponded with 125 youth-led marches (called the 'iMatter Marches') in cities across the US and in countries around the world, from Kuwait to Nepal, to Pakistan, to Bangladesh. The youth marches were organized by a non-profit organization, Kids vs Global Warming.[251] Subsequent youth marches were planned for summer of 2011 in India, Ghana, Egypt, Mississippi, The Netherlands, England, Mexico City, and New York City, among many other locations.[252] Alec Loorz, the 16-year-old lead plaintiff in the federal ATL lawsuit and founder of *Kids v Global Warming*, explained the youth's legal and social protest campaign in the following terms:

> [A]s youth, we are the last group of people in the US who don't have any official political rights. We can't vote, [and] we certainly can't compete with rich corporate lobbyists ... . So we are forced to simply trust our government to make good decisions on our behalf ... . However, it's become clear that our government has failed us, by not protecting the resources on this planet we need to survive. Even though scientists overwhelmingly agree that CO_2 emissions are totally messing up the balance of our atmosphere, our leaders continue to turn their backs on this crisis.

> The time has come for the youngest generation to hold our leaders accountable for their actions ... . Today, I and other fellow young people are suing the government, for handing over our future to unjust fossil fuel industries, and ignoring the right of our children to inherit the planet that has sustained all of civilization. I will join with youth and attorneys in every state in the US to demand that our leaders live and govern as if our future matters.

> The government has a legal responsibility to protect the future for our children. So we are demanding that they recognize the atmosphere as a commons that needs to be preserved, and commit to a plan to reduce emissions to a safe level.

> The plaintiffs and petitioners on all the cases are young people. We are standing up for our future ... . But we will not only stand up in the courts. We will stand up in the streets as well ... .

> This is a movement. A mass movement of young people standing up with a unified voice to tell the ruling generation that we will no longer just sit idly by as they make decisions that threaten our future. We matter. Our future matters.[253]

---

251 See http://imattermarch.org/.

252 See http://imattermarch.org/.

253 Why One 16-Year-Old Is Suing the US Government Over Climate Change (2011).

References

(2007) 'Britain Proposes Bold Environmental Legislation That Could Pave Way for Post-Kyoto Pact', *International Herald Tribune*, 13 March, available at http://www.iht.com/articles/ap/2007/03/13/europe/EU-GEN-Britain-Climate-Change.php.

(2008) *You Want Loopholes With That?*, available at http://www.ecoequity. org/2010/08/you-want-loopholes-with-that-2/#more-812.

(2009) 'UN Scientist Backs "350" Target for CO_2 Reduction', *Yahoo! News*, 25 August.

(2010) 'The Changing Climate For Environmental Legislation', *Newsweek*, available at http://www.newsweek.com/blogs/the-gaggle/2010/08/31/the-changing-climate-for-environmental-legislation.html.

(2011) 'Why One 16-year-old is Sueing the US Government Over Climate Change', *Guardian Environmental Network*, 5 May 2011, available at http://www.guardian.co.uk/environment/2011/may/05/sueing-us-government-climate.

David Adam (2008) 'World Carbon Dioxide Levels Highest for 650,000 Years, Says US Report', *The Guardian*, 13 May 2008, at 16, available at http://www.guardian.co.uk/environment/2008/may/13/carbonemissions/climatechange.

Adell Amos (2008) 'Freshwater Conservation in the Context of Energy and Climate Policy: Assessing Progress and Identifying Challenges in Oregon and the Western United States', 12 *U. Denver. Water L. Rev.* 1, 52, 65 (2008).

Thomas T. Ankersen (2003) 'Shared Knowledge, Shared Jurisprudence: Learning to Speak Environmental Law Creole (Criollo)', 16 *Tulane Env. L. J.* 807.

Hope M. Babcock (2009) 'The Public Trust Doctrine: What a Tall Tale They Tell', 61 *S. C. L. Rev.* 393.

Paul Baer, Tom Athanasiou, Sivan Kartha and Eric Kemp-Benedict (2008) *The Greenhouse Development Rights Framework 2d*, available at http://www.ecoequity.org/docs/TheGDRsFramework.pdf.

Paul Baer, Tom Athanasiou and Sivan Kartha (2009) *A 350 ppm Emergency Pathway: A Greenhouse Development Rights Brief*, available at http://gdrights.org/wp-content/uploads/2009/11/a-350-ppm-emergency-pathway-v2.pdf.

Ban Ki-moon (2009) 'Statement of, Secretary-General of the United Nations on the United Nations Environment Programme', *Climate Change Science Compendium II*, available at http://www.unep.org/pdf/ccScienceCompendium2009/cc_ScienceCompendium2009_full_en.pdf.

Peter Barnes (2006) Who Owns the Sky: Our common assets and the future of capitalism, Island Press.

(2004) *Black's Law Dictionary* 1477 (8th edn).

Michael C. Blumm and Brett M. Swift (1998) 'The Indian Treaty Piscary Profit and Habitat Protection in the Pacific Northwest: A Property Rights Approach', *Uni. Of Colorado Law Review*, available at http://papers.ssrn.com/sol3/papers.cfm?abstract_id=871518.

Michael C. Blumm and Jane G. Steadman (2009) 'Indian Treaty Fishing Rights and Habitat Protection: The Martinez Decision Supplies a Resounding Judicial Affirmation', 49 *Nat. Res. J.* 653.

George T. Bogert (1987) *Trusts*, 6th edn, West Pub. Co.

John Boitnott (2008) 'Berkeley Scientists: World in 'Mass Extinction Spasm'– Scientists: Humans to Blame', *News Report NBC*, 12 August, available at http://www.nbc11.com/news/17171725/detail.html.

City of Seattle (2006) *A Climate of Change: Meeting the Kyoto Challenge, Climate Action Plan Highlights*, available at http://www.seattle.gov/climate/docs/SeaCAP_summary.pdf.

Craig Collins (2010) *Toxic Loopholes: Failures and Future Prospects for Environmental Law*, Cambridge University Press.

Steve Connor (2007) 'The Earth Today Stands in Imminent Peril', *The Independent*, 19 June, available at http://environment.independent.co.uk/climate_change/article2675747.ece.

Karl S. Coplan (2010) 'Public Trust Limits on Greenhouse Gas Trading Schemes: A Sustainable Middle Ground?' 35 *Columbia J. of Env. L.* 287.

Robin Kundis Craig (2010) 'Adapting to Climate Change: The Potential Role of State Public Trust Doctrines', 34 *Vermont L. Rev..*

John Cronin and Robert F. Kennedy (1999) *The Riverkeepers*, Scribner.

Timothy P. Duane (2010) 'Greening the Grid: Implementing Climate Change Policy Through Energy Efficiency, Renewable Portfolio Standards, and Strategic Transmission Systems', 34 *Vermont L. Rev.* 711.

Harrison Dunning (1989) 'The Public Trust: A Fundamental Doctrine of American Property Law', 19 *Envtl. L.* 515.

Paul Finn (1995) 'A Sovereign People, A Public Trust, in P. Finn (ed.), *Essays on Law and Government*, Law Book Co.

Lisa Friedman (2010) 'US Bound by Obama's Copenhagen Emissions Pledge – U.N. Official', *Greenwire,* 10 January 2010, available at http://www.nytimes.com/gwire/2010/01/20/20greenwire-us-bound-by-obamas-copenhagen-emissions-pledge-17687.html.

Ross Gelbspan (2004) *Boiling Point: How Politicians, Big Oil and Coal, Journalists, and Activists Have Fueled a Climate Crisis–And What We Can Do to Avert Disaster*, Basic Books.

Ross Gelbspan (2007) 'Beyond the Point of No Return', 11 December, *Grist*, available at http://www.grist.org/article/beyond-the-point-of-no-return/.

Jennifer Gleason and Bern Johnson (1995) 'Environment Laws Across Borders', 10 *J. Envtl. L. & Litig.* 67.

Robert Glicksman (2008) 'Sustainable Federal Land Management, Protecting Ecological Integrity and Preserving Environmental Principle', 44 *Tulsa L. J.* 147.

Dale D. Goble and Eric T. Freyfogle (2002) *Wildlife Law, Cases and Materials*, West.

Suzanne Goldenberg, John Vidal and Jonathan Watts (2009) 'Leaked UN Report Shows Cuts Offered at Copenhagen Would Lead to 3C Rise', *The Guardian*, available at http://www.guardian.co.uk/environment/2009/dec/17/un-leaked-report-copenhagen-3c.

Jeff Goodell (2006*) Big Coal: The Dirty Secret Behind America's Energy Future*, Houghton Mifflin Harcourt.

Al Gore (2007) 'Moving Beyond Kyoto', *N.Y. Times*, 1 July.

Matt Gouras (2008) 'Judge: Ag Undersecretary Avoids Jail Time', *Associated Press*, available at http://hosted.ap.org/dynamic/stories/b/bush_official_contempt?site=ap&section=home&template=default&ctime=2008-02-28-00-41-37.

Douglas L. Grant (2001) 'Underpinnings of the Public Trust Doctrine: Lessons from Illinois Central Railroad', 48 *Ariz. St. L.J.* 849.

James Hansen (2006a) 'Climate Change: On the Edge', *The Independent*, 17 February, available at http://environmentindependent.co.uk/article345926.ece.

James Hansen (2006b), 'The Threat to the Planet', *The N.Y. Rev. of Books* 12, 13 July, available at http://www.nybooks.com/articles/19131.

James Hansen (2007a) 'Why We Can't Wait', *The Nation*, 7 May.

James Hansen (2007b), 'Dangerous Human-Made Interference with Climate – Testimony Before Select Committee on Energy Independence and Global Warming', US House of Representatives 5, 26 April, available at http://www.columbia.edu/~jeh1/testimony_26april2007.pdf.

James Hansen (2008a) 'Twenty Years Later: Tipping Points Near on Global Warming', *The Huffington Post*, 23 June, available at http://www.huffingtonpost.com/dr-james-hansen/twenty-years-later-tippin_b_108766.html.

James Hansen (2008b) 'Tipping Point: A Perspective of a Climatologist', *2008–09 State of the Wild*, available at http://www.columbia.edu/~jeh1/2008/StateOfWild_20080428.pdf.

James Hansen (2009) *Storms of My Grandchildren: The Truth About the Coming Climate Catastrophe and Our Last Chance to Save Humanity*, Bloomsbury.

James Hansen et al. (2007a) 'Climate Change and Trace Gases', 365 *Phil. Trans. R. Soc. A*, 1925, 1949, *available at* http://www.planetwork.net/climate/Hansen2007.pdf.

James Hansen et al. (2007b) 'Dangerous Human-Made Interference With Climate: A GISS Model Study', 7 *Atmos. -. Phys.* 2287, 2303, available at http://www.atmos-chem-phys.net/7/2287/2007/acp-7-2287-2007.pdf.

James Hansen et al. (2008) 'Target Atmospheric CO_2: Where Should Humanity Aim?', 2 *The Open Atmospheric Sciences Journal* 217, available at http://arxiv.org/abs/0804.1126, and at http://www.columbia.edu/~jeh1/2008/TargetCO2_20080407.pdf.

James Hansen et al. (2011) 'The Case for Young People and Nature: A Path to a Healthy, Natural, Prosperous Future', draft available at http://www.columbia.edu/~jeh1/mailings/2011/20110505_CaseForYoungPeople.pdf.

Richard G. Hildreth, David R. Hodas, Nicholas A. Robinson, James Gustave Speth (2010) *Climate Change Law: Mitigation and Adaptation*, West.

O. W. Holmes, (1881) *The Common Law*, Dover Books.

E. Hopkins (1896) *Handbook on the Law of Real Property*, Law Book Exchange.

David Hunter, Chris Wold and Melissa Powers (2009) *Climate Change and the Law*, Matthew Bender.

Robert F. Kennedy Jr (2005) *Crimes Against Nature: How George W. Bush and His Corporate Pals Are Plundering the Country and Hijacking Our Democracy*, Harper Perennial.

Jennifer Koons (2008) 'Following Pa. Mining Town's Example, Ecuador OKs Constitution Giving Rights to Nature', *Greenwire*, 30 September, available at http://celdf.org/article.php?id=185.

Laura H. Kosloff and Mark C. Trexler (2007) *Consideration of Climate Change in Facility Permitting*, in Michael B. Gerrard (ed.), *Global Climate Change*, American Bar Association.

Hanno Kube (1997), 'Private Property in Natural Resources and the Public Weal in German Law – Latent Similarities to the Public Trust Doctrine?', 37 *Natural Resources Journal* 857.

Jan G. Laitos, Sandra B. Zellmer, Mary C. Wood and Dan H. Cole (2006) *Natural Resources Law*, Thompson West.

Vinay Lal (1949) *Dandi, Salt March: History and Politics*, Manas, available at http://www.sscnet.ucla.edu/southasia/History/Gandhi/Dandi.html.

Gregory M. Lamb (2007) 'A Key Threshold Crossed', *Christian Sci. Monitor*, 11 October, available at http://www.csmonitor.com/2007/1011/p11s01-wogi.html (quoting climate scientist Tim Flannery.

Geoffrey Lean (2007) *A World Dying, But Can We Unite to Save It?*, *The Independent*, UK, 18 November, available at http://www.independent.co.uk/environment/climate-change/a-world-dying-but-can-we-unite-to-save-it-400847.html.

Jeffery J. Matthews (1999) 'Clean Water Act Citizen Suit Requests for Municipal Moratoria: Anatomy of a Sewer Hookup Moratorium Law Suit', 14 J. *Envtl. L. & Litig.* 25.

Michael McCarthy (2009) 'Carbon cuts Only Give 50/50 Chance of Saving Planet', *The Independent*, UK, 9 March.

Bill McKibben (2007) 'Remember This: 350 Parts Per Million', *Washington Post*, 28 December, available at http://www.washingtonpost.com/wp-dyn/content/article/2007/12/27/AR2007122701942.html.

Thomas W. Merrill (2005) 'Global Warming as a Public Nuisance', 30 *Colum. J. Env. L.* 293.

Vincent Mulier (2006) 'Recognizing the Full Scope of the Right to Take Fish Under the Stevens Treaties: The History of Fishing Rights Litigation in the Pacific Northwest', 31 *Am. Indian L. Rev.* 41.

Ved P. Nanda and William K. Ris, Jr (1976) 'The Public Trust Doctrine: A Viable Approach to International Environmental Protection', 5 *Ecol. L. Q.* 291.

President Nasheed (2009) 'UN Scientist Backs "350" Target for CO_2 Reduction', *High Level Conference on Climate Change: Technology*

Development and Transfer, New Delhi, India, 22 October, available at http://www.newdelhicctechconference.com/InauguralSession/Speech-PresidentofMaldives.pdf.

Patrick Parenteau (2010) *Come Hell And High Water: Coping with the Unavoidable Consequences of Climate Disruption*, available at: http://www.vjel.org/docs/Parenteau_Water_Draft.pdf.

Fred Pearce (2007) *With Speed and Violence: Why Scientists Fear Tipping Points in Climate Change*, Beacon Press 2007.

Jona Razzaque (2001) 'Case Law Analysis: Application of Public Trust Doctrine in Indian Environmental Cases', 13 *J. Envtl. L.* 221.

Restatement (Second) of Torts §821B (1959).

Restatement (Second) of Trusts § 177 (1959).

J. W. Rockström, W. Steffen, K. Noone, Å. Persson, F. S. Chapin III, E. Lambin et al. (2009a) 'Planetary Boundaries: Exploring the Safe Operating Space for Humanity', 14 *Ecology and Society* 2.

J. W. Rockstrom, W. Steffen, K. Noone et al. (2009b) 'A Safe Operating Space for Humanity', *Nature 461*, September.

Joseph Romm (2008a) 'Study: Water-Vapor Feedback is "Strong and Positive", So We Face "Warming of Several Degrees Celsius"', *Climate Progress Blog*, available at http://climateprogress.org/2008/10/26/study-water-vapor-feedback-is-strong-and-positive-so-we-face-warming-of-several-degrees-celsius.

Joseph Romm (2008b) *Is 450 ppm (or less) Politically Possible? Part 0: The Alternative is Humanity's Self-Destruction*, available at http://climateprogress.org/2008/04/26/is-450-ppm-or-lesspolitically-possible-part-0-the-alternative-is-humanitys-self-destruction.

Elizabeth Rosenthal (2007) 'UN Chief Seeks More Climate Change Leadership', *The New York Times*, 18 November, available at http://www.nytimes.com/2007/11/18/science/earth/18climatenew.html?

Elizabeth Rosenthal (2010) 'Britain Curbing Airport Growth to Aid Climate', *The New York Times*, 1 July, available at http://www.nytimes.com/2010/07/02/science/earth/02runway.html?emc=eta1.

Peter H. Sand (2004) 'Sovereignty Bounded: Public Trusteeship for Common Pool Resources?', 4 *Global Envtl. Politics* 47.

Joseph L. Sax (1970) 'The Public Trust Doctrine in Natural Resource Law: Effective Judicial Intervention', 68 *Mich. L. Rev.* 471.

Joseph William Singer (1997) *Property Law: Rules, Policies, and Practices*, 2nd edn, Little, Brown and Co.

Joseph William Singer (2006) *Property Law: Rules, Policies, and Practices*, 4th edn, Little, Brown and Co.

George P. Smith II and Michael W. Sweeney (2006) 'The Public Trust Doctrine and Natural Law: Emanations Within a Penumbra', 33 *B.C. Envtl. Aff. L. Rev.* 307.

Susan Solomona et al. (2009) 'Irreversible Climate Change Due to Carbon Dioxide Emissions', 106 *Proc Nat'l. Acad. Sci. U.S.* 1704 (Jan. 28, 2009), available at http://www.pnas.org/content/106/6/1704.full.pdf+html?sid=819c1042-fab1-4dce-88c7-e2c118f0f904.

James Gustave Speth (2008) *The Bridge at the End of the World: Capitalism, the Environment, and Crossing from Crisis to Sustainability*, Yale University Press.

David Spratt and Philip Sutton (2008) 'Climate Code Red: The Case for a Sustainability Emergency', *Friends of the Earth*, available at http://www.climatecodered.net/.

Jan S. Stevens (1980) 'The Public Trust: A Sovereign's Ancient Prerogative Becomes the People's Environmental Right', 14 *U.C. Davis. L. Rev.* 195.

William B. Stoebuck and Dale A. Whitman (2000) *The Law of Property*, 3rd edn.

Joseph Story (1836) *Commentaries on Equity Jurisprudence: as Administered in England and America*, Hilliard Gray and Co.

David Takacs (2008) 'The Public Trust Doctrine, Environmental Human Rights, and the Future of Private Property', 16 *NYU Envtl. L. J.* 711.

The University of New South Wales Climate Change Research Centre (2009) *The Copenhagen Diagnosis: Updating the World on the latest Climate Science*, available at http://www.copenhagendiagnosis.org/.

Gerald Torres (2002) 'Who Owns the Sky?', 19 *Pace Envtl. L. Rev* 515.

Mary Turnipseed, Raphael Sagarin, Peter Barnes, Michael C. Blumm, Patrick Parenteau and Peter H. Sand (2010) 'Reinvigorating the Public Trust Doctrine: Expert Opinion on the Potential of a Public Trust Mandate in US and International Environmental Law', 52 *Environment Magazine* 5.

United States Department of Commerce (NOAA, Earth System Research Laboratory) (2008) *Annual Greenhouse Gas Index 2008*, available at http://www.esrl.noaa.gov/news/quarterly/summer2009/2008_greenhouse_gas_index.html.

Union of Concerned Scientists (2007) 'How to Avoid Dangerous Climate Change: A Target for US Emissions Reduction', available at http://www.ucsusa.org/assets/documents/global_warming/emissions-target-report.pdf.

United States Global Change Research Program (2009) *Global Climate Change Impacts in the United States*, available at http://www.globalchange.gov/publications/reports/scientific-assessments/us-impacts/full-report.

John Vidal, Allegra Stratton and Suzanne Goldenberg (2009) 'Low Targets, Goals Dropped: Copenhagen Ends in Failure', *The Guardian*, 19 December, available at http://www.guardian.co.uk/environment/2009/dec/18/copenhagen-deal.

John Vidal (2011) 'Bolivia Enshrines Natural World's Rights with Equal Status for Mother Earth', *The Guardian*, 10 April, available at http://www.guardian.co.uk/environment/2011/apr/10/bolivia-enshrines-natural-worlds-rights.

W. Walsh (1947) *Commentaries on the Law of Real Property*, M. Bender.

Jonathan Watts (2010) 'India discloses Carbon Emissions for First Time Since More Than a Decade', *The Guardian*, 25 May, available at http://www.guardian.co.uk/environment/2010/may/25/india-carbon-emissions.

Edith Brown Weiss (1984) 'The Planetary Trust: Conservation and Intergenerational Equity', 11 *Ecology. L.Q.* 495.

John Willard (1863) *A Treatise of Equity Jurisprudence*, Platt Potter.

Charles F. Wilkinson (1989) 'The Headwaters of the Public Trust: Some of the Traditional Doctrine', 19 *Envtl. L. Rev.* 425.

M. C. Wood (1998) 'Reclaiming the Natural Rivers: The Endangered Species Act Applied to Endangered River Ecosystems', 40 *Ariz. L. Rev.* 197.

M. C. Wood (2000) 'The Tribal Property Right to Wildlife Capital (Part I): Applying Principles of Sovereignty to Protect Imperilled Wildlife Populations', 37 *Idaho L. Rev.* 1.

M. C. Wood (2006) 'Restoring the Abundant Trust: Tribal Litigation in Pacific Northwest Salmon Recovery', 36 *Envtl. L. Rep.* 10163.

M. C. Wood (2007) 'Nature's Trust: A Legal, Political and Moral Frame for Global Warming', 34 *Boston College Envt'l Affairs L. Rev.* (May 2007), available at http://www.law.uoregon.edu/faculty/mwood/docs/legal.pdf.

M. C. Wood (2008), 'A Framework of China-US Partnership to Address Global Warming', 3 *China Environmental and Resource Law Review*, Ocean University (Renmin Press), available at http://www.law.uoregon.edu/faculty/mwood/docs/china08.pdf.

M. C. Wood (2009a) 'Atmospheric Trust Litigation', in William C. G. Burns and Hari M. Osofsky (eds), *Adjudicating Climate Change: Sub-national, National, and Supranational Approaches*, Cambridge University Press.

M. C. Wood (2009b) 'Advancing the Sovereign Trust of Government to Safeguard the Environment for Present and Future Generations (Part I): Ecological Realism and the Need for a Paradigm Shift', 39 *Envt'l. L.* 43.

M. C. Wood (2010a) '"You Can't Negotiate With a Beetle": Environmental Law for a New Ecological Age', 50 *NAT. RES. L. J.*

M. C. Wood (2011 forthcoming) *Nature's Trust: Environmental Law for a New Ecological Age*, Cambridge University Press.

World Resources Institute (2005) *Climate Analysis Indicator Tool*, available at http://cait.wri.org/.

Victor John Yannacone Jr (1975) 'Agricultural Lands, Fertile Soils, Popular Sovereignty, The Trust Doctrine, Environmental Impact Assessment and the Natural Law', 51 *North Dakota L. Rev.* 615.

Alan Zarembo (2007) 'Kyoto's Failures Haunt New UN Talks', *L.A. Times*, 3 December, available at http://articles.latimes.com/2007/dec/03/science/sci-kyoto3.

Cases

1.58 Acres of Land, 523 F Supp 120 (D Mass 1981)

Action Mfg Co v Simon Wrecking Co, 428 F Supp 2d 28 (ED Penn 2006)

Alaska Ctr for the Env't v Browner, 20 F3d 981 (9th Cir 1994)

Alliance to Protect Nantucket Sound Inc v Energy Facilities Siting Board, 457 Mass. 663 (Mass S Ct 2010)

American Motorcyclist Ass'n v Watt, 543 F Supp 789 (C D Cal, 1982)

Anders v Meredith, 1839 WL 525 (NC 1839)

Arizona Center For Law In Public Interest v Hassell, 837 P2d 158 (Ariz App Div 1 1991)

Arnold v Mundy, 6 NJL 1 (NJ 1821)

Baxley v Alaska, 958 P2d 422 (Alaska 1998)

British Columbia v Canadian Forest Products, 2004 SCC 38

Burlington Northern & Santa Fe Ry v United States, 129 S Ct 1870 (2009)

California v. General Motors Corp., 2007 WL 2726871 (ND Cal 2007) (settled on appeal)

Center for Biological Diversity v FPL Group, 2008 WL 4255789 (Cal App 1 Dist, Sept 18, 2008)

Chosar Corp v Owens, 370 SE2d 305 (Va 1988)

Cobell v Kempthorne, 455 F3d 317 (DC Cir 2006)

Cobell v Norton, 240 F3d 1081 (DC Cir 2001)

Cobell v Norton, 283 F Supp 2d 66 (D DC 2003)

Coeur D'Alene Tribe v Asarco Inc, 280 F Supp 2d 1094 (D Id 2003)

Coleman v Schwarzenegger, 2009 WL 2430820 (ED Cal 2009)

Comer v Murphy Oil Co., No. 05-CV-436L Q (SD Miss, 20 Aug., 2007), rev'd in part, 585 F3d 855 (5th Cir 2009), judgement vacated, rehearing en banc granted, subsequently dismissed (for lack of quorum), 607 F.3d 1049 (5th Cir 2010)

Connecticut v American Electric Power, 406 F Supp 2d 265 (SD NY 2005), rev'd, 582 F3d 309 (2nd Cir 2009), cert granted, 131 S Ct 813 (2010)

Ctr. for Biological Diversity v Nat'l Highway Traffic Safety Admin, 508 F.3d 508 (9th Cir 2008)

Daubert v Merrill Dow Pharmaceuticals, Inc, 509 US 579 (1993)

Evans v Little, 271 SE 2d 138 (Ga 1980)

Fomento Resorts and Hotels Ltd v Minguel Martins, Civil Appeal Nos 4154 (S Ct India 2000)

Fraser v Southeast First Bank of Jacksonville, 417 So2d 707 (Fla App 1982)

Garber v Whittaker, 174 A 34 (Super Ct Del 1934)

Geer v Connecticut, 161 US 519 (1896)

Green Mountain Chrysler v Crombie, 508 F. Supp 2d 295, 313–17 (D Vermont 2007)

Her Majesty v City of Detroit, 874 F2d 332 (6th Cir 1989)

Hill v Ground, 114 Mo App 8, 343, 9 SW 343 (Ct App Mo 1905)

Idaho ex rel Evans v Oregon, 462 US 1017 (1983)

Th. Majra Singh v Indian Oil Corporation, AIR 1999 J&K 81

Thomas v Peterson, 753 F2d 754 (9th Cir 1975)

United States v Metropolitan Dist Comm'n, 757 F Supp 121 (D Mass 1991)

United States v Washington, 2007 US Dist LEXIS 61850 *23 (WD Wash. 2007)

United States v Washington, 384 F Supp 312 (WD Wash 1974)

United States v Washington, 520 F2d 685 (9th Cir 1975)

United States v White Mountain Apache Tribe, 537 US 465, (2003)

United States v Washington, 384 F Supp 312, (WD Wash 1974)

US v 1.58 Acres of Land, 523 F Supp 120, (D Mass 1981)

US v Oregon, 913 F 2d 576 (9th Cir 1990)

US v Washington, 384 F Supp312 (WD Wash 1974)

Waihole Ditch 94 Haw 97, 130–31 (Haw 2000)

Washington v Washington State Commercial Passenger Fishing Vessel Ass'n, 443 US 658 (1979)

Waweru v Republic, Misc. Civil Application No. 118 of 2004, at 689 (High Court, at Nairobi, March 2, 2006), available at http://www.chr.up.ac.za/index.php/browse-by-subject/339-kenya-waweru-v-republic-2007-ahrlr-149-kehc-2006-.html.

Weinberger v Romero-Barcelo, 456 US 305 (1982)

White v Smyth, 214 SW2d 967 (Tex 1948)

Willmon v Koyer, 143 P 694 (Cal 1914)

Zuch v Conn. Bank & Trust Co, 500 A2d 565 (Conn App 1985)

Legislation

106 Proc. Nat'l Acad. Sci. U.S. 1704.

Comprehensive Environmental Response, Compensation, and Liability Act (CERCLA), 42 U.S.C. § 9601 (2006).

Haw. Const., art. XI, §1.

India Const. Part III, art. 21.

Kenya Const.

Kyoto Protocol to the United Nationals Framework Convention on Climate Change, Dec. 10, 1997, 37 I.L.M. 32.

Louisiana. Const., art. IX, §1.

Oregon. Rev. Stat. Ann. § 540.140.

R.I. Const., art. I, §16.

S. Afr. Const. 1996 §24.

UNFCCC, S. Treaty Doc. No. 102–38, Art. 3.

Const. of Ukraine, art. 13.

United Nations Framework Convention on Climate Change, Copenhagen Accord, FCCC/CP/2009/L.7 (18 Dec., 2009).

Chapter 7

Fiduciary Principles and International Organizations

Donald Feaver

Introduction

A growing number of legal scholars, practitioners and jurists in the Anglo-American tradition acknowledge that 'there is something deeply fiduciary about the interaction between a state and its subjects'.[1] 'The most fundamental fiduciary relationship in our society is manifestly that which exists between the community (the people) and the state, its agencies and officials'.[2] Even though the notion that a fiduciary relationship should exist between a state and its people is appealing and intuitive, the fiduciary principle has traditionally been regarded as a private rather than public law concept.[3] Within national states, constitutions define the limits of state power while common law and statute provide standards that regulate public officials who exercise public decision-making powers. The same, however, cannot be said of public international organizations and officials. There is no body of international constitutional and administrative law that regulates public international decision-making processes. Hence, the question that is investigated in this chapter is whether and how the fiduciary principle might be applied as a standard to regulate the exercise of governance powers beyond the nation state?

Even though legal scholars haven't examined this question in great depth,[4] political economy and international relations scholars have been using 'fiduciary-like' concepts to investigate the behaviour of international organizations for

---

1 Fox-Decent (2005), p. 261.
2 Finn (1994), pp. 225–44.
3 Glover (2004).
4 In his excellent book, Prof Sarooshi discards the use of private law concepts as a method of regulating the behaviour of international organizations in favour of extending administrative law concepts from the domestic into the international sphere. This suggestion is not entirely convincing, as Sarooshi concedes, because of problems in the variation among and uneven application between the different administrative law principles between nation states. See: Sarooshi (2005).

several decades.[5] Proponents of the analytical approach, known as 'principal-agency theory', use fiduciary-like concepts as the foundation for political theories that explain how nation states *control* international organizations through the use of *political pressure*.[6] International lawyers have been reluctant to apply the fiduciary principle because, in large part, the legal relationship between nation states and international organizations is far more complex than agency theorists generally acknowledge.[7] Indeed, the most prominent international organizations do not conform to the legal definition of what constitutes an 'agent' and are not 'black boxes' created to act upon the instructions of nation states. As will be discussed later, a specific class of international organization, exemplified by the United Nations, the World Trade Organization and the International Court of Justice, possess and exercise a high degree of operational independence and decision-making autonomy.[8] These organizations routinely make decisions that are contrary to the political preferences of the most powerful nation states.[9]

The International Court of Justice (ICJ) recognized the autonomous and independent status of the United Nations in the *Reparations Case* handed down in 1949. In it, the ICJ unequivocally stated that international organizations are entities that possess a 'will' and 'legal personality' independent of member nation states.[10] This decision was followed by the House of Lords in *Watson v International Tin Council*, in which the Court states that an international organization with

---

5   Agency theory is a concept that has its origins in economic thought. For a useful survey article summarizing the principal-agent problem as perceived in political economy, see Eisenhardt (1989), p. 57.

6   Alter (2008), pp. 33–63: 'A number of scholars have used the ideas of Principal-Agent theory (P-A) to argue that states are actually controlling what merely appear to be independent International Courts. P-A theory focuses on the unique tools of political control that states have by virtue of being part of the 'Principal' body that writes, and thus can re-write, the Agent's 'delegation contract'. P-A theory posits that the ability of the Principal to 'sanction' an Agent by changing the contract (firing or not reappointing the Agent, rewriting contractual terms to undercut the Agent's realm of authority, or cutting the Agent's budget) provides states with significant political leverage that they can use to reign in Agents who go astray. See also: *Hawkins et al.* (2006).

7   Pollack (2007).

8   Specific bodies, or organs, within international organizations such as the International Court of Justice, UN Security Council and the Appellate Body of the WTO possess high degrees of decision-making autonomy and independence. For a thorough explanation, see Posner (2005), pp. 1–74.

9   A second problem with Principal-Agent theory is that it is extremely difficult, if not virtually impossible, to 're-negotiate' the terms of multilateral treaties once ratified by numerous Member states.

10   Reparation for Injuries Suffered in Service of United Nations, Advisory Opinion, 1949 I.C.J. 174 (Apr. 11)

separate legal personality acts on its own behalf ... when exercising power under a constituent treaty.[11]

As a consequence, the idea that certain types of public international organizations are independent and autonomous governance entities has led other international relations theorists to argue that these organizations operate more in the nature of trustees – not as agents of nation states.[12] This view, however, raises a further range of questions, such as: of what are public international organizations trustees?[13]

The main argument made in this chapter is that the transfer of sovereign governance powers by nation states to public international organizations creates a fiduciary relationship between those organizations, nation states and the citizens of those nation states. In developing this argument, several relevant contextual and theoretical issues are discussed. This discussion explains the theoretical underpinnings of the relationship between the right to exercise sovereign governance powers and the fiduciary principle. Then, the legal characteristics of international organizations are explained in relation to different types of conferrals of sovereign governance powers and their implications for the application of the fiduciary principle. In a later section, a specific form of transfer of sovereign governance powers leading to the characterization of governance powers as 'political property rights' is discussed in relation to trust concepts. Finally, the notion of a general international fiduciary obligation is distinguished from 'public trust' concepts and their potential application in international environmental law discourse.

Theoretical Considerations

The number and influence of public international organizations that exercise governance functions and powers has proliferated since the end of World War II.[14] The new level of public decision-making apparatus that has emerged over top nation states conforms to what is described as a 'regulatory model' of governance. As Kirsch and Kingsbury observe:

---

11   *MacLaine Watson & Co. Ltd v International Tin Council*, decision of 26 October 1989, House of Lords, in 81 ILR 670.

12   Abbott and Snidal (1998), pp. 3–32.

13   It is unclear, at law, whether the fiduciary duty owed is by the international organization to Member States or whether it is Member States that owe an obligation to an organization? In Case 22/70, *EC Commission v Council* (ERTA), (1971) E.C.R. 263, the European Court of Justice found that nation states owe a duty not to impinge upon or impede the work of an organization. A further question is who should be responsible for breaches of international law that may occur as a result of the organization's exercise of conferred powers – the Member States or the international organization or both?

14   Multilateral treaties, or treaties between multiple nation states agreeing to abide by a single common set of rules are a relatively new phenomenon. For a brief history, see: Ruggie (1982), pp. 379–415; and Ruggie (1992), pp. 561–98.

> much of global governance can be understood as regulation and administration, and that we are witnessing the growth of a 'global administrative space': a space in which... administrative functions are performed in often complex interplays between officials and institutions on different levels, and in which regulation may be highly effective, despite its predominantly non-binding forms.[15]

However, concerns surrounding the accountability and control of this regulatory space are increasingly being raised. A major reason for this growing concern is that nation states, "who constitute the main members of the international community, have never consciously come together to establish a constitution regulating the international public order and setting forth the guiding principles for the main functions of international governance".[16]

The fiduciary principle provides one means of addressing some of the deficiencies in international governance by providing a body of rules that set a standard of conduct for decision-makers. However, the application of fiduciary concepts to international organizations poses several complex legal questions, not least of which is whether there is a robust and coherent juridical theory that provides a solid foundation for the imposition of such an obligation? Public international organizations are organizational constructs created pursuant to a contractual agreement made between nation states. In spite of the decision in the *Reparations case*, which was made only in relation to the United Nations, the legal relationship between nation states and international organizations under international law is by no means clear. If a fiduciary relationship does exist, who owes a fiduciary obligation to whom? Do international organizations owe a fiduciary duty to nation states, or vice versa?[17] One approach to answering this question is to not focus on the legal relationship between nation states and public international organizations as the relevant units of analysis. Instead, the analytical focus should be directed towards the source, transfer and exercise of *sovereign governance powers* as the basis for applying the fiduciary principle. In using this conceptual approach, a distinction must first be made between the general concept of 'governance' and the more specific notion of 'governance powers'. By focusing on the *exercise of governance powers* (and the authority to use them) as the central unit of analysis, a juridical framework is presented that maps the source of the fiduciary obligation, as applied in a public law context, along a chain of delegation up through the nation state and beyond to the international organization.

---

15 Krisch and Kingsbury (2006), pp. 1–13.
16 De Wet, Erika (2006); Tomuschat (1993).
17 Case 22/70, *EC Commission v Council* (ERTA), (1971) E.C.R. 263.

Sovereign Governance Powers as the Source of Fiduciary Obligations

Governance is a term that has many different meanings and definitions.[18] A large body of literature from several branches of political science and sociology examines its many dimensions. In spite of the multitude of definitions, a central theme underlying all is that the creation, administration and enforcement of 'rules' promotes and contributes to social stability. More contemporary definitions of governance consider the relationship between public and private sources of rules and advocate that the terms 'governance' and 'government' are not synonymous. Numerous non-governmental governance arrangements have evolved through history.[19] Similarly, public-private partnership arrangements have been created to perform governance functions.[20] A critical feature of private governance arrangements is that they do not rely upon state-sanctioned compliance mechanisms. Instead, compliance is motivated through, for example, religious, cultural and contractual relationships.[21]

In spite of the growing scholarly consensus that governance need not fall within the exclusive domain of governments, an important concept that distinguishes public from private governance arrangements is the concept of sovereignty. Governance by a sovereign is a special category of governance arrangement.[22] Although acknowledging that the concept of sovereignty is both enormous and contested,[23] the foregoing analysis draws upon *legal* conceptualizations of sovereignty, which define the delimitation of power and authority of nation states within their national boundaries (internal sovereignty) as well as the status of, and relationship between, nation states beyond national boundaries (external sovereignty).[24]

Sovereignty is the legal principle that gives rise to the *right* of a state to exercise public governance powers in relation to a group of people within a specific geographical territory (which includes the power to exercise coercive force to achieve certain ends).[25] It is this *legal concept of governance powers*, as distinguished from the more *general concept of governance*, that can be sub-divided into the three specific legal powers expressed in the constitutional doctrine of the separation of powers: i) the setting of rules (legislative power), the

---

18   *Kjaer* (2004).

19   Gessner (2009).

20   Osborne (2000).

21   Richman (2006), p. 31.

22   *'Sovereignty means many things, some essential, some insignificant, some agreed, some controversial, some that are not warranted and should not be accepted.'* Louis Henkin, cited in Nagan and Hammer (2005) p. 142.

23   Sarooshi, D. (2004) pp. 1107–40.

24   *The Case of the S.S. 'Lotus'*, 7 September, 1927 Twelfth (Ordinary) Session of the Permanent Court of International Justice, Series A – No. 10 Collection of Judgements.

25   Dugger (1996), p. 2.

administration and enforcement of rules (executive power) and the adjudication of rules (judicial power).[26]The right to exercise sovereign governance powers within a state is no longer considered to be absolute. It is widely recognized in domestic constitutional and international law that states are constrained in their exercise of internal sovereignty and, more controversially, that a state's right to exercise governance powers vested in it by the people is mediated by a corresponding duty to exercise those powers responsibly.[27] The fiduciary principle provides a substantive standard of conduct that informs the public duty to exercise governance powers responsibly.

Mapping the Transfer of Sovereign Governance Powers and Fiduciary Obligations from the Individual to International Organizations

As mentioned above, sovereignty, defined as the right to exercise sovereign governance powers, is a highly contested.[28] The classical notion that these rights are devolved from a Monarch and vested solely and absolutely in the parliament is no longer orthodoxy. The constructivist argument that parliaments, states and international organizations are abstract social constructs and, therefore, cannot possess rights unless people grant them rights implies that the source of sovereign rights must ultimately stem from humanity. The notion that all human beings are born with certain inalienable rights attests to this notion. It is then through an intricate network of social contracts and international treaties that governance power is delegated upwards to the appropriate social construct that humans create to organize their societies. Sovereignty, therefore, can be viewed as an organizing principle that is neither absolute nor indivisible.

Further, given that sovereignty is divisible, 'state sovereignty, popular sovereignty and individual sovereignty' need not be competing concepts.[29] Each describes different levels, dimensions, or spheres of decision-making responsibility, beginning with the individual, extending to the citizen and the state and ultimately, as will be argued below, beyond the nation state to international organizations. In all cases, with the exception of individual sovereignty, the right to exercise of sovereign governance powers requires a legal instrument evidencing the terms and conditions under which governance powers are transferred.[30]

Individual sovereignty is the newest dimension of sovereignty. Yet, like most interesting ideas, it is not new at all and its origins can be traced back to Kantian notions of human dignity. Kant wrote that 'the dignity of humanity in us' provides that we should not 'suffer [our] rights to be trampled underfoot by

---

26   Feeny (1993).

27   Evans and Sahnoun (2002), pp. 99–110.

28   Wendt (1992), p. 391.

29   Petersmann (2006).

30   The conceptualization described here broadly conforms with the classical principle of subsidiarity. For an historical account of the principle's origins, see Carozza (2003), p. 38.

others with impunity'.[31] A careful echoing of these sentiments can be found in the definition given by former UN Secretary-General Kofi Annan, who describes individual sovereignty as the 'fundamental freedom of each individual, enshrined in the charter of the UN and subsequent international treaties … is enhanced by a renewed and spreading consciousness of individual rights'.[32] Although still largely a political concept, individual sovereignty is gradually finding its way into legal discourse.

The idea that some individual rights are so deeply embedded that they cannot be challenged by any level of authority was raised by Justice Cooke in *Taylor v New Zealand Poultry Board*. In *Taylor*, Justice Cooke explored the ultimate limits of statutory validity in New Zealand opining that:

> I do not think that literal compulsion, by torture for instance, would be within the lawful powers of Parliament. Some common law rights presumably lie so deep that even Parliament could not over-ride them. [33]

At its core, individual sovereignty advances the notion that every human possesses the right to make decisions fundamental to their personal dignity and well-being as argued by Hart:

> Any adult human being capable of choice (1) has the right to forbearance on the part of all others from the use of coercion or restraint against him save to hinder coercion or restraint and (2) is at liberty to do (i.e., is under no obligation to abstain from) any action which is not one coercing or restraining or designed to injure other persons.[34]

Yet, human beings do not live in a 'splendid isolation'.[35] Rather, they live in communities as part of a social system. The conversion from 'human being' to 'citizen' corresponds with the legal transition from 'individual sovereignty' to 'popular or constitutional sovereignty'. 'Human beings' become political 'citizens' by being members of a political society. Governance powers in legally constituted societies are granted to the legal construct known as the state by way of a Constitution. Constitutions are social contracts that set jurisdictional boundaries and authoritative limits upon the use of those sovereign governance powers. The notion that these governance powers are granted to the state by the 'people' (not a Monarch) is evident in the words of Chief Justice Mason in the *Australian Capital Television* case in which he stated that:

---

31   Kant (1797) [Hacket 1983].

32   Annan (1999).

33   *Taylor v New Zealand Poultry Board* [1984] 1 NZLR 394 (CA).

34   Hart (1955), pp. 175–91.

35   A phrase attributed to Canadian politician George Foster as noted in Hamilton (1952).

> The *Australia Act* 1986 (UK) marked the end of the legal sovereignty of the
> Imperial Parliament and recognized that ultimate sovereignty resided in the
> Australian people.[36]

This notion echoes the sentiments of Forsey made in relation to the source of
governance authority within the Canadian polity. Stated succinctly, every 'act of
government is done in the name of the Queen, but the authority for every act flows
from the ... people'.[37]

In the same way, a constitution is a legal instrument that transfers sovereign
governance powers from the people to the state, international treaties are legal
instruments that 'may' contain provisions that transfer sovereign governance
powers from nation states to international organizations. The law underlying the
empowering of international organizations is contentious and complex. In brief,
however, the constituent treaty represents an agreement to transfer (or confer)
sovereign powers belonging to nation states to international organizations by
prescribing an international organization's:

1. organizational structure;
2. jurisdictional scope of authority;
3. governance functions; and
4. depth of discretionary powers in respect of those functions.[38]

In combination, these four elements establish the composition of an international
organization. However, it is the latter two elements that largely determine whether
an international organization will owe a fiduciary obligation in the conduct of its
activities as discussed further below.

Attaching Fiduciary Obligations to International Transfers

A fiduciary relationship is one that, according to Justice Wilson in *Frame v Smith*,
has three characteristics. First, the fiduciary must be in the position of one who is
responsible for exercising a discretionary power. Second, the fiduciary must be
able to unilaterally exercise that power in a manner that affects the beneficiary's
legal or practical interests. Third, the beneficiary must be vulnerable or at the mercy
of the fiduciary holding the discretion or power. Once a fiduciary relationship has
been found to exist, 'equity will then supervise the relationship by holding [the
fiduciary] to the fiduciary's strict standard of conduct'.[39]

---

36 *Australian Capital Television v Commonwealth* (1992) 177 CLR 106.
37 Forsey (2005).
38 Feaver (2010).
39 *Frame v Smith* [1987] 2 S.C.R. 99. Prior to the application of fiduciary concepts
in a growing range of areas of public action, the unlawful or inappropriate actions of public

The question whether the fiduciary principle applies to the exercise of sovereign governance powers in an international context is summarized by Judge Higgins, of the International Court of Justice, who observes that:

> issues of fiduciary duty and international responsibility depend for their proper resolution upon an understanding of precisely what type of powers have been transferred and the character of their exercise by the organization concerned.[40]

Stated another way, international organizations take many different forms. Not all possess conferrals of sovereign governance powers. Whilst there are thousands of private non-governmental organizations (more frequently referred to now as transnational organizations), which are not granted and cannot possess sovereign governance powers,[41] there are comparatively few public international organizations that possess conferrals of power. Public international organizations that are created to perform governance functions requiring the exercise of governance powers can be identified by the following legal characteristics. Public international organizations:

- are established by agreements made between nation states (and are open to all) which take the form of a 'constituent instrument';
- possess an organizational structure that is separate from its members;
- are recognized under international law as having international legal personality;
- have an exclusive membership of nation states.[42]

The constituent instrument is an international treaty that is the primary source of law governing nearly all aspects of an international organization's functions, powers and operational activities.[43] It is recognized that constituent treaties have

---

decision-makers were subject only to common law and statutory standards of judicial review familiar to administrative lawyers. Any obligation that public decision-makers should be held to a higher standard of conduct and, hence, accountability, did not hitherto exist.

40   Higgins (2007).

41   Public international organizations are generally classified as being multilateral (such as the United Nations or the World Trade Organization), supranational (such as the European Union) and regional (such as NAFTA and ASEAN).

42   Amerasignhe (2005).

43   As Klabbers (2003) notes: 'The constitutive [instruments] of international organizations are strange creatures, often said to occupy a special place in international law. On one hand, they are treaties, concluded between duly authorized representatives of states, and as such no different than other treaties. Thus, one would expect, they are simply subject to the general law of treaties. Yet, such constituent documents are not ordinary treaties: they establish an international organization, and, for that reason, most authors appear inclined to grant those treaties a separate status, from which follows the applicability of some special rules … .'

a special status under international law giving them what has been described as having an organic quality.[44] Such a special status stems from the fact that a central purpose of a constituent instrument is to articulate the nature of the functions and powers contained in the conferral.[45] The form of a conferral was considered in the *WHO Advisory Opinion* case in which the ICJ stated that a conferral is:

> normally the subject of an express statement in their constituent instruments. Nevertheless, the necessities of international life may point to the need for organizations, in order to achieve their objectives, to possess subsidiary powers which are not expressly provided for in the basic instruments which govern their activities. It is generally accepted that international organizations can exercise such powers, known as 'implied' powers.[46]

Whether an international organization is granted sovereign governance powers (and hence, will owe a fiduciary duty in respect of their exercise) depends on the 'type' of conferral contained in a constituent instrument. Professor Sarooshi identifies three different types of conferral which he classifies as: i) agency relationships, ii) delegations, and iii) transfers. Each type of conferral can be distinguished as to whether: a conferral is revocable; the extent of control over the exercise of those powers is retained by states; and, finally, whether an international organization has the sole right to exercise a conferred power.[47] Agency relationships and delegations are conferrals that are revocable and have a considerable degree of the exercise of those powers retained by states and can be exercised concordantly with international organizations. In the case of nation states that transfer powers to an international organization, a nation state:

> does not confer its powers *in toto* on the organization. It retains the powers as part of its sovereignty, but has agreed to limit its right to exercise those powers in favour of an exclusive right of the organization to exercise the conferred powers. Put differently, the organization is the sole place for the lawful exercise of transferred powers.[48]

---

44   Zacklin (1969), p. 8.

45   In the *WHO Advisory Opinion case*, the International Court of Justice confirmed that a constituent treaty can act as a mechanism for conferrals by States of express powers on an international organization. In that case, the ICJ observed that 'the powers conferred on international organizations are normally the subject of an express statement in their constituent instruments'. However, as well as being express, the powers conferred may also be ascribed under the doctrine of implied powers.

46   *WHO Advisory Opinion case*, p. XX.

47   Sarooshi (2005).

48   Sarooshi (2005), p. 47.

In following Sarooshi's typology, the argument can be made that the higher the degree of independence and autonomy granted to an international organization arising from the type of conferral of powers contained in its constituent instrument, the higher the likelihood that it and its officials are bound by the fiduciary principle. However, it must be noted that few international organizations are created on the basis of transfers of powers. Relatively few international organizations are designed to be independent and autonomous in their exercise of governance powers to give rise to a fiduciary obligation within the meaning articulated by Justice Wilson in *Frame v Smith*.[49]

Governance Powers as Political Property Rights Exercised in Trust

The transfer of sovereign governance powers to independent and autonomous international organizations raises the question as to whether international organizations are more accurately described as 'agents' created to act upon the instruction of nation states or whether they occupy an altogether different place in international governance? It is clear from the foregoing discussion that international organizations that receive 'transfers' of sovereign governance powers cannot be agents within the legal meaning of the concept because of their independent and autonomous status as decision-makers. Several scholars such as Alter, Keohane and Grant and Majone have turned to trust-like concepts as a means of explaining the compositions of transfers as well as the exercise of sovereign powers by international organizations.[50] This perspective has resulted in a corresponding analytical shift away from examining issues surrounding the control of international organizations to, instead, analysing the characteristics of those organizations' decision-making processes. In particular, this line of analysis focuses on authority and legitimacy in the exercise of governance powers. In characterizing 'public power' and its exercise as a form of fiduciary and political trusteeship, Majone explains how:

> [John] Locke thought in terms of a contract of society, establishing a political (or 'civil') society, followed by the creation of a fiduciary sovereign under and by a trust-deed. In chapter XIII of the *Second Treatise of Government*, he writes that the legislative, even though it is the supreme power, is 'only a Fiduciary Power to act for certain ends'. Hence:
>
> there remains still *in the People a Supreme Power* to remove or *alter the Legislative*, when they find the *Legislative* act contrary to the trust reposed in them. For all *Power given with trust* for the attaining an *end*, being limited by

---

49 *Frame v Smith* [1987] 2 S.C.R. 99.

50 Alter (2008), pp. 33–63; Barnett and Finnemore (2004); Keohane and Grant (2005), pp. 29–43: Majone (2001), pp. 103–22.

that end, whenever that *end* is manifestly neglected, or opposed, the *trust* must necessarily be *forfeited*, and the Power devolve into the hands of those that gave it, who may place it anew where they shall think best for their safety and security. (emphases in the original)[51]

Majone contends that Locke's conceptualization of the government as trustee and of the community as both trustor and beneficiary was not metaphorical. In support of his argument, he comments that, according to Barker, Locke's notion of 'power given in trust' is consistent with concepts derived from the English law of trusts as it existed in his time.[52]

Locke's notion that governance powers are given 'in trust' can be extended beyond the nation state and applied to their conferral by means of a 'transfer' to international organizations. In building upon Moe's contention that the 'exercise of authority' should be considered a political property right,[53] Majone distinguishes the notion of agency from that of a trust relationship:

> The agent is not ordinarily the owner of property for the benefit of the principal. Strictly speaking, when property is transferred to a person who is supposed to manage it for the benefit of a third person, we have not an agency but a special type of fiduciary relation known as trusteeship. In Anglo-American law a trust is a situation where the owner of some property, the 'trustor' (or 'settlor'), transfers it to a 'trustee' with the stipulation that the trustee should not treat it as her own but manage it for the benefit of the 'beneficiary' – who could be the trustor himself. Now, since agency may possess the element of trust and confidence of a fiduciary relation, both agents and trustees can be classed together for many purposes, but the two concepts are distinct. In the legal literature this distinction is sometimes expressed by saying that all trustees are agents but not all agents are trustees: a trustee is an agent and something more (Bogert 1987: 36). The trustee's fiduciary duty is not simply a personal obligation but is attached to a piece of property – the trust assets.[54]

If sovereign governance powers that are the subject of a transfer to international organizations are considered to be 'political property right' of nation states, a strong argument can be made that international organizations, therefore, stand in the position of a fiduciary in respect of the exercise of those powers.

---

51   Majone (2001), pp. 103–22.
52   Barker (1962), pp. vii–xliv.
53   Moe (1990), pp. 213–53. A political property right, as defined by Majone, is a 'right to exercise public authority in a given policy area'.
54   Majone (1990), p. 23.

The Doctrine of Public Trust and International Environmental Law

In the previous sections, it is argued that international organizations that are conferred governance powers by means of a constituent transfer, owe a *general* fiduciary obligation in respect of the exercise of those powers. However, a general fiduciary obligation owed in respect of an exercise of discretionary powers is different from the notion of a *positive fiduciary duty* to act to preserve trust assets for the benefit of trustors/settlors that underlies the notion of a 'public trust'.

The public trust doctrine is said to be the cornerstone of contemporary environmental law and is based on the notion that certain classes of cultural and natural resources are 'held in trust' by the sovereign on behalf of all citizens.[55] The origins of the public trust doctrine has been traced back to the sixth century Justinian code doctrine of *res communes*, which provides that certain resources are 'common to mankind – the air, running water, the sea, and consequently the shores of the sea [and] the right of fishing in a port, or in rivers, [are] common to all men'.[56] Ownership of this class of natural resources is said to be vested in the state, as the sovereign, in trust for the people. *Res communes* is considered to be excluded from private control and a trustee is charged with a positive duty to act to preserve the resources in order that they can be made available for the benefit of the whole of society.[57]

Although it is clear that the public trust doctrine is embedded within the domestic law of a number of nation states, there is some question as to whether it also forms part of international law.[58] Legal scholars have used various means to argue that the public trust doctrine is part of international law governing resources that fall outside the jurisdiction of nation states. Even those resources that fall within the internal sovereignty of nation states, it is argued, may be subject to the public trust doctrine with various forms of trusteeship and guardianship being suggested in relation to issues such as the common heritage of mankind, law of the sea and human rights law, or a combination thereof.[59] For example, the notion that environmental rights should be inextricably linked to fundamental human rights concerns has long been argued. Toepfer, an early director of the UN Environment Program, states that:

> Human rights cannot be secured in a degraded or polluted environment. The fundamental right to life is threatened by soil degradation and deforestation and by exposures to toxic chemicals, hazardous wastes and contaminated drinking water. ... Environmental conditions clearly help to determine the extent to

---

55 Sax (1970), pp. 471–556; Sax (1980), 185–94.

56 Jarman (1986), p. 1.

57 *Illinois Central Railroad vs People of the State of Illinois* (1892) US Supreme Court (5 December 1892), 146 U.S. 387.

58 Kameri-Mbote (2007), p. 195.

59 Sands (1989), pp. 393–420.

which people enjoy their basic rights to life, health, adequate food and housing, and traditional livelihood and culture. It is time to recognize that those who pollute or destroy the natural environment are not just committing a crime against nature, but are violating human rights as well.[60]

Similarly, in the Danube Dams case, Justice Weeramantry of the ICJ stated that:

The protection of the environment is likewise a vital part of contemporary human rights doctrine, for it is a *sine qua non* for numerous human rights such as the right to health and the right to life itself. It is scarcely necessary to elaborate on this, as damage to the environment can impair and undermine all the human rights spoken of in the Universal Declaration and other human rights instruments.[61]

According to Sands (consistent with the argument made in previous sections of this chapter), it is the 'exercise of sovereign rights' to which the duty attaches:

In spite of the irritant amount of rhetoric surrounding it, the concept of public trusteeship is *not* a mere figure of speech or a utopian scenario, as some commentators and orators seem to assume. ... the trusteeship status of a resource is not at all incompatible with the legitimate exercise of sovereign rights by a host state, just as – and here the analogy from trust law seems perfectly appropriate – a common law trustee has legitimate property rights over the corpus, always provided those rights are exercised in accordance with the interests of the beneficiary and the terms of the trust.[62]

The difficulty and challenge is not one of finding a juridical foundation for a positive duty to act to preserve the object of a public trust obligation, and hence a positive duty to act to preserve the common property of mankind. Rather it is the enforcement of this duty in an international context. The international community is not a single, coherent entity. It is composed of numerous state and non-state actors, the most powerful of which are nation states. 'Given the decentralized structure of the world community, securing international compromise and consensus often requires adopting vague declarations and ineffective institutional arrangements, thereby avoiding the establishment of the strong enforcement mechanisms needed.'[63] The ideal solution is an international organization that is granted the broad jurisdictional authority and governance powers to enable it to compel nation states to recognize their collective duty to combat climate change.

---

60   Toepfer (2001), quoted in Shelton (2002), p. 2.
61   Case Concerning the Gabcikovo-Nagymaros Project between Hungary and Slovakia ('*Danube Dam Case*') 37 ILM 162 at p. 217.
62   Sands (1989), pp. 393–420.
63   Nanda and Ris (1975).

Although the Preamble of the UNFCCC provides that States 'have the responsibility to ensure that activities within their jurisdiction or control do not cause damage to the environment of other States or of areas beyond the limits of national jurisdiction', the UNFCCC does not contain any general enforcement powers against those countries primarily responsible for atmospheric warming, i.e., the Western industrialized countries. Rather than an enforcement power, the Kyoto Protocol to the UNFCCC includes a mechanism designed to encourage and assist Parties to comply with the obligations contained in the agreement.[64] A potential compliance problem (referred to as a 'question of implementation') can be raised by an 'expert review team' or by a Party about its own compliance or by a Party raising questions about the actions of another Party. After a preliminary examination by the Compliance Committee, the 'question of implementation' is allocated to the appropriate branch of the Compliance Committee for further investigation and consideration.[65] In the event of an adverse finding by the Compliance Committee, the Compliance Committee only has the power to recommend that the Party either abandon or amend its practices to ensure compliance.

Even in the absence of a multilateral mechanism capable of enforcing public trust obligations, the public trust doctrine is still relevant in two respects. It first arises in respect of the question whether the governments of nation states, as sovereign trustees of trust assets harmed by atmospheric warming caused by other nation states (such as vulnerable coastal states), owe a fiduciary obligation to their citizens to protect those trust assets from damage? In this regard, the decision of Brown J. in *State v City of Bowling Green*, provides some guidance in that 'where the state is deemed to be the trustee of property for the benefit of the public it has the obligation to bring suit not only to protect the corpus of the trust property but to recoup the public's loss ... '.[66] Accordingly, it may well be that nation states that have already been harmed by and are vulnerable to future atmospheric warming have an obligation to pursue compensation from those nation states primarily responsible for the harm. One method is to protect and recoup losses by pursuing legal proceedings against those nation states responsible for the damage in the International Court of Justice (ICJ).

The related second aspect arises in respect of those nation states causing the harm. Can states be held responsible under international law for current or future climate change damages? Is there an obligation under public international law to prevent and to compensate for such damages? Might the ICJ recognize the public trust doctrine as a basis upon which to find in favour of those nation states harmed by atmospheric warming? The answers to the first two questions are closely linked to the law of State responsibility. In an environmental context, 'the law of State

64 Mitchell (2005), p. 73.

65 Marrakesh Accords, UN Doc. FCCC/CP/2001/13/Add.3 (2002) *Report of the Conference of the Parties on its Seventh Session*; Marrakesh Accords, Decision 24/CP.7 'Procedures and Mechanisms relating to Compliance under the Kyoto Protocol'.

66 *State v City of Bowling Green*, 313 N.E.2d 409, 411 (Ohio 1974).

responsibility has two functions: first, to support primary rules established by treaties or customary law which aim at preventing environmental damages and, second, to provide injured states with a right to restitution and compensation'.[67] Under customary international law, there is a long-standing norm that States are obliged not to inflict damage on or violate the rights of other States as established in the arbitral decision in the *Trail Smelter* case. The *Trail Smelter* case provides the basis of the 'no-harm rule' central to the law of state responsibility. In it, the tribunal concluded:

> Under the principle of international law … no State has the right to use or permit the use of its territory in such a manner as to cause injury by fumes in or to the territory of another state or the properties or persons therein, when the case is of serious consequence and the injury is established by clear and convincing evidence.[68]

The origins of the 'no harm rule' can be traced to the principle of good neighbourliness between States which are equal under international law. As Verheyen notes, 'Principle 2 of the 1992 Rio Declaration, which echoes Principle 21 of the 1972 Stockholm Declaration, reiterates this rule of customary international law, outlawing transboundary environmental injury':

> States have, the sovereign right to exploit their own resources pursuant to their own environmental and developmental policies, and the responsibility to ensure that activities within their jurisdiction or control do not cause damage to the environment of other States or of areas beyond the limits of national jurisdiction.

Under international customary law, in order for nation states to be held responsible for climate change damage, it is necessary to identify a legally significant behaviour by a state or to attribute the damaging behaviour of private persons to the state. State behaviour having damaging effects would include, for example, allowing emissions of greenhouse gases *per se* or during a certain time, and/or not having put in place the regulatory means to arrest emissions over and above a certain threshold: both clearly legally relevant state actions or omissions.'[69] Further, a claim for damages under international law requires the following elements to be satisfied: (i) identifying a damaging activity attributable to a state, (ii) establishing a causal link between the activity and the damage, (iii) determining either a violation of international law or a violation of a duty of care

---

67   Voight (2008).
68   Trail Smelter Arbitration of 1941, Reports of International Arbitration Awards (RIAA) III.
69   Tol and Verheyan (2004).

(due diligence), which is (iv) owed to the damaged state. Step (v) in a court of law would be to quantify the damage caused and relate this back to the activity.[70]

In addition to the more conventional 'no harm rule', an additional and more novel approach to establishing a breach of international law as per (iii) above flows from the equitable rule imposing a duty upon all nation states to preserve 'common property of mankind'. As mentioned above, common heritage equity, assumes that certain resources are the patrimony of all humankind. 'A number of principles proceed from this assumption, namely: non-ownership of the heritage, shared management, shared benefits, use exclusively for peaceful purposes, and conservation for future generations.'[71] The legal status of the atmosphere as falling within the scope of 'common heritage property' was clarified by UN General Assembly Resolution 43/53 in 1988. Paragraph 1 of Resolution 43/53 provides that 'climate change is a common concern of mankind, since climate is an essential condition which sustains life on earth'.[72] According to Baslar, 'under the peaceful purpose clause of the common heritage of mankind concept, the global atmosphere should not be used in such a way that threatens the quality of the resource itself as well as the survival of mankind. Acting contrary to this should give rise to *locus standi*, that is, either a right of appearance before the International Court of Justice, or a right to be heard before an 'International Environmental Court'.[73]

Accordingly, as a first step towards prompting the creation of adequate international institutional machinery to deal with issues relating to climate change related harm, it may be necessary for those nation states to take the initiative and assert their rights under international law. If nothing else, such a challenge before the ICJ would give the Court an opportunity to clarify the status of the atmosphere and the international law relating to the common heritage of mankind.[74]

Summary and Conclusions

A growing number of legal scholars, practitioners and jurists in the Anglo-American tradition acknowledge that a fiduciary relationship exists between a state and its subjects. Although the constitutional and administrative law of nation states provide standards regulating public officials who exercise public decision-making powers, the same cannot be said of public international organizations and officials. There is no body of international constitutional and administrative law that regulates public international decision-making processes. Hence, the question

---

70 Tol and Verheyan (2004), p. 1112.

71 Verheyen (2005).

72 General Assembly Resolution 43/53 (6 December 1988).

73 Baslar (1999), p. 305.

74 There is a concern that the law surrounding the international law of the common heritage of mankind is ambiguous and unclear, although other authors believe this view to be overstating the concern. See Shakelford (2008).

that is investigated in this Chapter is whether and how the fiduciary principle might be applied as a standard to regulate the exercise of governance powers beyond the nation state?

The application of fiduciary concepts to international organizations poses several complex legal questions and challenges, not least of which is whether there is, or can be developed, a robust and coherent juridical theory that provides a solid foundation for the imposition of such an obligation? By focusing on the *exercise of governance powers* (and the authority to use them) as the central unit of analysis, a juridical framework maps how the fiduciary obligation travels with or along a chain of delegation from people up through the nation state and beyond to the international organization. The argument is made that the higher the degree of independence and autonomy granted to an international organization arising from the type of conferral of powers contained in its constituent instrument, the higher the likelihood that it and its officials are bound by the fiduciary principle. However, it must be noted that few international organizations are created on the basis of transfers of powers. Relatively few international organizations are designed to be independent and autonomous in their exercise of governance powers to give rise to a fiduciary obligation. A general fiduciary obligation owed in respect of an exercise of discretionary powers is very different from the notion of a positive fiduciary duty to act to preserve trust assets for the benefit of trustees that underlies the notion of a 'public trust'. Although there is a strong implication arising under the public trust doctrine that there is a positive duty to act to preserve the common property of mankind, a whole range of difficulties exist as to the enforcement of this duty in an international context.

References

K. W. Abbott and D. Snidal (1998) 'Why States Act Through Formal International Organizations', 42(1) *The Journal of Conflict Resolution* 3–32.
Karen Alter (2008) 'Agents or Trustees', 14 (1) *European Journal of International Relations* 33–63.
C. Amerasignhe (2005) *Principles of the Institutional Law of International Organizations*, Cambridge University Press.
K. Annan (1999) 'Two Concepts of Sovereignty', *The Economist*, 18 September.
E. Barker (1962) 'Introduction', in E. Barker (ed.), *Social Contract: Locke Hume Rousseau*, Oxford University Press, pp. vii–xliv.
M. Barnett and M. Finnemore (2004) *Rules for the World: International Organizations in Global Politics*, Cornell University Press.
K. Baslar (1999) *The Concept of the Common Heritage of Mankind*, Martinus Nijhoff.
P. Carozza (2003) 'Subsidiarity As a Structural Principle Of International Human Rights Law', 97 *American Journal of International Law* 38.

Erika De Wet (2006) 'The International Constitutional Order', 55 *International and Comparative Law Quarterly* 51–76.

W. Dugger (1996) 'Sovereignty in Transaction Cost Economics: John R. Commons and Oliver E. Williamson', 30 *Journal of Economic Perspectives* 2.

M. K. Eisenhardt (1989) 'Agency theory: An Assessment and Review', *Academy of Management Review* 14(1), 57.

G. Evans and M. Sahnoun (2002) 'The Responsibility to Protect', 81 *Foreign Affairs* 99–110.

D. Feaver (2010) 'Architectures of Transnational Organizations', *Working Paper Series*, Melbourne: Centre for Work Governance and Technology, RMIT University.

David Feeny (1993) 'The Demand for and Supply of Institutional Arrangements', in Vincent Ostrom et al. (eds), *Rethinking Institutional Analysis and Development: Issues, Alternatives and Choices*, ICS Press.

Paul Finn (1994) 'Public Trust and Public Accountability', 3(2) *Griffith Law Review* 225–44.

E. Forsey (2005) *How Canadians Govern Themselves*, 6th edn, Minister of Public Works and Government Services, Canada.

Evan Fox-Decent (2005) 'The Fiduciary Nature of State Legal Authority', 31 *Queen's Law Journal* 259.

V. Gessner (2009) *Contractual Certainty in International Trade*, Hart.

John Glover (2004) *Equity, Restitution and Fraud*, LexisNexis Butterworths.

R. Hamilton (1952) *Canadian Quotations and Phrases: Literary and Historical*, McClelland and Stewart.

H. L. A. Hart (1955) 'Are there any natural rights?, 64 *The Philosophical Review* 175.

Darren G. Hawkins, David A. Lake, Daniel L. Nielson and Michael J. Tierney (eds) (2006) *Delegation and Agency in International Organizations*, Cambridge.

C. Jarman (1986) 'Public Trust Doctrine in the Exclusive Economic Zone', 65 *Oregon Law Review* 1.

P. Kameri-Mbote (2007) 'The Use of the Public Trust Doctrine in Environmental Law', 3(2) *Law, Environment and Development Journal* 195.

Immanuel Kant (1797)[1983] *The Metaphysical Principles of Virtue*, trans. J. W. Ellington, Hacket.

R. Keohane and Grant, R (2005) 'Accountability and Abuses of Power in World Politics', 99(1) *American Political Science Review* 29.

A. M. Kjaer (2004) *Governance*, The Polity Press.

J. Klabbers (2003) *An Introduction to International Institutional Law*, Cambridge University Press.

N. Krisch and B. Kingsbury (2006) 'Introduction: Global Governance and Global Administrative Law', 17(1) *European Journal of International Law* 13.

Giandomenico Majone (2001) Two Logics of Delegation: Agency and Fiduciary Relations in EU Governance, 2(1) *European Union Politics* 103.

Ronald B. Mitchell (2005) 'Flexibility, Compliance and Norm Development in the Climate Regime', 65/82, in Olav Schram Strokke, Jon Hovi and Geir Ulfstein (eds), *Implementing the Climate Regime: International Compliance*, Earthscan.

T. M. Moe (1990) 'Political Institutions: The Neglected Side of the Story', 6 *Journal of Law, Economics and Organisation* 213.

W. Nagan and C. Hammer (2005) 'The Changing Character of Sovereignty in International Law and International Relations', 43 *Columbia Journal of Transnational Law* 142.

V. P. Nanda and W. Ris (1975) 'The Public Trust Doctrine: A Viable Approach to International Environmental Protection', 5 *Ecology Law Quarterly* 291.

S. P. Osborne (2000) *Public Private Partnerships: Theory and Practice in International Perspective*, London: Routledge.

E. Petersmann (2006) 'State Sovereignty, Popular Sovereignty and Individual Sovereignty: From Constitutional Nationalism to Multilevel Constitutionalism in International Economic Law?', available at SSRN: http://ssrn.com/abstract=964147.

Mark A. Pollack (2007) 'Principal-Agent Analysis and International Delegation: Red Herrings, Theoretical Clarifications and Empirical Disputes (February 2007)', available at SSRN: http://ssrn.com/abstract=1011324.

E. A. Posner and J. C. Yoo (2005) 'Judicial Independence in International Tribunals', 93(1) *California Law Review* 1.

B. Richman (2006) 'How Communities Create Economic Advantage: Jewish Diamond Merchants in New York', *Law and Social Inquiry* 31.

John G. Ruggie (1982) 'International Regimes, Transactions, and Change: Embedded Liberalism in the Postwar Economic Order', 36(2) *International Organization* 379.

John G. Ruggie (1992), 'Multilateralism: The Anatomy of an Institution', 46(3) *International Organization* 561.

P. Sands (1989) 'The Environment, Community and International Law', 30 *Harvard International Law Journal* 393.

D. Sarooshi (2004) 'The Essentially Contested Nature of the Concept of Sovereignty: Implications for the Exercise by International Organizations of Delegated Powers of Government', 25 *Michigan Journal of International Law* 1107.

D. Sarooshi (2005) *International Organizations and Their Exercise of Sovereign Powers*, Oxford University Press.

J. L. Sax (1970) 'The Public Trust Doctrine in Natural Resources Law: Effective Judicial Intervention', 68 *Michigan Law Review* 471.

J. L. Sax (1980) 'Liberating the Public Trust Doctrine from its Historical Shackles', 14 *University of California-Davis Law Review* 185.

S. Shakelford (2008) 'The Tragedy of the Common Heritage of Mankind', 27 *Stanford Environmental Law Journal* 101.

D. Shelton (2002) 'Health and Human Rights Working Paper 1', *Human Rights, Health and Environmental Protection: Linkages in Law and Practice*, WHO, available at *www.who.int/.../Series_1%20%20Human_Rights_Health_ Environmental%20Protection_Shelton.pdf.*

R. Tol and R. Verheyan (2004) 'State Responsibility and Compensation for Climate Damages', 32 *Energy Policy* 1109.

Christian Tomuschat (1993) 'Obligations Arising for States Without or Against their Will', *Recueil des Cours* 235.

K. Voight (2008) 'State Responsibility for Climate Change Damages', 77 *Nordic Journal of International Law* 1.

A. Wendt (1992) 'Anarchy Is what States Make of It: The Social Construction of Power Politics, 46 *International Organization* 391.

Ralph Zacklin (1969) *The Amendment of Constitutive Instruments of the United Nations and Specialized Agencies*, Leydon.

Cases

Australian Capital Television v Commonwealth (1992) 177 CLR 106

Case 22/70, *EC Commission v Council* (ERTA), (1971) E.C.R. 263

Case Concerning the Gabcikovo-Nagymaros Project between Hungary and Slovakia ('*Danube Dam Case*') 37 ILM 162

Frame v Smith [1987] 2 S.C.R. 99

Illinois Central Railroad vs People of the State of Illinois (1892) US Supreme Court (5 December 1892), 146 US. 387

MacLaine Watson & Co. Ltd v International Tin Council, decision of 26 October 1989, House of Lords, in 81 ILR 670

Marrakesh Accords, UN Doc. FCCC/CP/2001/13/Add.3 (2002) *Report of the Conference of the Parties on its Seventh Session*; Marrakesh Accords, Decision 24/CP.7 'Procedures and Mechanisms relating to Compliance under the Kyoto Protocol'

Reparation for Injuries Suffered in Service of United Nations, Advisory Opinion, 1949 I.C.J. 174 (Apr. 11)

State v City of Bowling Green, 313 N.E.2d 409, 411 (Ohio 1974)

Taylor v New Zealand Poultry Board [1984] 1 NZLR 394 (CA)

The Case of the S.S. 'Lotus', 7th September 1927 Twelfth (Ordinary) Session of the Permanent Court of International Justice, Series A – No. 10 Collection of Judgements

Trail Smelter Arbitration of 1941, Reports of International Arbitration Awards (RIAA) III.

WHO Advisory Opinion case, p. XX

Legislation

General Assembly Resolution 43/53 (6 December 1988)

Chapter 8

High Court of Australia on Fiduciary Theory

Rosemary Teele Langford*

Introduction

The application of the fiduciary principle to oblige governments to take action to protect our natural environment and to reduce climate change has strong initial appeal. Given the importance of the environment to citizens' lives and well-being, it seems fitting to impose duties on governments and other authorities to safeguard natural resources and actively prevent further deterioration.[1] The fiduciary standard is strict and imposes a high level of loyalty. It would therefore appear a suitable mechanism for the imposition of duties in relation to climate change.

In light of calls for the imposition of such fiduciary duties, it is opportune to explore whether such implementation is possible in Australia. This chapter will examine the nature of fiduciary duties in Australia and the direction the High Court of Australia has been taking in relation to fiduciary theory. The difficulties of applying fiduciary duties in relation to climate change (*climate change fiduciary duties*) in Australia will then be discussed. The High Court appears to have significantly restricted the scope of the fiduciary principle and warned against expanding the fiduciary concept to recognise new types of fiduciary duties. The introduction of positive fiduciary duties to actively protect the environment or to take steps to reduce climate change appears no longer to be achievable in Australia.

The focus of this chapter is on whether climate change fiduciary duties are possible under Australian general law.[2] Issues such as whether a requisite fiduciary relationship could be made out between government and citizens (or between

---

* The author is grateful to Dr Michael Bryan for very helpful comments and discussion.

1 A number of fiduciary obligations in relation to climate change were discussed at the *Fiduciary Duty, Public Trust and the Governance of Climate Change* workshop (the *Workshop*). These included duties to protect and restore natural resources (particularly a duty to bring down atmospheric concentrations of greenhouse gas pollution) and duties of loyalty – for a detailed proposal see Wood (2009). Obligations attaching to the exercise of discretionary power were also discussed – see generally Fox-Decent (2005), pp. 268, 276, 280. The imposition of directors' duties in relation to climate change was also raised – in this respect note Australian Government Corporations and Markets Advisory Committee (2006), pp. 7–8.

2 This chapter will therefore not discuss the legislative imposition of fiduciary-like climate change duties, which may in fact be the best method of imposing the suggested duties.

sovereign states) and whether a trust model could instead be used to implement climate change fiduciary duties in Australia are left to other chapters.[3] The application of obligations attaching to the exercise of discretionary powers by fiduciaries will also not be considered.[4] The focus of this chapter is on the types of fiduciary duties that could be recognised in light of the High Court's more recent statements on the fiduciary principle.

The Nature of Fiduciary Duties in Australia

General Principles

The fiduciary concept has traditionally imposed a very high standard of loyalty on those to whom it applies. As the most demanding of the standards imposed by the law of equity,[5] fiduciary duties have compelled fiduciaries to act in the interests of their principal. Unsurprisingly, these duties have been reserved for relationships in which such a high degree of loyalty is appropriate. Examples include trustee and beneficiary, agent and principal, solicitor and client, director and company and partners.[6]

Outside of these accepted categories fiduciary duties may be imposed if the criteria for establishing a fiduciary relationship are satisfied. For example, fiduciary duties have been imposed on bankers, stockbrokers and certain employees.[7] The factors necessary to establish a fiduciary relationship are not settled – in fact there is a vast literature on this issue. In *Hospital Products Ltd v United States Surgical Corporation*[8] Mason J observed that the critical feature of well-recognised fiduciary relationships is the undertaking or agreement by the fiduciary to act for or on behalf of or in the interests of another person in the exercise of power or discretion which will affect in a legal or practical sense the interests of that other person.[9]

---

3   For details of public trust models see Finn (1995), pp. 140–41; Sax (1970), p. 471; Wood (2009).

4   See Chapter 5; see also Teele Langford (2011).

5   'Equity' refers to that body of cases, maxims, doctrines, rules, principles and remedies which derive ultimately from the specific jurisdiction established by the original English High Court of Chancery': Heydon and Loughlan (1997), p. 3.

6   See *Breen v Williams* (1996) 186 CLR 71 at 92 per Dawson and Toohey JJ. These relationships are known as status-based fiduciary relationships.

7   See generally Finn (1989), pp. 49–54.

8   (1984) 156 CLR 41 at 97 quoted in *Pilmer v Duke Group Limited (in liq)* (2001) 207 CLR 165, [2001] HCA 31 at [70] per McHugh, Gummow, Hayne and Callinan JJ.

9   For other approaches see Fox-Decent (2005); Criddle and Fox-Decent (2009); Glover (2004), chapter 2; Finn (1989), p. 46; *Breen v Williams* (1996) 186 CLR 71 at 107 per Gaudron and McHugh JJ; Flannigan (2004a); Meagher, Heydon and Leeming (2002) at [5–005]; Conaglen (2010), chapter 9.

Given that a fiduciary is expected to act in the interests of the principal, the fiduciary concept prohibits fiduciaries from pursuing a conflict between their personal interests and their duty to act in the interests of the principal or between another duty and their duty to the principal (the *conflicts rule*). The fiduciary principle also prohibits fiduciaries from making a profit from their position (the *profits rule*). If a fiduciary wishes to pursue a conflict or obtain a profit from their position they must obtain the informed consent of their principal, which involves full disclosure.[10]

The fiduciary principle has not traditionally been limited to the conflicts and profits rules. For example, company directors have for many years been subject to additional fiduciary duties. These include fiduciary duties to act bona fide in the interests of the company, to act for proper purposes and to retain discretions,[11] as well as a fiduciary duty of disclosure in certain circumstances.[12] Other fiduciary obligations have been imposed outside the company law sphere.[13]

Because of the high standard to which fiduciaries are held, fiduciary duties are supported by wide-ranging remedies in the event of breach of duty. These include injunctions, rescission, account of profits, constructive trust and equitable compensation.[14] The potentially significant consequences following a finding of breach of fiduciary duty serve to deter such breaches and to reinforce the high standard required of fiduciaries.[15] Another advantage of proving breach of fiduciary duty is that liability may be imposed on third parties. A claim can be brought against a third party who receives property knowing of a breach of fiduciary duty or who knowingly assists in a breach of fiduciary duty.[16]

---

10   See, for example, *Chan v Zacharia* (1984) 154 CLR 178, 198–199 per Deane J.

11   See Teele Langford (2009); Teele Langford (2011).

12   See Teele Langford (2008).

13   See *Lawfund Australia Pty Ltd v Lawfund Leasing Pty Ltd* (2008) 66 ACSR 1, [2008] NSWSC 144 at [26]–[28] and *Ambridge Investments Pty Ltd (in liq) v Baker* [2010] VSC 59 at [563] in relation to a fiduciary duty of good faith in the context of joint ventures. In relation to a duty to consider the exercise of powers (which applies to fiduciary power-holders but not to non-fiduciary power-holders) see Finn (1977) at [74]–[75]; Thomas (1998) at [1–40], [6–02], [6–06]; Hardingham and Baxt (1984) at [204]; Hayton and Mitchell (2005) at [3–103]; Meagher, Heydon and Leeming (2002) at [5–030]; cf Conaglen (2010), pp. 52–53. In relation to a fiduciary duty to act in the interests of another party see *Chan v Zacharia* (1984) 154 CLR 178 at 197 per Deane J and *Mabo v Queensland (No2)* (1992) 175 CLR 1 at 199–201 per Toohey J.

14   See, for example, Meagher, Heydon and Leeming (2002) at [5–235]–[5–280]; Conaglen (2005); Stafford and Ritchie (2008), chapter 9.

15   The availability of these remedies has in fact led to the abuse of the fiduciary principle in some cases – see Finn (2003), p. 5; Teele (1996); Stafford and Ritchie (2008) at [1.13].

16   This liability is based on *Barnes v Addy* (1873–74) L.R. 9, Ch. App. 244. For recent authority see *Farah Constructions Pty Ltd v Say-Dee Pty Ltd* (2007) 230 CLR 89, [2007] HCA 22. See also Glover (2004) at [8.6], [8.25].

Recent Fiduciary Theory

In the last fifteen years the High Court has delivered two judgements which appear to have considerably curtailed the content of the fiduciary principle. The scope of fiduciary duties is unclear as a result of these cases.

The first of these judgements is *Breen v Williams*.[17] The issue in that case was whether a patient (Ms Breen) could access her files held by a medical practitioner (Dr Williams). One of the alleged bases for access to the files was that a fiduciary relationship existed between Dr Williams and Ms Breen, which gave rise to a duty of disclosure. The imposition of a fiduciary duty of disclosure in these circumstances was rejected by all the members of the High Court.

In a celebrated passage, Gaudron and McHugh JJ held:

> In this country, fiduciary obligations arise because a person has come under an obligation to act in another's interests. As a result, equity imposes on the fiduciary proscriptive obligations – not to obtain any unauthorised benefit from the relationship and not to be in a position of conflict ... . But the law of this country does not otherwise impose positive legal duties on the fiduciary to act in the interests of the person to whom the duty is owed. If there was a general fiduciary duty to act in the best interests of the patient, it would necessarily follow that a doctor has a duty to inform the patient that he or she has breached their contract or has been guilty of negligence in dealings with the patient. That is not the law of this country.[18]

Dawson and Toohey JJ disapproved a tendency (which they found to be discernible in the United States and to a lesser extent Canada) to view a fiduciary relationship as imposing obligations which go beyond the exaction of loyalty – the fiduciary principle should not displace the role so far played by the law of contract and tort by becoming an independent source of positive obligations and creating new forms of civil wrong.[19]

---

17   (1996) 186 CLR 71 (*Breen*).

18   (1996) 186 CLR 71 at 113. Brennan CJ held that there was no fiduciary relationship which gave rise to a duty to give access to or to permit the copying of Ms Breen's records – see *Breen* (1996) 186 CLR 71 at 83.

19   (1996) 186 CLR 71 at 95. Gummow J stated: 'Fiduciary obligations arise (albeit not exclusively) in various situations where it may be seen that one person is under an obligation to act in the interests of another. Equitable remedies are available where the fiduciary places interest in conflict with duty or derives an unauthorised profit from abuse of duty. It would be to stand established principle on its head to reason that because equity considers the defendant to be a fiduciary, therefore the defendant has a legal obligation to act in the interests of the plaintiff so that failure to fulfil that positive obligation represents a breach of fiduciary duty' – (1996) 186 CLR 71 at 137–138.

The approach taken by Gaudron and McHugh JJ in *Breen* was followed by the majority of the High Court in *Pilmer v The Duke Group Ltd (in liq)*.[20] At issue in that case was whether members of a partnership of accountants were in breach of fiduciary duty to a company in preparing an allegedly inaccurate report in relation to a proposed takeover that the company was making of another company. McHugh, Gummow, Hayne and Callinan JJ applied *Breen* in holding that fiduciary obligations are proscriptive rather than prescriptive in nature; there is not imposed on fiduciaries a quasi-tortious duty to act solely in the best interests of their principals.[21] Kirby J dissented, questioning the dichotomy between prescriptive and proscriptive duties.[22] This point is taken up below.

The principles laid down by the High Court in *Breen* and *Pilmer* have been applied in a number of subsequent cases[23] and favoured in many academic commentaries.[24] A popular view is that the conflicts and profits rules are now the only recognised fiduciary duties.[25] Any positive duties to which a fiduciary is subject are said to be sourced in other areas of law such as tort and contract.

A fiduciary duty that is often rejected in cases applying the principles in *Breen* and *Pilmer* is the duty of disclosure.[26] Disclosure is now often seen as a step

---

20   (2001) 207 CLR 165, [2001] HCA 31 (*Pilmer*).

21   Ibid at [73]; see also Finn (1977), p. 3; *Cubillo v Commonwealth of Australia* (2001) 112 FCR 455, [2001] FCA 1213; *Williams v Minister, Aboriginal Land Rights Act 1983* (1999) 25 Fam LR 86, [2000] NSWCA 255; *Paramasivam v Flynn* (1998) 90 FCR 489.

22   See (2001) 207 CLR 165 at [128].

23   Notable examples include *Aequitas Ltd v Sparad No 100 Ltd* (2001) 19 ACLC 1006, [2001] NSWSC 14 at [283]–[285]; *Dresna Pty Ltd v Linknarf Management Services Pty Ltd* (2006) 156 FCR 474, [2006] FCAFC 193 at [132]; *Gibson Motorsport Pty Ltd v Forbes* (2006) 149 FCR 569, [2006] FCAC 44; *Australian Securities and Investments Commission v Citigroup Global Markets Australia Pty Ltd (No 4)* (2007) 160 FCR 35, [2007] FCA 963 at [290]; *P&V Industries Pty Ltd v Porto* (2006) 14 VR 1, [2006] VSC 131 at [23].

24   See, for example, Glover (2004) at [4.1]; Conaglen (2005, 2010); Stafford and Ritchie (2008) at [1.10]; Dempsey and Greinke (2004); Harding (2007); Flannigan (2004a); cf Hepburn (1996); Smith (2003); Lee (2009); Teele Langford (2011, 2009, 2008).

25   See, for example, *Australian Securities and Investments Commission v Citigroup Global Markets Australia Pty Ltd (No 4)* (2007) 160 FCR 35, [2007] FCA 963; *Gibson Motorsport Pty Ltd v Forbes* (2006) 149 FCR 569, [2006] FCAC 44 at [22] per Finn J; *P & V Industries Pty Ltd v Porto* (2006) 14 VR 1, [2006] VSC 131 at [23]; *Kenneth Gordon Webber v New South Wales* (2004) 31 Fam LR 425, [2003] NSWSC 1263; Conaglen (2005, 2010); Stafford and Ritchie (2008) at [2–39]; Dempsey and Greinke (2004); Vann (2005), p. 67; Harding (2007), cf Teele Langford (2011, 2009, 2008); Burrows (2002); Heydon (2005); Edelman (2010), p. 316; Nolan (2009), p. 315; Lee (2009).

26   See *P&V Industries Pty Ltd v Porto* (2006) 14 VR 1, [2006] VSC 131; *National Mutual Property Services (Australia) Pty Ltd v Citibank Savings Ltd* [1998] FCA 564; *Compaq Computer Australia Pty Ltd v Merry* (1998) 157 ALR 1 at 21; *Fitzwood Pty Ltd v Unique Coal Pty Ltd (in liq)* (2001) 188 ALR 566, [2001] FCA 1628; *Dresna Pty Ltd*

towards obtaining consent to a conflict of interest or profit from position, operating akin to a defence to breach of fiduciary duty rather than as an independent duty.[27] However, a fiduciary duty of disclosure continues to be recognised in the corporate law context.[28] In fact, if narrowly interpreted, the High Court's statements on fiduciary theory in *Breen* and *Pilmer* do not appear to fit well with the way in which fiduciary duties have been applied to company directors.[29]

The High Court's statements in *Breen* and *Pilmer* represent a shift from previous High Court cases on fiduciary principles, which do not limit the fiduciary principle to the same extent.[30] In this respect, the acknowledgement by Austin J in *Aequitas Ltd v Sparad No 100 Ltd*[31] that 'judicial thinking about the content of fiduciary duties has changed significantly over the last decade' is significant. There is now doubt as to whether duties that have previously been classified as fiduciary can continue to be so classified. Moreover, as will be shown below, the distinction between prescriptive and proscriptive duties is causing difficulties in practice.

Issues Arising from Breen and Pilmer

Despite the need for further clarification from the High Court, it is important to examine the scope and nature of possible fiduciary obligations following *Breen* and *Pilmer* in order to assess the potential applicability of these obligations in the area of climate change.

---

v Linknarf Management Services Pty Ltd (in liq) (2006) 156 FCR 474, [2006] FCAFC 193; *Gibson Motorsport Pty Ltd v Forbes* (2006) 149 FCR 569, [2006] FCAC 44 at [22]; *Australian Securities and Investments Commission v Citigroup Global Markets Australia Pty Ltd (No 4)* (2007) 160 FCR 35, [2007] FCA 963; Stafford and Ritchie (2008) at [1.10], [2.1], [2.6], [2.39]. See also Harding (2007), p. 3; Flannigan (2004b), p. 294.

27   This follows comments by Gummow J in *Breen* that 'informed consent' is an answer to circumstances which otherwise indicate disloyalty, not a mainspring of equitable liability: see *Breen* (1996) 186 CLR 71 at 125; see also *P&V Industries Pty Ltd v Porto (No 2)* [2007] VSC 64 at [22]; *Dresna Pty Ltd v Linknarf Management Services Pty Ltd (in liq)* (2006) 156 FCR 474, [2006] FCAC 193; *Wilden Pty Ltd v Green* [2009] WASCA 38 at [106], [241]; Flannigan (2004b), p. 294; Vann (2005), p. 74.

28   See Teele Langford (2008). See also *Michael Wilson and Partners Ltd v Nichols* [2009] NSWSC 1033 at [180]; *Southern Real Estate Pty Ltd v Dellow* (2003) 87 SASR 1, [2003] SASC 318 at [29]; *Fitzwood Pty Ltd v Unique Goal Pty Ltd* (2001) 188 ALR 566, [2001] FCA 1628 at [33].

29   See Teele Langford (2011). It is in fact arguable that the High Court's comments on fiduciary matters in *Breen* and *Pilmer* do not apply to status-based fiduciary relationships – see for example *Breen* (1996) 186 CLR 71 at 112 (per Gaudron and McHugh JJ); Svehla (2006); Heydon (2005), p. 233; cf *P & V Industries Pty Ltd v Porto* (2006) 14 VR 1, [2006] VSC 131 at [21]; *Aequitas Ltd v Sparad No 100 Ltd* (2001) 19 ACLC 1006, [2001] NSWSC 14 at [286]–[287].

30   See generally Teele Langford (2009), p. 329.

31   (2001) 19 ACLC 1006, [2001] NSWSC 14 at [283].

There are in fact several ways of interpreting the ambit of the fiduciary principle in light of *Breen* and *Pilmer*. A prevalent interpretation is that the conflicts and profits rules are now the only fiduciary duties recognised in Australia. A second interpretation is that the High Court is requiring fiduciary duties to be proscriptive in nature but not restricting fiduciary duties beyond this requirement. A third interpretation is that the only fiduciary duty rejected by the High Court is a broad prescriptive fiduciary duty to act in the interests of the principal. These interpretations will now be examined.

First Interpretation: Conflicts and Profits as the Exclusive Fiduciary Duties

A number of commentators and judges have expressed the view that Australian fiduciary duties are now limited to the conflicts and profits rules.[32] This is because the High Court in *Breen* and *Pilmer* appeared to single out the conflicts and profits rules as accepted proscriptive fiduciary duties.[33] For example, Hollingworth J in *P & V Industries Pty Ltd v Porto* stated:

> The decisions in *Breen* and *Pilmer* clearly confirm that, in Australia, fiduciary duties are limited to proscriptive duties of loyalty ... . This means that the no conflict and no profit rules encompass the whole content of fiduciary obligations and the duty of loyalty imposed on a fiduciary is promoted by prohibiting disloyalty rather than by prescribing some positive duty.[34]

If this interpretation is correct then fiduciary duties in Australia are very limited indeed.

This interpretation does not, however, clearly follow from the High Court's judgements in *Breen* and *Pilmer*. This can be seen from the fact that the majority judges in *Pilmer* cite the conflicts and profits rules as particular examples of fiduciary duties:

> The words of Frankfurter J in *Securities and Exchange Commission v Chenery Corp* bear repetition. His Honour said:

> 'But to say that a man is a fiduciary only begins analysis; it gives direction to further inquiry. To whom is he a fiduciary? What obligations does he owe as a fiduciary? In what respect has he failed to discharge these obligations? And what are the consequences of his deviation from duty?'

---

32   See above n 25.

33   See Gaudron and McHugh JJ in *Breen* (1996) 186 CLR 71 at 113 (extracted above); Gummow J in *Breen* (1996) 186 CLR 71 at 135. See also *Youyang Pty Ltd v Minter Ellison Morris Fletcher* (2003) 212 CLR 484, [2003] HCA 15 at [41].

34   (2006) 14 VR 1, [2006] VSC 131 at [23] citing Nolan (1997), p. 222.

In particular, the fiduciary is under an obligation, without informed consent, not to promote the personal interests of the fiduciary by making or pursuing a gain in circumstances in which there is 'a conflict or a real or substantial possibility of a conflict' between personal interests of the fiduciary and those to whom the duty is owed. That is how the matter was put by Mason J in *Hospital Products*. Similar reasoning applies where the alleged conflict is between competing duties, for example where a solicitor acts on both sides of a transaction (emphasis added).[35]

The issue of whether the conflicts and profits rules are the only remaining Australian fiduciary duties was considered recently by Owen J in *Bell Group Ltd (in liq) v Westpac Banking Corp (No 9)*.[36] After a detailed examination of the relevant case law, Owen J concluded:

Breen stands for the proposition that Australian law only recognises as fiduciary those duties that stem from the fundamental obligation of loyalty and which are proscriptive. It does not necessarily follow that their Honours intended to suggest that the duties to avoid conflicts and not to profit were, in every species of fiduciary relationship (the categories for which are not closed), the only duties that could possibly qualify.[37]

In other words, the conflicts and profits rules are underlying fiduciary duties that are applied in all fiduciary relationships but other fiduciary duties may also be imposed depending on the particular fiduciary relationship at issue. In the context of the relationship between director and company, Owen J upheld the fiduciary

---

35   See *Pilmer* (2001) 207 CLR 165 at [77] – [78] (footnotes omitted). The two most recent decisions of the High Court concerning fiduciary duties arguably support the view that fiduciary duties are not limited to the conflicts and profits rules, although this was not directly in issue in these decisions. For example, in *Friend v Brooker* [2009] 239 CLR 129, [2009] HCA 21 the High Court rejected the imposition of a fiduciary duty between co-directors 'to be equally and personally liable to each other for losses flowing from personal borrowings.' The Court did not reject this duty because it was not the conflicts or profits rule but because it was not proscriptive and because the parties had made a commercial decision to adopt a corporate structure (in which they would owe duties to the company) – see [2009] 239 CLR 129, [2009] HCA 21 at [84], [86] per French CJ, Gummow, Hayne and Bell JJ (Heydon J concurring). In its most recent judgment on fiduciary duties, the High Court refers to an article that recognises the fiduciary status of a number of duties – see *John Alexander's Clubs Pty Ltd v White City Tennis Club Ltd* (2010) 241 CLR 1, [2010] HCA 19 at [44] per French CJ, Gummow, Hayne, Heydon and Kiefel JJ citing Edelman (2010).

36   (2008) 22 FLR 1, [2008] WASC 239 (*Bell*). Note that this decision has been appealed.

37   Ibid at [4569]. See also Parkinson (1995), p. 444; Parkinson (2003) at [1013]; Teele Langford (2009), p. 329.

classification of duties to act bona fide in the interests of the company and for proper purposes.[38]

Although the preferable view is that Australian fiduciary duties are not limited to the conflicts and profits rules, it is improbable that Australian courts will accord fiduciary status to fiduciary duties that have not previously been accepted as fiduciary (in contrast to the duties recognised by Owen J in *Bell*, which his Honour found to have been classified as fiduciary for 'eons').[39] This is because the High Court in *Breen* has indicated that the declaration of any such new fiduciary duties would be a matter for Parliament rather than the Court:

> Advances in the common law must begin from a baseline of accepted principle and proceed by conventional methods of legal reasoning. Judges have no authority to invent legal doctrine that distorts or does not extend or modify accepted legal rules and principles. Any changes in legal doctrine, brought about by judicial creativity, must 'fit' within the body of accepted rules and principles. The judges of Australia cannot, so to speak, 'make it up' as they go along. It is a serious constitutional mistake to think that the common law courts have authority to 'provide a solvent' for every social, political or economic problem. The role of the common law courts is a far more modest one.
>
> In a democratic society, changes in the law that cannot logically or analogically be related to existing common law rules and principles are the province of the legislature. From time to time it is necessary for the common law courts to reformulate existing legal rules and principles to take account of changing social conditions. Less frequently, the courts may even reject the continuing operation of an established rule or principle. Such steps can be taken only when it can be seen that the 'new' rule or principle that has been created has been derived logically or analogically from other legal principles, rules and institutions.[40]

The recognition of climate change fiduciary duties in Australia is therefore unlikely.

---

38 In this respect, the following comments of Professor Austin are also significant: 'Perhaps the duty to act lawfully and within their powers and authority is the most obvious duty of company directors... There may be room for debate as to whether these duties are strictly fiduciary duties, given the narrow definition of the concept in *Breen v Williams* ...Assuming it is correct, notwithstanding *Breen v Williams*, to regard the positive duty of a fiduciary to act in good faith in the interests of the principal as part of the fiduciary responsibility rather than an extraneous duty of good faith, it seems appropriate to treat the duty to act within the terms of the constitutive instrument for the fiduciary relationship as part of that positive duty, and therefore itself a fiduciary requirement. Indeed, properly analysed, it is a fiduciary duty of fundamental importance '– see Austin, Ford and Ramsay (2005) at [11.1] (footnotes omitted), see also at [11.9], [11.12] and [11.13].

39 *Bell* (2008) FLR 1 at [4570].

40 *Breen* (1996) 186 CLR 71 at 115 per Gaudron and McHugh JJ (footnotes omitted).

Second Interpretation: Fiduciary Duties as Proscriptive Duties

It will be recalled that the second possible interpretation of the cases of *Breen* and *Pilmer* is that the High Court is requiring fiduciary duties to be proscriptive (but not restricting the fiduciary principle to the conflicts and profits rules). The requirement that fiduciary duties be proscriptive was reaffirmed by the High Court in the recent case of *Friend v Brooker*.[41]

The distinction between prescriptive and proscriptive duties plays an important part in the judgements in *Breen* and *Pilmer*. Unfortunately, however, the distinction between prescriptive and proscriptive obligations is not precise.

The wording used by Gaudron and McHugh JJ in *Breen* indicates that the term 'proscriptive' is being used in contrast to positive or affirmative duties. This is confirmed by the wording used by Kirby J in *Pilmer:*

> ... *Breen* upholds the principle stated in the aphorism that fiduciary obligations are 'proscriptive' and not 'prescriptive'. This, in my view, is the fundamental reason why all members of this court in *Breen* rejected Ms Breen's claim of a fiduciary obligation. Whatever the differing views which the justices held concerning the character of the relationship in question there and whether it was, or was not, a fiduciary one for some or all purposes, there was agreement that Ms Breen's claim failed because it would have involved imposing on the suggested fiduciary positive obligations to act. It would have burdened him with an affirmative obligation to grant access to his notes to a patient ('prescriptive' duties). It would thus have gone further than the conventional ('proscriptive') duties of loyalty, of avoiding conflicts of interest or of misusing one's power, such as fiduciary duties have traditionally upheld.[42]

In other words the simplest meaning of the term 'proscriptive' as applied to duties is to require restraint – a proscriptive duty spells out what a fiduciary cannot do. By contrast, a prescriptive duty mandates positive action.[43] In basic terms a prescriptive duty could be described as positive and a proscriptive duty as negative, although the distinction is not so simple in practice.[44] The conflicts

---

41   (2009) 239 CLR 129, [2009] HCA 21 at [84] per French CJ, Gummow, Hayne and Bell JJ (Heydon J concurring).

42   (2001) 207 CLR 165 at [127] (footnotes omitted).

43   Ford and Lee state: 'A prescriptive duty requires action by the person owing it; a proscriptive duty requires restraint' – see Ford and Lee (2010) (looseleaf accessed August 2011) at [9.100]; see also Vann (2005), p. 67. Glover (2004) at [4.1] writes: 'Fiduciary duties stipulate certain things which fiduciaries *cannot* be influenced by and *cannot* do. Actions are outlawed by fiduciary duties. Nothing is mandated or prescribed'. Harpum describes fiduciary obligations as 'negative principles' – see Harpum (1997), pp. 145, 147. See also Nolan (1998), p. 12; Sealy (1989), p. 268.

44   Kirby J in *Pilmer* states: 'Whilst, for my own part, I question the viability of this supposed dichotomy (because omissions quite frequently shade into commissions) I

and profits rules are said to be proscriptive because they require avoidance of unauthorised conflicts and profits.

The requirement that fiduciary duties be proscriptive is resulting in duties being reworded in negative terms. This can be seen in the case of *Motor Trades Association of Australia Superannuation Fund Pty Ltd v Rickus (No 3)*.[45] In that case a former chairman (Mr Rickus) of a superannuation trustee company (the *Trustee*) had given documents to the Australian Prudential Regulation Authority at its request. He later refused to produce those documents to the rest of the board. At issue was whether Mr Rickus was under a duty to produce such documents in light of cases such as *Breen* and *Pilmer*. Flick J found that Mr Rickus was under a duty to produce a copy of the documents, at least when so requested by the Trustee. His Honour considered that this was not a 'positive obligation', but rather a duty to place the board in the position whereby it could respond to an investigation being pursued by the regulator and respond to that information:

> Rather than the obligation being the imposition of a *'positive obligation'*, it was his obligation not to withhold information from the board. The *'obligation'* was to accede to the request made of him to provide the documents ... . Replying to a request properly made of him cannot properly be characterised as the imposition of any *'positive obligation'*.[46]

Essentially what Flick J did was to rephrase a positive obligation to disclose as a negative obligation not to withhold information. This rephrasing was perhaps more plausible in this case because the information had previously been disclosed. Nevertheless, the actual requirement to disclose the information was arguably still a positive one.[47]

However, such attempts to reformulate positive obligations in negative terms are not always necessarily successful. For example, in *Loxias Technologies Pty Ltd v Curacel International Pty Ltd*,[48] Moore J rejected an attempt to couch a

---

must accept that *Breen* embraces the distinction...': (2001) 207 CLR 165 at [128]. For more detail on the prescriptive-proscriptive distinction and an examination of the operation of this distinction in the context of obligations relating to discretionary powers see Teele Langford (2011).

45   (2008) 69 ACSR 264, [2008] FCA 1986 (*Motor Trades*).

46   Ibid at [71] and [72].

47   See also *Bell* in which the duties to act bona fide in the interests of the company and for proper purposes were rephrased proscriptively but nevertheless imposed positive requirements. These included active consideration and investigation and ensuring the existence of appropriate corporate benefit – see (2008) 22 FLR 1 at [4619], [5605], [6033], [6040], [6043], [6047], [6050]–[6052], [6060], [6064], [6065], [6083], [6088], [6089], [6091], [6096], [6100]. In some circumstances these duties require positive action to prevent a transaction from going ahead – see Teele Langford (2011).

48   [2002] FCA 1473; [2002] FCA 753 (*Loxias*). See also *Friend v Brooker* (2009) 239 CLR 129, [2009] HCA 21 at [84]–[85].

duty of disclosure (in this case in relation to what a director knew about certain products manufactured by another company of which he was also a director) in terms of a duty to avoid a conflict of interest.[49]

The prescriptive-proscriptive distinction is not always straightforward.[50] A proscriptive duty may in fact sometimes include positive requirements.[51] A fiduciary may be required to take positive action in order to meet the requirements of the conflicts and profits rules.[52] For example, a director faced with a conflict of duties may have to take a positive step (such as resigning from the board of one company) in order to avoid breach of the conflicts rule.

Despite the need for further clarification and reconsideration of the prescriptive-proscriptive distinction, it appears that any climate change fiduciary duties would need to be proscriptive to be recognised in Australia.

Third Interpretation: Rejection of Overarching Prescriptive Duty Only

An alternative view of the prescriptive-proscriptive distinction laid down in *Breen* and *Pilmer* is that the only fiduciary duty rejected by the High Court is a broad prescriptive fiduciary duty to act in the interests of the principal.[53] Some of the problems with such a broad prescriptive duty are said to be the difficulty of assessing compliance and the likely inhibition of entrepreneurial

---

49   Moore J found that the pleading was in form an allegation that a position of conflict arose which required the relevant director to advise the company not to distribute the relevant products (to meet the requirement in *Breen* that fiduciary duties be proscriptive rather than prescriptive). However in substance it was a pleading that the relevant director was obliged to take steps to disclose to the company the information he had about relevant products. The company would not suffer damage or other consequences from the director breaching this obligation. Moore J therefore found there was no real prospect of this aspect of the claim succeeding – see *Loxias* [2002] FCA 1473 at [2].

50   See Conaglen (2010), p. 203; Teele Langford (2011); DalPont, Chalmers and Maxton (2007), p. 94.

51   See *Motor Trades; Bell,* above n 47. For more detail see Teele Langford (2011).

52   Watt (2008), p. 330 writes: 'Some consider the duty of good faith to be the positive aspect of the fiduciary duty. They contrast it with negative aspects of the fiduciary duty, such as the duty *not* to put oneself in a position of potential conflict and the duty *not* to make an unauthorised gain. However, this distinction is not convincing. Even the so-called 'negative' fiduciary rules have their positive counterparts. Hence a fiduciary has a positive duty, unless he is authorised to the contrary, to extricate himself from any position of actual conflict and to disgorge any gain made from his position of trust.' See also *Loxias* [2002] FCA 753 at [13]; cf Nolan (1997), p. 223.

53   The foundation of the claim of a fiduciary obligation to provide access to records was a fiduciary duty that Dr Williams would act in Ms Breen's interests – see *Breen* (1996) 186 CLR 71 at 109, 110, 113 per Gaudron and McHugh JJ; see also *Breen* (1996) 186 CLR 71 at 137 per Gummow J; *Pilmer* (2001) 207 CLR 165 at [74] per McHugh, Gummow, Hayne and Callinan JJ. See also Young, Croft and Smith (2009) at [7.300].

activity. [54] If this view is correct, then fiduciary duties can be phrased in positive terms or impose positive requirements provided that these requirements do not amount to a prescriptive duty to act in the interests of the principal. Although this interpretation would give welcome flexibility and scope to the fiduciary concept, it does not appear to be the interpretation favoured by most decisions since *Breen*. [55]

The High Court's rejection of a prescriptive fiduciary duty to act in the interests of the principal means that it will not be possible to impose a prescriptive fiduciary duty to act in the interests of citizens as concerns climate change. The imposition of such a broad duty, incorporating various more specific duties, [56] might have been an effective way of implementing effective climate change duties.

Summary of the Australian Position on Fiduciary Duties

Given the arguable ambiguity of the High Court's recent fiduciary statements, it is hoped that the High Court will re-examine and qualify its recent fiduciary statements. However, pending such re-examination and possible qualification, Australian fiduciary law is in a state of flux. The reach of Australian fiduciary duties is unclear, with several interpretations possible.

The law on fiduciary obligations in Australia can be summarised as follows:

1. The High Court appears to have rejected the existence of a prescriptive fiduciary duty to act in the interests of the principal, at least in the context of non-status based fiduciary relationships.
2. The preferable view is that the conflicts and profits rules are not the only remaining accepted fiduciary duties. Certain duties that have previously been recognised as fiduciary can continue to be so recognised. The High Court has, however, cautioned against the expansion of fiduciary duties to new categories.

---

54   See Nolan (2005), p. 423; cf Pey-Woan Lee and Lusina Ho (2007). Professor (now Justice) Finn states: 'If a fiduciary's liability was to be determined by reference to whether or not the beneficiary's interests had in fact been served, an often impossible inquiry, more than curious consequences would follow. Much of the law of trusts, of agency and of companies would, for example, be rendered superfluous. The law of torts and of contracts would be displaced from their now accepted roles in many relationships' – see Finn (1989), p. 28; see also Finn (1977) at [31]; Worthington (2001), p. 448.

55   See, for example, the cases in n 26 above. See also *Pilmer* (2001) 207 CLR 165 at [127] per Kirby J (extracted above); *Fico v O'Leary* [2004] WASC 215 at [155]; *Kenneth Gordon Webber v New South Wales* (2004) 31 Fam LR 425, [2003] NSWSC 1263 at [35], [47]; *Youyang Pty Ltd v Minter Ellison Morris Fletcher* (2003) 212 CLR 484, [2003] HCA 15 at [41].

56   See, for example, Thomas (2008).

3. It appears that fiduciary obligations are required to be proscriptive, although the application of this requirement has been shown to be unclear.

Application of Climate Change Fiduciary Duties in Australia

Having analysed the current state of fiduciary theory in Australia, it is appropriate to test the application of that theory to proposed climate change fiduciary duties. Particular reference will be made to the climate change fiduciary duties propounded by Professor Mary Wood in her seminal articles proposing a sovereign trust model to safeguard the environment.[57]

The suggested climate change fiduciary duties can be divided into two broad categories. The first category comprises proscriptive 'duties of loyalty'. These duties broadly correspond to the conflicts and profits rules. In this respect Professor Wood notes problems such as bribery, abuse of discretion, undue political influence and the favouring of industry friends.[58] She states:

> The duty of loyalty reaches its pinnacle with respect to natural assets necessary for public survival – like the atmosphere. Because such assets are crucial and irreplaceable, breaching the strict duty of loyalty may bring irreversible damage to society and future generations. Thus, the inquiry into fiduciary loyalty must be particularly demanding with respect to issues such as global warming. While it is true that government sometimes must balance competing public interests in managing the natural trust, that situation is much different than making a trade-off of public interests to benefit private singular interests.[59]

The second category of climate change fiduciary duties comprises prescriptive duties. The following are examples of suggested prescriptive fiduciary duties:

a. a duty to protect and restore public assets;[60]
b. a duty to maintain the ability of public assets to provide a steady abundance of environmental services for future generations;[61] and

---

57   See Wood (2009). Note that the duties proposed by Professor Wood are based on a public trust model. Not all duties owed by trustees are fiduciary duties.

58   Wood (2009), pp. 58–59, 63, 100.

59   Wood (2009), p. 100 (footnotes omitted). This quote raises the issue (also raised in general discussion at the Workshop) of dealing with competing interests of beneficiaries – see generally Finn (1977), chapters 12 and 13; Thomas (1998) at [6–165]; Glover (2004) at [4.4].

60   See Wood (2009), p. 104. Note that the specific fiduciary obligations vary according to the nature and needs of the particular asset – see Wood (2009), p. 111.

61   Wood (2009), p. 95.

    c. a duty to protect the atmosphere, by following measured targets to reduce atmospheric concentrations of greenhouse gas pollution.[62]

An issue also arises as to whether fiduciary duties could be imposed on the Australian Government to take action to realise the goals of the Copenhagen Accord in relation to matters such as limiting global warming, voluntarily reducing greenhouse gas emissions and the provision of financial aid. Future climate change summits are also likely to give rise to other obligations or voluntary undertakings. Whether such obligations or undertakings could be enforced by citizens by way of the imposition of fiduciary obligations on the Australian government involves analysis similar to that relating to the duties in category two.

The imposition of these fiduciary duties on the Australian government or governmental agencies is dependent on the establishment of a fiduciary relationship.[63] Provided that the requisite fiduciary relationship between state and citizens could be made out, the following could be said about the application of climate change fiduciary duties in Australia:

1. The proscriptive duties of loyalty outlined in category one above could be imposed in Australia, given the High Court's explicit approval of the conflicts and profits rules as fiduciary duties. The fiduciary concept could therefore be used, for example, to prevent governments and authorities from using money earmarked for environmental protection or prevention of climate change for other purposes or from pursuing conflicts when undertaking environmental or climate change initiatives.

2. Perhaps the simplest way of implementing the prescriptive duties in category two (including obligations arising from the Copenhagen Accord) would be to impose a broad prescriptive fiduciary duty on the Government and governmental agencies to act in the interests of present and future citizens as concerns climate change. The specific duties outlined in category two or framed pursuant to the Copenhagen Accord could be imposed as aspects of this overarching duty. However, it was shown above that acceptance of such an overarching fiduciary duty to act in the interests of citizens as concerns climate change is very unlikely. The High Court has rejected the imposition of such a prescriptive fiduciary duty in *Breen* and *Pilmer* and has also cautioned against extending the fiduciary concept.

---

    62   Wood (2009), pp. 96, 102. A general fiduciary duty to protect the public and to cause the burden (in relation to carbon allocation) to be shared equally (as suggested during the Workshop) would also be included in category two, as would the positive fiduciary duties suggested by Will McGoldrick, Donald Feaver and Andrew Maver in Chapter 2.

    63   This issue is dealt with in Chapter 5. See also Finn (1977), p. 10; *Habib v Commonwealth (No2)* (2009) 175 FCR 350, [2009] FCA 228; Young et al (2009) at [7.520], [7.530], [7.540]; Keane (2010).

3. If a narrow view of the High Court's statements on fiduciary theory is taken and the conflicts and profits rules are the only recognised fiduciary duties then there is no scope for the recognition of individual climate change fiduciary duties as outlined in category two (or based on the terms of the Copenhagen Accord). However, as shown above, the better view is that Australian fiduciary duties are not limited to the conflicts and profits rules.

4. Given the apparent requirement that fiduciary duties be proscriptive, prescriptive climate change fiduciary duties are unlikely to be upheld by Australian courts. Any Australian climate change fiduciary duties would need to be proscriptive (or at least phrased proscriptively). This means that it is doubtful that the prescriptive fiduciary duties outlined under category two (or based on the Copenhagen Accord) would be recognised as fiduciary duties in Australia.

5. It might be possible to rephrase the duties in category two in negative terms to render them proscriptive, although this process is arguably somewhat artificial and may be overturned. For example, a fiduciary duty to protect natural resources could perhaps be rephrased as a fiduciary duty to avoid further deterioration of such resources. A duty to bring down atmospheric concentrations of greenhouse gas pollution could possibly be brought in as a requirement of a proscriptive duty to avoid climate change. However, it must be remembered that the High Court has warned against the expansion of the fiduciary concept. This means that even the recognition of such proscriptive fiduciary duties in relation to climate change is improbable.

The imposition of effective climate change fiduciary duties in Australia is therefore unlikely to be possible in light of the High Court's statements in *Breen* and *Pilmer.*

The narrow approach to fiduciary duties taken in Australia can be contrasted with the more expansive approach taken in Canada and the United States.[64] Courts in these jurisdictions recognise prescriptive fiduciary duties.[65] The prescriptive fiduciary duties proposed by Professor Wood and other commentators are in keeping with the more expansive prescriptive fiduciary model in force in Canada. However, the Australian High Court has cautioned against following this

---

64 Fiduciary law in England is less restrictive than in Australia but not as expansive as in the United States or Canada – see *Bristol & West Building Society v Mothew* [1998] Ch 1; Teele Langford (2011).

65 See, for example, Finn (1989), p. 25; Parkinson (1995); Ellis (2009) (looseleaf accessed February 2010) particularly at 1–4, 1–22.1, 5–8(1), 9–8.2(1), 9–8.5, 10–4 – 10–6, 15–3, 19–4(6), 19–14.1, 19–14.7, 19–24.16, 19–20, 19–20.3, 19–20.5, 19–26.4; Rosen (2007); Milnes (2009); Berryman (2009); *McInerney v McDonald* [1992] 2 SCR 138 at [20], [21], [22], [28]; *Strother v 3469420* [2007] SCC 24; Dorsett (1996); Finn (2003), p. 5; Flannigan (2004a); *P&V Industries Pty Ltd v Porto* (2006) 14 VR 1, [2006] VSC 131 at [42].

approach.[66] Gaudron and McHugh JJ (with whom Gummow J agreed) in *Breen* stated that:

> ... the Canadian law on fiduciary duties is very different from the law in this country with respect to [access to records]... . One significant difference is the tendency of Canadian courts to apply fiduciary principles in an expansive manner so as to supplement tort law and provide a basis for the creation of new forms of civil wrongs ... . With great respect to the Canadian courts, however, many cases in that jurisdiction pay insufficient regard to the effect that the imposition of fiduciary duties on particular relationships has on the law of negligence, contract, agency, trusts and companies in their application to those relationships. Further, many of the Canadian cases pay insufficient, if any, regard to the fact that the imposition of fiduciary duties often gives rise to proprietary remedies that affect the distribution of assets in bankruptcies and insolvencies.[67]

Conclusion

The use of the fiduciary model to impose positive duties to protect the environment and prevent climate change is likely to be difficult in Australia due to the High Court's recent statements on fiduciary theory. If a narrow view of these statements is taken, the fiduciary principle is limited to preventing unauthorised conflicts and profits. The fiduciary concept could therefore not be used to impose positive duties to protect the environment or to reduce climate change. It would not even allow more negative duties to avoid climate change or harm to the environment.

The preferable view is that fiduciary duties are not so limited. Particularly in the context of status-based fiduciary relationships, fiduciary duties other than the conflicts and profits rules can be recognised. However, even on this view the imposition of positive duties to protect the environment or to reduce climate change is dubious. Perhaps duties framed in more negative terms might be upheld. However, this too is doubtful given the High Court's warning against expansion of the fiduciary concept.

There have been calls for a new paradigm for holding governments accountable in protecting natural wealth.[68] This chapter has shown that the fiduciary principle is unlikely to constitute that paradigm as concerns Australia. Further clarification from the High Court in relation to its more recent statements on the fiduciary principle would be welcome, not just as concerns the imposition of climate change fiduciary duties but also as concerns fiduciary loyalty in general.

---

66 See also *Breen* (1996) 186 CLR 71 at 95 per Dawson and Toohey JJ; *Pilmer* (2001) 207 CLR 165 at [124] per Kirby J.

67 (1996) 186 CLR 71 at 112; see also (1996) 186 CLR 71 at 137 per Gummow J; see also Parkinson (1995); Dorsett (1996).

68 See Wood (2009), p. 93.

References

R. P. Austin, H. A. J. Ford and I. M. Ramsay (2005) *Company Directors – Principles of Law and Corporate Governance*, LexisNexis Butterworths.

Australian Government Corporations and Markets Advisory Committee (2006) *The Social Responsibility of Corporations*.

Jeff Berryman (2009) 'Fact-Based Fiduciary Duties and Breaches of Confidence: An Overview of Their Imposition and Remedies for Breach', 15 *New Zealand Business Law Quarterly* 35.

Andrew Burrows (2002) 'We Do This at Common Law but That in Equity', 22 *Oxford Journal of Legal Studies* 1.

Matthew Conaglen (2005) 'The Nature and Function of Fiduciary Loyalty', 121 *Law Quarterly Review* 452.

Matthew Conaglen (2010) *Fiduciary Loyalty – Protecting the Due Performance of Non-Fiduciary Duties*, Hart Publishing.

Evan J. Criddle and Evan Fox-Decent (2009) 'A Fiduciary Theory of Jus Cogens', 34 *Yale Journal of International Law* 331.

G. E. DalPont, D. R. C. Chalmers and J. K. Maxton (2007) *Equity and Trusts: Commentary and Materials*, 4th edn, Lawbook Co.

Gillian Dempsey and Andrew Greinke (2004) 'Proscriptive fiduciary duties in Australia', 25 *Australian Bar Review* 1.

Shaunnagh Dorsett (1996) 'Comparing Apples and Oranges: The Fiduciary Principle in Australia and Canada after *Breen v Williams*', 8 *Bond Law Review* 158.

James Edelman (2010) 'When do fiduciary duties arise?', 126 *Law Quarterly Review* 302.

Mark Vincent Ellis (2009) *Fiduciary Duties in Canada*, Carswell.

P. D. Finn (1977) *Fiduciary Obligations*, Law Book Co.

P. D. Finn (1989) 'The Fiduciary Principle', in T. G. Youdan (ed.), *Equity, Fiduciary and Trust*, Carswell.

Paul Finn (1995) 'The Forgotten "Trust": The People and the State', in Malcolm Cope (ed.), *Equity: Issues and Trends*, The Federation Press.

Paul Finn (2003) 'Fiduciary Reflections', *13th Commonwealth Law Conference*, 13–17 April, Melbourne.

Robert Flannigan (2004a) 'The Boundaries of Fiduciary Accountability', 83 *Canadian Bar Review* 35.

Robert Flannigan (2004b) 'Fiduciary Duties of Shareholders and Directors', *Journal of Business Law* 277.

H. A. J. Ford and W. A. Lee (2010), *The Law of Trusts,* 4th edn (Update 72), Lawbook Co.

Evan Fox-Decent (2005) 'The Fiduciary Nature of State Legal Authority' 31 *Queen's Law Journal* 259.

John Glover (2004) *Equity, Restitution and Fraud*, LexisNexis Butterworths.

Matthew Harding (2007) 'Two Fiduciary Fallacies', *Journal of Equity* 1.

I. J. Hardingham and R. Baxt (1984) *Discretionary Trusts*, 2nd edn, Butterworths.

Charles Harpum (1997) 'Fiduciary Obligations and Fiduciary Powers: Where Are We Going?', in P. Birks (ed.), *Privacy and Loyalty*, Clarendon Press.

David Hayton and Charles Mitchell (2005) *Hayton and Marshall Commentary and Cases on the Law of Trusts and Equitable Remedies*, 12th edn, Sweet & Maxwell.

Samantha Hepburn (1996) '*Breen v Williams*', 20 *Melbourne University Law Review* 1207.

J. D. Heydon and P. L. Loughlan (1997) *Cases and Materials on Equity & Trusts*, 5th edn, Butterworths.

J. D. Heydon (2005) 'Are the Duties of Company Directors to Exercise Care and Skill Fiduciary?', in Simone Degeling and James Edelman (eds), *Equity in Commercial Law*, Lawbook Co.

P. A. Keane (2010) 'The Conscience of Equity', 10(1) *Queensland University of Technology Law and Justice Journal* 106.

Rosemary Teele Langford (2008) '*ENT Pty Ltd v Sunraysia Television Ltd:* A Positive Fiduciary Duty of Disclosure', 26 *Company and Securities Law Journal* 470.

Rosemary Teele Langford (2009) 'The Fiduciary Nature of the Bona Fide and Proper Purposes Duties of Company Directors: *Bell Group Ltd (in liq) v Westpac Banking Corp (No 9)*', 31 *Australian Bar Review* 326.

Rosemary Teele Langford (2011) 'The Duty of Directors to Act Bona Fide in the Interests of the Company: A Positive Fiduciary Duty? Australia and England Compared', 11 *Journal of Corporate Law Studies* 215.

Pey-Woan Lee and Lusina Ho (2007) 'A Director's Liability to Compete', *Journal of Business Law* 98.

Rebecca Lee (2009) 'Rethinking the Content of the Fiduciary Obligation', *The Conveyancer and Property Lawyer* 236.

R. P. Meagher, J. D. Heydon and M. J. Leeming (2002) *Meagher Gummow and Lehane's Equity Doctrines and Remedies*, 4th edn, Butterworths LexisNexis.

Robert E. Milnes (2009) 'Acting in the Best Interests of the Corporation: To Whom Is This Duty Owed by Canadian Directors? The Supreme Court of Canada in the BCE Case Clarifies the Duty', 24 *Banking & Finance Law Review* 601.

Richard Nolan (1997) 'A Fiduciary Duty to Disclose?', 113 *Law Quarterly Review* 220.

R. C. Nolan (1998) 'The Proper Purposes Doctrine and Company Directors', in Barry A. K. Rider (ed.), *The Realm of Company Law*, Kluwer Law International.

Richard C. Nolan (2005) 'The Legal Control of Directors' Conflicts of Interest in the United Kingdom: Non-Executive Directors Following the Higgs Report', 6 *Theoretical Inquiries in Law* 413.

R. C. Nolan (2009) 'Controlling Fiduciary Power', 68 *Cambridge Law Journal* 293.

Patrick Parkinson (1995) 'Fiduciary Law and Access to Medical Records: *Breen v Williams*', 17 *Sydney Law Review* 433.

Patrick Parkinson (2003) *The Principles of Equity*, Lawbook Co.

Kenneth M. Rosen (2007) 'Fiduciaries', 58 *Alabama Law Review* 1041.

Joseph L. Sax (1970) 'The Public Trust Doctrine in Natural Resource Law: Effective Judicial Intervention', 68 *Michigan Law Review* 471.

L. S. Sealy (1989) '"Bona Fides" and "Proper Purposes" in Corporate Decisions' 15 *Monash University Law Review* 265.

Lionel Smith (2003) 'The Motive, Not the Deed', in J. Getzler (ed.), *Rationalizing Property, Equity and Trusts – Essays in Honour of Edward Burn*, LexisNexis Butterworths.

Andrew Stafford and Stuart Ritchie (2008) *Fiduciary Duties – Directors and Employees*, Jordan Publishing Ltd.

Julian Svehla (2006) 'Director's Fiduciary Duties', 27 *Australian Bar Review* 192.

Rosemary Teele (1996) 'The Search for the Fiduciary Principle: A Rescue Operation', 24 *Australian Business Law Review* 110.

Geraint Thomas (1998) *Thomas on Powers*, 1st edn, Sweet & Maxwell.

Geraint W. Thomas (2008) 'The duty of trustees to act in the "best interests" of their beneficiaries', 2 *Journal of Equity* 17.

Vicki J. Vann (2005) 'Solicitors and the duty to disclose: *Hilton v Barker Booth & Eastwood (A Firm)*', 2 *University of New England Law Journal* 67.

Gary Watt (2008) *Trusts and Equity*, 3rd edn, Oxford University Press.

Mary Christina Wood (2009) 'Advancing the Sovereign Trust of Government to Safeguard the Environment For Present and Future Generations (Parts I and II): Ecological Realism and the Need For a Paradigm Shift', 39 *Environmental Law* 43.

Sarah Worthington (2001) 'Reforming Directors' Duties', 64 *Modern Law Review* 439.

Peter Young, Clyde Croft and Megan Louise Smith (2009) *On Equity*, LawBook Co.

Cases

Aequitas Ltd v Sparad No 100 Ltd (2001) 19 ACLC 1006, [2001] NSWSC 14

Ambridge Investments Pty Ltd (in liq) v Baker [2010] VSC 59

Australian Securities and Investments Commission v Citigroup Global Markets Australia Pty Ltd (No 4) (2007) 160 FCR 35, [2007] FCA 963

Barnes v Addy (1873–74) L.R. 9, Ch. App. 244

Bell Group Ltd (in liq) v Westpac Banking Group (No 9) (2008) 22 FLR 1, [2008] WASC 239

Breen v Williams (1996) 186 CLR 71

Bristol & West Building Society v Mothew [1998] Ch 1

Chan v Zacharia (1984) 154 CLR 178

Compaq Computer Australia Pty Ltd v Merry (1998) 157 ALR 1

Cubillo v Commonwealth of Australia (2001) 112 FCR 455, [2001] FCA 1213

Chapter 9

Applying Fiduciary Duty in Real Politik[1]

Andrew Murray[2]

Introduction

The Rudd ALP Government's Carbon Pollution Reduction Scheme Bill 2009 (CPRS) and ten related bills were supposed to significantly reduce harmful greenhouse gas emissions in Australia.

The Australian Greens say those bills do nothing of the sort, even amended as some Liberals proposed. The Nationals[3] and some Liberals[4] say that economically the cost far outweighs the benefit. If those legislators are right, it is their public duty to vote against bills they believe do little good environmentally and will be economically harmful.

Others say they should vote in accordance with the wishes of the media and the public, but the Australian people are neither engaged with nor understand these specific bills; en masse, the public are ignorant of the nature, cost and effect of the CPRS. If they have a view, it is that something must be done, and trust the Government to do it.

If the CPRS legislation had passed and was effective, then well and good; but if, later, the public found it to be of little environmental value and it had a large personal cost, there would have been a reckoning.

---

1   The brief: Long-term policy issues and administration affecting the public interest, such as climate change, raise major concerns about how parliaments and governments are held to account. Beyond the crude, blunt instrument of elections, what principles and practices can ensure that longer-term and intergenerational interests predominate over the usual media and electoral political cycles? This chapter is based on a paper prepared prior to the UNFCCC COP15 conference in Copenhagen, December 2009 and the defeat of the Australian Carbon Pollution Reduction Scheme Bill 2009.

2   Andrew Murray was a Senator for Western Australia from July 1996 to June 2008. He is best known in politics for his work on finance, economic, business, industrial relations and tax issues; on accountability and electoral reform; and for his work on institutionalised children.

3   Members of the National Party of Australia.

4   Members of the Liberal Party of Australia.

Fiduciary Duty

The Workshop topic was 'Fiduciary Duty, Public Trust and the Governance of Climate Change' and my part in it was to consider 'Applying Fiduciary Duty in Real Politik'.

That is a false juxtaposition as fiduciary duty and realpolitik cannot be either/or. 'Fiduciary' means 'held or given in trust',[5] and 'duty' means 'a moral or legal obligation, a responsibility';[6] while Realpolitik is 'politics based on realities and material needs, rather than on morals or ideals'.[7]

A nineteenth century concept, 'realpolitik' was a defensive strategy, designed to better accommodate the national interest in a threatening world, but realpolitik recognizes the moral or legal behaviour expected of a nation, either by outsiders or insiders. Realpolitik is more a question of political emphasis than a political alternative.

Politics based on realities rather than morals suggests substitution. Except in the most brutal of national circumstances, morality is always in play, but emphasis can change. While realpolitik certainly means the rejection of dogma and the embrace of pragmatism, political reality is usually presented within a moral framework, and almost always within a legal framework. There were, and are, things that practitioners of realpolitik cannot do, although the murkier depths of history indicate evils that can attach to the national or partisan interest.

But this is Australia, where considerations of power have to play against the background of an open society and a liberal parliamentary democracy. Morals and ideals will, and do, come into play, even when periodically mugged by reality.

What is the ongoing governance framework in which climate change policy will sit? There are existing strengths and weaknesses. First, the strengths.

There is already a substantial, formidable and impressive Commonwealth architecture supporting the concepts of budget transparency and financial accountability.[8] Budget transparency and financial accountability are part of the rule of law, mechanisms which deliver integrity and a real underpinning to our political economy, and which enable law to operate effectively and affordably. Budget transparency and financial accountability are not only ethically, morally, and managerially sound concepts with positive and beneficial consequences; they are not only the natural accompaniment of parliamentary democracy; but they are legal requirements that flow from the higher law of the Australian Constitution, as supplemented by statute.

The Commonwealth's power to tax and spend is arguably its most important power. It is fundamental to the Commonwealth's ability to achieve its policy

5   (1997) *The Australian Concise Oxford Dictionary of Current English*, p. 488.

6   (1997) *The Australian Concise Oxford Dictionary of Current English*, p. 412.

7   (1997) *The Australian Concise Oxford Dictionary of Current English*, p. 1123.

8   For further material on the transparency of government budgets, see Murray (2008) from which some of this draws.

priorities and objectives. Legitimacy in the exercise of these powers is provided by Australia's highest legal authority, the Australian Constitution. These constitutional provisions must not be avoided or evaded; they constitute a command – a constitutional imperative.

Our budget transparency and financial accountability systems are entrenched in law and practice because parliaments have understood that if you want high standards, accountability and good governance, you have to institutionalize and legislate those standards.

Nevertheless, reality dictates that governments and bureaucracies always have 'wriggle room', which is why the Auditor-General, the media, the Opposition and the Senate are so important.

The fundamental characteristics of robust and useful information for provision to governments, parliaments and citizens is for it to include relevance and timeliness, reliability and representational faithfulness, comparability, and understandability. The Government must be judged by results. The current government has said it is willing to be measured through KPIs, targets and benchmarks. They must be held to that, and the measurement should be brutally honest.

The reality of attempted greenhouse gas reduction can mug principle and best practice. How and on what is the government to be measured? If there is a tonne less of greenhouse gas is it caused naturally, or by a policy measure, and if a policy measure, which one?

Perhaps, one could argue, expenditure is at least easily measured, what was budgeted versus what was spent; maybe so with direct outlays, but tracking the money can get murky depending on the accounting devices used.

Compensation was a big feature of the CPRS. The major one to watch would have been tax concessions, especially for the polluters.

Treasury defines a tax expenditure as a tax concession that provides a benefit to a specified activity or class of taxpayer. A tax expenditure can be provided in many forms, including a tax exemption, tax deduction, tax offset, concessional tax rate or deferral of a tax liability.[9] Tax expenditures provide what is, in effect, a subsidy through income foregone for certain activities or categories of persons. Tax expenditures incur an opportunity cost; they represent revenue that, if collected, would have been available to fund spending programs to meet similar objectives.[10] In the case of energy, a withheld tax concession may have effectively resulted in a higher carbon price, which was the point of it all, was it not?

Tax expenditures operate under the radar. Treasury has observed that the cost of tax expenditures is generally not directly observable as it does not arise from a direct transaction with Government.[11] Compared to direct outlays, tax expenditures are subject to a less comprehensive management and reporting framework. This process hampers the effective monitoring and scrutiny of individual tax

---

9   Department of the Treasury (2006), p. 1.
10   Australian National Audit Office (2008), p. 10.
11   Department of the Treasury (2005), p. 2.

expenditures. In many cases, it is not possible to show whether objectives are being achieved and whether the actual benefits are proportionate to the costs.[12]

Of particular note has been the significant underestimation by departments of tax expenditures, which have been growing markedly.[13] Tax expenditures have a long way to go before they catch up to the accounting and reporting standards that apply to direct outlays. There is a need for a settled, nationally applicable and comprehensive reporting framework and set of benchmarks and accounting standards for tax expenditures.

In its May 2008 audit of tax expenditures, the ANAO found that many previous reviews of tax expenditure have identified problems such as poor or non-existent data collection, inadequate reporting and ineffective merits assessment.[14]

Then there is the longer term. The Intergenerational Report (IGR) was a great Costello[15] innovation, but it has weaknesses. The IGR states that 'the wellbeing of successive generations requires sustainable economic, social and environmental conditions'.[16] The IGR does not recognize and account for the impacts of environmental and social costs and benefits, influences that are increasingly important in the overall assessment of wellbeing of the Australian people. The Government must include key environmental and social impacts, such as climate change, in future intergenerational reports. Climate change does not respect political demarcations. It is an issue of shared inter-governmental responsibility. In order to evaluate the performance of Australian national, state and territory governments, Council of Australian Governments (COAG) cross-governmental climate change reports are needed.

Now to the nub of the problem: did anyone really have a clue as to what the CPRS would have cost in direct or indirect outlays or in deadweight costs, or what its environmental effects would have been?[17] The great fear is that the economic cost would have been greater than the environmental benefit.

Treasury has the brightest minds, yet they cannot get their figures for revenue, for employment, for growth, right for a few months, never mind years. Treasury officers have 'Buckley's'[18] of accurately modelling the economic or environmental effects of the CPRS. They are not to blame; it is an impossible task.

---

12   Australian National Audit Office (2008), p. 14.

13   Underestimations of expenditure are moving towards the $100 billion mark.

14   Think CPRS permits and the difficulties that would have attached to data collection on those financial derivatives!

15   Hon Peter Costello was Treasurer of Australia during the period of the Howard Coalition Government.

16   Commonwealth of Australia (2007), p. 1, available at http://www.treasury.gov.au/documents/1239/PDF/IGR_2007_final_report.pdf.

17   There is also the matter of 'first movers'. 'First movers suffer losses in competitiveness compared with late movers or no movers'. Carmody (2009), p. 63.

18   An Australian colloquialism meaning: no chance.

Scientists already quarrel over climate change calculations; government statisticians would have struggled to calculate real costs and real carbon effects resulting from a CPRS. Post-facto measurement and reporting would have been very difficult. Pre-facto it is impossible to get it right. That is just reality.

There is also a giant misconception – the idea that the CPRS would have been the most important mechanism in changing behaviour and reducing greenhouse gas emissions. It would not have been because it could not have been. The price impost per tonne of carbon in Australia would have to be so high to change behaviour significantly by force of cost that it would have been politically, economically and socially unacceptable.[19]

The CPRS at the modest price per tonne anticipated was a policy adjunct, and a very costly one at that. It is the other policies that would have had to do the heavy lifting, primarily investment in efficient renewable and lower carbon energy, coupled with the voluntary efforts and commitments of households and businesses. Personally, I have always expected coordinated and extensive energy efficiency measures, greater consumer awareness and new technology to be the biggest change-drivers.

Realpolitik

On *realpolitik* here is my line of thinking. While there are politicians who think climate change is wholly or partly man-made, and others who think it is not, few dispute that climate change is occurring that has dangers for the Australian economy and society.

A majority of Australians want a *domestic* plan for carbon reduction. Globally, they also want Australia to do its bit. Logically, this means a higher price for carbon and indications are that a majority of Australians will accept a 'reasonable' price consequence on goods and services they consume.

In meeting this voter demand, the alternative policy responses have to be tested for their effects on jobs and the economy, and for simplicity, efficiency and cost. Fiduciary duty says these have to be honestly laid out for voters in an understandable form. They have not been.

A carbon policy can have four main components: more efficient energy use meaning lower carbon use; substituting lower-carbon renewable energy for higher-carbon energy; an emissions trading scheme (which can only really be production based); a carbon tax (which can be production or consumption based).

There was no likelihood at all that Copenhagen would have signed off on a binding internationally harmonized greenhouse gas policy. So if Australia was going to act unilaterally, any Australian carbon policy needed to be designed with

---

19 As at October 2010, Western Australia had seen electricity prices rise 45 per cent in the last two years, but with little apparent effect on energy consumption to date. Which raises the question – what carbon price on electricity will really curb consumption?

Australia's best economic and environmental interests in mind; a difficult balance to achieve.

The Government's ETS – the CPRS – failed on domestic and international grounds; domestically because every other Australian political party rejected it; internationally because it was not part of a global system or a standard design.

In the wider community, many had grave doubts as to whether the CPRS stacked up environmentally; many others did not think it stacked up economically and contended that it was unnecessarily costly and profoundly inefficient. If they are right, *realpolitik* means that the Australian Labor Party (ALP) needed a partner to share the blame for its own failure; there was only one candidate, the Liberals.

Unlike their detailed and well-documented case for a GST, the Howard Coalition Government's climate change policy was weakly prepared. As a Cabinet, they had never moved into policy belief and were instead focused on political tactics. The Howard Cabinet and Party Room never debated the ETS properly because it opted for an ETS as it was facing an election and because other countries were going down this path.

Realpolitik is reality politics. I thought it through like this: neither the Liberals, the Greens, the Nationals, Family First nor Senator Xenophon supported the CPRS. Most thought that with respect to the CPRS, nothing would have been better than something. It was judged severely flawed on policy grounds and would not have passed without amendment.

The only alternative then seemed to have been an ALP/Liberal ETS. What was thought might happen if the Liberal Party Room rejected that concept? What would the Liberals be expected to propose instead? At the time, I thought 'nothing' was unlikely. These were big assumptions, but assuming there was not a final ALP/Liberal deal and no double-dissolution to wave through the Government's CPRS, what then? Having invested so much policy capital in their CPRS, the ALP would have been expected to fight an election with it, and on then-current polls, probably win.

The Liberals could then have been expected to fight the 2010 election with an ETS policy of their own or with a consumption-based carbon tax policy. However, in that scenario, after the 2010 election, an ALP Government would still not have had the support of the Greens and Nationals in the Senate for the CPRS. So, it would have been back to square one for them and the Liberals.

We now know that a quite different chain of events unfolded. Opposition Leader Turnbull did agree to an ALP/Coalition deal but was quickly ousted from the leadership by the ETS-opposed Abbott. The Coalition then turned its back on any ETS or carbon tax. The ALP Government narrowly survived the following election, returning as a minority government.

The Coalition remains resolutely opposed to any price on carbon emissions. It does, however, retain a commitment to 'direct action', meaning targeted interventionist policy. The direct action approach has long been part of Australian Government carbon-reduction policy, for instance with renewable energy programs.

The variants on direct action offered by the Coalition need to be analysed from a cost-benefit perspective, no differently from any other carbon reduction policy.

I cannot help thinking that the whole debate nationally and internationally suffers from the political fear of anything called a tax. Yet the CPRS has the same effect as a tax – it raises prices. There are good reasons many economists and others support an ETS tax. There are good reasons many support a carbon tax. Much more has been written on an ETS tax than on a carbon tax.

Arguments for a carbon tax include these. A carbon tax will be easier understood by voters, consumers and businesses. An ETS tax will remain incomprehensible to most. The effects of a carbon tax are immediate and much more easily measured than an ETS tax. A carbon tax can be more easily calibrated than an ETS tax – it can be set low and rise as needed. An ETS is indirect, a derivative; a carbon tax is direct. An ETS is a complex unpredictable uncertain market-based system with erratic price effects – its European trial has hardly been encouraging. A carbon tax is a simple, certain price signal.

An ETS can have disadvantageous market arbitrage effects if permits are internationally traded. An ETS tax hides the true cost to the consumers, such as the wholesale sales tax. A carbon tax shows the true cost to the consumer, such as the GST. An ETS encourages special deals and compensation for polluters. A carbon tax can compensate low-income consumers. An ETS tax taxes production. A carbon tax taxes consumption. An ETS tax advantages imports. An ETS tax cannot exempt exports (but it can compensate for them in a complicated costly churning manner); a carbon tax can exempt exports. An ETS tax discourages individual action because the cumulative voluntary carbon savings of concerned consumers gives polluters more room to pollute.

Will the CPRS or another ETS triumph? Has the carbon tax bird flown? Is the CPRS *realpolitik*? – designed more for a political need than an environmental end?

All I am sure of is that transparency, reporting, understandability and measurement would have been a lot easier with a carbon tax. Whatever the final result, there will eventually be an accounting, and the truth will out.

References

The Australian Concise Oxford Dictionary of Current English, 3rd edn (1997), Oxford University Press.

Australian National Audit Office (2008) 'Preparation of the Tax Expenditures Statement', Audit Report 32, Australian Government.

Geoff Carmody (2009) 'Flawed ETS model needs major rethink', *The Australian Financial Review*, 19 November, p. 63.

Commonwealth of Australia (2007) Intergenerational Report, available at http://www.treasury.gov.au/documents/1239/PDF/IGR_2007_final_report.pdf.

Department of the Treasury (2005) Tax Expenditures Statement, Australian Government.

Department of the Treasury (2006) Tax Expenditures Statement, Australian Government.

Andrew Murray (2008) 'Review of Operation Sunlight: Overhauling Budgetary Transparency', Report for the Commonwealth Minister of Finance, available at http://www.finance.gov.au/financial-framework/financial-management-policy-guidance/operation-sunlight/docs/budget-transparency-report.pdf.

Legislation

Carbon Pollution Reduction Scheme Bill 2009 (CPRS) (Cth).

Chapter 10

Fiduciary Duty, Democracy and the Rule of Law

Robert Clark

A lot has been written about the notion of fiduciary duty in relation to preserving our environment, and in particular in relation to climate change. This paper makes some observations from the perspective of a currently serving Member of Parliament about the broader implications for law and government of trying to establish such duty in legally binding terms. In particular, this chapter argues that:

- universal pessimism about existing institutions is not justified, and attempting to tackle institutional problems directly is likely to produce better outcomes than trying to circumvent those problems through the notion of a legally binding fiduciary duty;
- attempting to create and impose a legally binding fiduciary duty is likely to do violence to the rule of law and respect for democratic decision-making, which will in turn prejudice achieving good environmental outcomes; and,
- the concept of a legally binding fiduciary duty is unlikely to work effectively in responding to climate change.

Professor Mary Wood[1] mounts the case for fiduciary or trust duty by arguing, in effect, that American public environmental policy has been a grave failure, that the regulatory agencies that are supposed to have addressed environmental problems have gone off the rails, and that the US government has proved ineffectual. As a result, the only way to get much needed environmental outcomes, Wood argues, is to impose on governments a legally binding fiduciary duty to secure those outcomes, or at least a fiduciary duty not to act in a way that would prejudice them.

Wood's grounds for universal pessimism about existing regimes, however, are not justified. There are many instances around the world where environmental problems have been resolved by government and community action over time. Urban air and water pollution, for example, has been dramatically reduced in many major cities compared with the levels of sixty years ago.

---

1 See, for example, M. C. Wood (2009b) 'Advancing the Sovereign Trust of Government to Safeguard the Environment for Present and Future Generations (Part I): Ecological Realism and the Need for a Paradigm Shift', 39 *Envt'l. L.* 43.

Internationally, cooperation and treaties have often successfully resolved issues to the point where we almost take the outcomes for granted. For example, the penny post was introduced in Great Britain in 1840. The international treaties on postage and mail established within the next few decades have been so successful that we rely on them today without giving them a second thought. Similar success has been achieved with international air navigation arrangements. Although other international treaties are under pressure, such on whaling or Antarctica, such treaties can still at least provide a framework within which nations try to resolve issues.

Professor Wood's description of the problems with existing regimes is given in the context of the United States of America. However, the USA has not been through the root and branch reform of the processes and administration of government that various other countries, such as Australia, have been through over recent decades. In Australia, there has been an often bipartisan reform process that can be traced back to the later years of the Fraser Government in the 1970s and early 1980s, and which was followed through by the reforms introduced by the Hawke, Keating and Howard Governments nationally and by the Kennett Government in Victoria.

While far from perfect, and arguably subject to serious backsliding in recent years, these reforms have made institutions of government far more responsive to and capable of implementing public policy decisions made by elected governments. Many jurisdictions in the United States are a long way behind Australia in terms of these reforms.

Arguably, a rigid executive/legislative separation of powers has also not served modern government as well as the Westminster model of responsible government. A rigid separation of powers has meant neither the President nor the Congress can take full ownership of problems nor have full control over delivering solutions. Furthermore, modern government requires a large amount of what is essentially executive action to be transacted in the form of legislation, making the old executive/legislative distinction less clear and less functional.

Thus it can be argued that Wood's frustration with existing regimes arises to a significant extent from problems specific to the jurisdiction with which she is most familiar. The better way forward may be to tackle those problems of politics and of government directly, rather than trying to by-pass them by asserting a theory of legally binding fiduciary duty.

Of course, Australian governmental structures are far from perfect. Wood identifies at least one systemic failure that also applies in Australia, namely independent agencies that cease to be truly independent and in fact become covert extensions of political government. This is currently occurring in the state of Victoria and perhaps at a national level as well.

For example, in Victoria in relation to freeway noise, the Environmental Protection Authority, which is supposed to be an independent environmental watchdog, ought to operate independently to set and apply appropriate standards. However, in practice, it avoids taking any action that would contradict the noise level policy of the roads agency, VicRoads.

The better solution to this problem would be to fix it directly, in this case by confronting the issue of whether the agency is expected to be truly independent or to be an arm of government. If an agency is explicitly an arm of government, at least the government is politically accountable for what it does or fails to do. However, a problem exists when the government pretends that an agency is an independent watchdog, but in fact has covertly turned it into a government lapdog.

Part of the solution may lie in the approach that the Howard Government took in Australia at a national level. If an agency is intended to be genuinely independent, it should have a corporate structure and governance separate from government and a charter that makes clear its independent responsibilities. However, if it is not intended to be genuinely independent, it should be integrated back into a government department where there is political accountability through to the Minister.

The next issue to be addressed is how decisions on environmental issues should best be made.

Most people want good environmental outcomes. Few want to see our environment trashed. However, there are many different views about the appropriate balance between human use of our natural surrounds and the preservation and conservation of those surrounds.

There are also many heated factual debates as to what the available trade-offs are; in other words, as to what the consequences of undertaking, or ceasing to undertake, various human activities would be.

The question we should be seeking to answer is: what regimes of factual determination, policy-making, law-making and administration are most likely to lead to the best resolutions of these issues and to the best outcomes?

The author's view is that the concept of a fiduciary duty imposed on government by judicial fiat as a way of tackling environmental problems, and in particular as a way of tackling climate change, is the wrong way to go. It is the wrong way to go both because it will be ineffectual and because it will threaten the rule of law and ultimately threaten democratic institutions.

The concept of fiduciary duty has a long antecedence. It is analogous to the notion of stewardship in the Christian faith. Each generation is a steward of resources for those who come after them. In that sense the concept is nothing new, but it has not previously been an issue about law. It has been about good policy or of good practice. In other words, it has been about the right thing to do. Arguing that it should regarded as a legal duty is radically different.

The rule of law is fundamental to the debate about fiduciary duty. The rule of law requires upholding a consensus about an agreed set of rules, including rules about rule-making, within which people can live together; a consensus which reflects generations of experience about what makes good law and what generally delivers justice between different people.

Once a group within a community starts asserting or imposing rules (that is, laws) in a way inconsistent with the agreed rules about law-making, they are abandoning the rule of law. Once impositions of rules start to occur other than by agreed processes, those who start the practice will not necessarily be able to continue to change the law in the direction they want. Rather, they will find others striving to impose changes to the law that work in entirely different directions. The rule of law will be replaced by the rule of the strongest.

This approach also has consequences for democratic institutions. It is possible to have the rule of law without democracy, but it is hard to have democracy without the rule of law.

When viewed in an historical perspective, democracy is not a guaranteed state of affairs. There are many past civilizations where democracy has collapsed from within. There have also been many democracies overborne from without. In the world at the moment there are a range of threats faced by democracies, and an increasing proportion of the world growing in wealth and influence under non-democratic traditions.

Failing to remedy fundamental problems with the institutions of government, and instead seeking to devise 'work-arounds' that make those institutions even more complex and dispute-ridden, is only going to further weaken democratic governments and undermine public confidence in them.

It is of course arguable that there is a code of morality or a natural law that commands a higher allegiance than the positive law of the nation. There are times when people have passionate views that the law of their nation is compelling them to do something that they regard as unjust and they therefore have a duty and a right to instead obey such higher morality. That approach poses very difficult issues. On the one hand, it is desirable to uphold the right of individuals to defy manifestly immoral laws. On the other hand, people can use assertions of a higher morality in many different ways to justify taking the law into their own hands.

In relation to climate change, much of the debate is about facts. Community members and policy-makers who are not scientists are often dependent on the scientific advice that they receive. The prevailing scientific assessment is that climate change is a very serious issue indeed, but others put a contrary view vociferously.

Climate change is an issue that creates for policy-making a dilemma in the classical sense of the word. There are two sharply divergent approaches. If one accepts the prevailing scientific assessment, the nation and the world need a very radical response indeed. If one does not accept the prevailing scientific assessment, to undertake that radical response would amount to expending a huge amount of resources on something that is completely unnecessary. Not much of a middle ground exists between those two diametrically divergent positions.

In this context, if one side of the argument tries to use a legal device to resolve what is essentially a dispute about facts and policy trade-offs, it is likely to provoke a vehement reaction from the other side of the argument, not just about the conclusion reached, but about the legitimacy of the way in which that conclusion is being attempted to be imposed. Such a rejection of legitimacy is likely to produce a divided society in which respect for institutions and decision-making processes is diminished.

Trying to develop and impose a concept of fiduciary duty is also likely to prove ineffectual. It will graft yet another layer of complexity on top of legislative, administrative and legal regimes that are already highly complex. If judges were to try to create law on what governments should do about environmental protection matters, they would simply be one more group trying to drag the public policy debate in a particular direction, without necessarily being particularly well qualified to do so.

Australia has, for some time, been debating in a political context the best way to limit greenhouse emissions. If the High Court were to enter into that debate and try to tell governments that they are under some sort of fiduciary duty to act to mitigate climate change, how would a court be capable of specifying what government was required to do? Any attempt to do so would not only be likely to be ineffectual, it would also be dangerous for democratic institutions, because it would be turning what should be debates about policy and politics, to be decided by democratic means, into issues of law.

There is an analogous issue in Victoria and Australia about a so-called charter or bill of human rights. Much of what is in the Victorian charter document[2] is attempting to use the law to impose policy decisions, and therefore to pre-empt the Parliament and those who elect the Parliament. The Victorian charter also uses a very broad notion of reasonableness which is in effect a disguise for asking for what are in fact policy decisions to be made by either a judge or a tribunal member.

There is a vast difference between 'reasonableness' in contexts such as these and, for example, the notion of 'reasonableness' associated with what started off being the 'reasonable man' or 'the man on the Clapham omnibus' in the genesis of the law of negligence.

In relation to the law of negligence, there was a fairly general consensus about what reasonableness was and what the 'ordinary' man or woman would think. In such a context, it is possible to use the concept of 'reasonableness' to apply a generally understood policy decision to individual factual situations. However, if one tries to use open-ended terms such as 'reasonableness' in relation to environmental policy or, to give another example, in relation to the appropriate modifications of a building for disability access, that is, in effect, transferring a policy-making responsibility from a Parliament to a court or tribunal. That action is likely both to result in bad policy outcomes and to undermine the democratic way of decision-making.

There is a further difficulty in seeking to apply the notion of fiduciary duty to international problems such as climate change, namely that there is no readily apparent supernational entity that can be made liable to fulfil that fiduciary duty.

So, by all means let us use the notion of fiduciary duty as a moral, political and/or ethical argument. There is no dispute that those who are in government from time to time, on behalf of the community, have a moral duty to conserve the

---

2   *Charter of Human Rights and Responsibilities Act 2006* (Victoria).

community's assets, to think about future generations, and to seek to make the world a better place.

But to try to declare that to be a principle of law is taking entirely the wrong direction. At least in terms of Australian jurisprudence, it is a huge stretch to argue that there is a fiduciary duty of a legal nature owed by governments of the sort that Wood has advocated.

Those who really want to follow a legal route should argue for a referendum to amend the constitution to impose a fiduciary duty on government. If such a referendum were carried, it would not violate the rule of law nor be an affront to democratic decision-making. Whether it was a good idea or not would be subject to public debate during the course of the referendum campaign. At least it would have the benefit that in framing the question to be put to referendum, careful attention would need to be given to defining the terms of the fiduciary duty which it was sought to impose. However, to argue that Australian courts should, without a referendum, simply invent a notion of fiduciary duty where one has not existed before, does violence to the rule of law and undermines our democratic system.

The best way forward in seeking effective action on climate change is to focus on the issues in hand in public policy debate – what needs to be done and what is the best way to get it agreed to and implemented?

To try to make this a matter of law, rather than a matter of public policy, is to head in a wrong and unproductive direction.

References

M. C. Wood (2009b) 'Advancing the Sovereign Trust of Government to Safeguard the Environment for Present and Future Generations (Part I): Ecological Realism and the Need for a Paradigm Shift', 39 *Envt'l. L.* 43.

Legislation

Charter of Human Rights and Responsibilities Act 2006 (Victoria)

Chapter 11

The Role of Fiduciary Duty in Safeguarding the Future

Kelvin Thomson

A fiduciary is said to be under a duty not to use his position or the power or opportunity it gives him to serve an interest other than his beneficiaries – his own or a third party's. A fiduciary cannot use his position, or knowledge or opportunity gained from it, to his own or someone else's advantage or to the beneficiary's disadvantage. Fiduciary duty is not a commonly applied concept to political or policy-making contexts. Courts have been traditionally reluctant to impose trust obligations on Government. It is fundamental to the principle of separation of powers that courts should not interfere in the governmental, administrative or executive functions of the Crown. Dal Pont and Chalmers describe the government's duty to the people as trustee in somewhat limited terms:

> the duty to comprehensively account faithfully and accurately to the people in
> a form that facilitates the public monitoring of the government's performance.
> This is required as a check upon misuse of the power conferred by the public
> to its representatives and as such, founds the basis of responsible government.[1]

Justice Paul Finn argues that Australian law has been different from case law in the United States concerning the public trust doctrine in that Australian courts 'seem for the moment far less ready to regard the types of rights and interests protected in the United States as being ones which would justify reading down legislation or in making them significant relevant considerations in decisions'.[2]

But despite the likely limited legal application of terms such as fiduciary duty and public trust to questions of politics and government, it seems to me that fiduciary duty and public trust can be useful political constructs.

Most Australian adults own private property or aspire to ownership. Most Australian adults also have children or plan to. Most Australian adults look after that property reasonably responsibly with a view to passing it on to their children.

I can attest to the strong resistance to any public policy measures, such as assets testing the pension, or aged care accommodation bonds, which threaten to

---

1 Pont and Chalmers (2004), pp. 619–20.
2 See Finn (2011), chapter 3 this volume.

run down these assets over the course of one's lifetime and prevent it being passed on to the children. The same is not true, however, for public property.

We say we love our children, but we are acting with much more concern for our material wealth than for their future well-being. We are not safeguarding our neighbourhood, our city, our country, or our planet. We are not meeting what I believe is an obligation to pass on to our children, and our grandchildren, a world, and an Australian way of life, in as good a condition as the one our parents and grandparents passed on to us. It may be helpful if parliamentarians or governments were to think of this obligation in terms of fiduciary duty or public trust.

When we were boys my brother and I explored an Australian bush which boasted kookaburras, koalas, lyrebirds, wombats and platypus. When she was a young girl, my partner and her sister would run through water sprinklers on the front lawn on hot days to keep cool. When I was 25, I put down a deposit and bought a house. Later on, I drove my children up the east coast of Australia to tropical Queensland. At night you could drive across the open spaces between the towns and savour the silence. The legacy we are leaving our children may well have none of these things.

In central Victoria's woodlands the number of kookaburras has declined by 30 per cent in just four years. The number of koalas in Queensland and New South Wales has declined dramatically, and it is possible that there are now fewer than 50,000 koalas left in the wild. The same study that found the 30 per cent decline in kookaburras also found a 30 per cent or greater decline in robins, and even greater declines for various honeyeaters, flycatchers and kingfishers. The study noted that eucalypt flowering had declined significantly over the past 12 years of drought, and detected virtually no bird breeding in the latest survey periods. Some 30 per cent of Australia's bird species are now regarded as threatened, and internationally the number of birds listed as critically endangered continues to rise.[3] A survey released by the International Union for Conservation of Nature in November 2009 found 17,291 species of plant and animal in some degree of danger – more than one third of 47,677 species surveyed.[4]

As for water, we live in an age of long-standing water restrictions, of rising water prices and long-term decline in water availability. The search is underway for alternative sources of water, such as desalination. It is clear that these alternatives are expensive – in 2009 Victoria's Water Minister observed that water prices would be doubling, and so parlous was our water situation that this observation scarcely raised an eyebrow. Disputes about water entitlement in the Murray Darling are commonplace, and controversy rages over water harvesting in Queensland through projects like the Cubbie Station and Traveston Dam.

Our levels of housing affordability are in very poor shape. My parents' generation and my generation could save for a house and aspire to home ownership. Today's generation of young people is confronted by the steep walls

3   Fischer, Lindenmayer and Montague-Drake (2008), pp. 38–46.
4   International Union for Conservation of Nature (2009).

of record high prices fuelled by competition from both an increasing population and from investors.

Now, instead of the open spaces, we have never-ending suburbs and traffic lights. We are behaving selfishly, with cavalier disregard for the interests of our children and future generations, and for the future of the Australian way of life.

Why is this happening? There is a variety of reasons, but one of them is the triumph of short-term over long-term thinking and policy-making. The pressures in the direction of short-termism are obvious and need little elaboration. Governments and Members of Parliament are up for re-election every three or four years. Company directors and Chief Executive Officers may have even shorter tenure. Their pay turns on annual profit and loss statements. Treasury Departments sweat on quarterly GDP numbers.

The best way I can think of to defeat short-termism is to persuade the community to look beyond the immediate and to consider the world they will be living in, in the years ahead. I also think it would be helpful if we developed new indicators of community performance and downgraded our obsession with GDP. Such performance indicators would place more focus on long-term economic, social and environmental outcomes.

Politicians and business leaders will take a more long-term view if the community demands it of them. That demand, however, seems a long way off. Politicians are geared to maximizing GDP. They expect that anything short of a 'full-throttle', 'flat to the floor' approach to GDP will be attacked by political opponents, and fear the electoral consequences of these attacks.

But does GPD really provide a good measure of living standards? If it does not, how are policy-makers to properly discharge their fiduciary obligation to safeguard the future? The Report by Nicolas Sarkozy's Commission on the Measurement of Economic Performance and Social Progress found that those attempting to guide the economy and our societies are 'like pilots trying to steer a course without a reliable compass'.[5] In early 2008, French President Nicolas Sarkozy, dissatisfied with the current state of statistical information about the economy and society, established the Commission on the Measurement of Economic Performance and Social Progress, composed of Professors Joseph Stiglitz, Amartya Sen and Jean-Paul Fitoussi. The Commission's report made an extensive critique of GDP. It commenced by noting that what we measure affects what we do, and that if our measurements are flawed, decisions may be distorted.

Commission members noted that neither the private nor the public accounting systems were able to deliver an early warning of the Global Financial Crisis. Commission members also found that we are facing a looming environmental crisis, especially associated with global warming. The Report stated:

> (m)arket prices are distorted by the fact that there is no charge imposed on carbon emissions; and no account is made of the cost of these emissions in

5   Stiglitz, Sen and Fitoussi (2009).

standard national income accounts. Clearly, measures of economic performance that reflected these environmental costs might look markedly different from standard measures.[6]

The Commission found that the time was ripe for our measurement system to shift emphasis from measuring economic production to measuring people's well-being, and measures of well-being should be put in a context of sustainability. It said that changing emphasis does not mean dismissing GDP and production measures. They reported:

> (t)hey emerged from concerns about market production and employment; they continue to provide answers to many important questions such as monitoring economic activity. But emphasising well-being is important because there appears to be an increasing gap between the information contained in aggregate GDP data and what counts for common people's well-being.[7]

The Commission made the obvious point that no single measure can capture something as complex as the well-being of members of society, and therefore a range of different measures were required. It made a number of recommendations about how to go about this. First it recommended that when evaluating material well-being, we should look at income and consumption rather than production.

Second, it emphasized the household perspective. In a number of countries real household income has grown quite differently from real GDP per capita, and typically at a lower rate. Third, it recommended that income and consumption be considered jointly with wealth. To construct proper balance sheets, we need comprehensive accounts of both assets and liabilities. Fourth, it recommended that more prominence be given to the distribution of income, consumption and wealth. Fifth, it recommended broadening income measures to non-market activities. The Commission observed that many of the services received from other family members in the past are now purchased on the market:

> This shift translates into a rise in income as measured in the national accounts and may give a false impression of a change in living standards, while it merely reflects a shift from non-market to market provision of services.[8]

The Commission also drew attention to the question of measuring leisure, making the valid point that consuming the same bundle of goods and services but working for 1500 hours a years instead of 2000 hours per year represents an increase in one's standard of living.

---

6   Stiglitz, Sen and Fitoussi (2009), p. 9.
7   Stiglitz, Sen and Fitoussi (2009), p. 12.
8   Stiglitz, Sen and Fitoussi (2009), p. 14.

The Commission also made recommendations aimed at dealing with the problem that GDP puts no value on the environment. It says that sustainability assessment requires a well-identified dashboard of indicators. The components of the dashboard should be interpretable as variations of some underlying 'stocks'. Finally, it recommends a separate follow-up of the environmental aspects of sustainability based on a well-chosen set of physical indicators. It said there was a need for a clear indicator of our proximity to dangerous levels of environmental damage, such as are associated with climate change or the depletion of fishing stocks. Members of the Commission said there was a need for a clear indicator of increased atmospheric concentration of greenhouse gases associated with proximity to dangerous climate change, or levels of emissions that might reasonably be expected to lead to such concentrations in future. Climate change was 'special in that it constitutes a truly global issue that cannot be measured with regard to national boundaries'.[9]

Politicians wrestle with the question of whether the electorate is selfish or altruistic. Clearly, there is no one size fits all answer to a question like that – it varies from person to person, it varies from issue to issue, and it varies across time and circumstance.

Voters are regularly assumed to be selfish, and to be principally concerned with economic self-interest. In the debate over global warming, for example, many of those wanting action, including me, make regular use of Nicholas Stern's conclusion that 'inaction will be far more costly than adaptation'.[10] We try to get across a message to people that they will be financially better off if we do act, a message addressed to those members of our audience who we fear are principally motivated by financial concerns.

My father, who shares my concern about the number of endangered species in the world, says to me: 'Get the evidence about those plant and animal species which have been the source of medicines, which restore our health and lengthen our lives, and point out the short-sightedness of allowing species whose potential value we are unaware of to become extinct'. He makes a good point, of course, but there is a part of me which is resistant to it because I feel it concedes too much ground to forces I disagree with – who believe that unless something has an economic value that it has no value at all.

What I feel is that other species have a place in the world too. We should be able to share the world with them. I know they do not vote, but I do not think this means they have no rights at all and can be cheerfully consigned to oblivion.

I think the world for my children and their children will be a poorer place if it does not have the cackle of the kookaburra or the whip crack of the whipbird, but for me it is more than simply about the quality of their lives. It is a moral question, a values question, about whether we have the right to steamroll other species into oblivion in our never-ending quest for material wealth.

---

9   Stiglitz, Sen and Fitoussi (2009), p. 18.
10   Stern (2006), p. 430.

So you can see I cast the fiduciary net pretty wide. Others cast it rather more narrowly. Given that I think our responsibilities as policy-makers extend to other species, you would be right in assuming that I think our responsibilities also extend to other peoples in other countries, and to a responsibility for the well-being of the world as a whole.

But this is not immediately clear from the concept of fiduciary duty. Many representatives see their responsibility as being limited to representing the interests of their voters, and no-one else. Many governments see their responsibility as being limited to looking after their country and no-one else. This assumption lies behind Ross Garnaut's description of climate change as a diabolical policy problem. The negotiations at Copenhagen made it perfectly clear that although there is widespread awareness that we are all in the same boat, everyone wants someone else to pick up the oar and start rowing. The climate crisis is the most spectacular example of Garret Hardin's 'tragedy of the Commons' – if we approach it in terms of economic self-interest, from a self-centred or selfish point of view, it made sense for every country to have gone to COP15 in Copenhagen and tried to get away with doing as little as possible, and tried to persuade everyone else to do as much as possible. Yet, if we all take that view, the planet will go to hell in a hand-basket. The boat we are all in will slowly gather momentum and end up sailing over Niagara Falls.

So I have a strong view that fiduciary duties involve concern about the welfare of people in other countries. I have advocated a lift in our country's refugee intake of 45 per cent, from 13,750 to 20,000, and a lift in our overseas aid to meet the United Nations target of 0.7 per cent of Gross National Income. These are compassionate, humanitarian, internationalist proposals, and I have little difficulty persuading people on the left to support them.

But I also seek to win support for these proposals from the right of politics by arguing that the compassionate approach is also the smart approach. I believe that the problems we read about in the news every day, and worry about as we go to sleep at night – terrorism, war, boat people – have at their heart conflicts over access to scarce resources – land, water, food, energy. Too many people; not enough land, water, food, energy, to go around. And if we do not get in there and try to help deal with these underlying problems, then they will rebound on us in the form of terrorism, war and uncontrolled arrival of asylum-seekers by sea – 'boat people'.

So the compassionate thing is also the smart thing, and there is good reason why both those on the left and those on the right should embrace this approach, and good reason why our fiduciary duties should not be defined as stopping at our electorate boundaries. So how should the concept of fiduciary duty be applied to the great policy challenge of our time, global warming?

Whatever the political pressures we are under, and no doubt they are significant, we have a duty to take a science-based, evidence-based approach. It has been claimed that, climate change, for the extreme left, provides the opportunity to do what they have always wanted to do, which is to de-industrialize the Western world.

Let me observe that I do not hear anyone proposing that we de-industrialize the Western world. Proposals to move to solar, wind and geo-thermal energy, or for greater energy efficiency or electric cars, do not represent an atom of de-industrialization. Proposals for solar panels and rainwater tanks promote the kind of household self-reliance one would expect champions of rugged individualism to be enthusiastic about.

Discharging a fiduciary duty means taking the blinkers off and being prepared to look at the facts. Second, I think fiduciary duty fits neatly with the definition of sustainability developed in 1983 by a United Nations Commission headed by Norwegian Prime Minister Brundtland: 'Meeting the needs of the present without compromising the ability of future generations to meet their own needs'.[11] Sustainability is a somewhat over-used word, having been fought over and prostituted by various vested interests. But the core idea of safeguarding the future remains as important as ever.

So how can we realistically tackle climate change and discharge that trust obligation to meet the needs of the present without compromising the ability of the future generations to meet their own needs?

One thing I think is particularly important, but in constant danger of being left out, is the issue of protecting tropical forests. According to Al Gore's book, *Our Choice – A Plan to Solve the Climate Crisis*,[12] the carbon dioxide emissions from deforestation are second only to the burning of fossil fuels for the production of electricity and heat as the largest source of global warming on the planet. An estimated 20–23 per cent annual CO_2 emissions – more than that of all of the cars and trucks in the world – result from the destruction and burning of forests. Almost 20 per cent of the Amazon forest has already been destroyed, and Brazil is responsible for 48 per cent of all deforestation in the world. African forests have been purchased by Chinese interests and are being chopped down for large-scale agriculture to grow food for import to China. Indonesia and Malaysia are clearing rainforests to produce palm oil plantations for export to North America and Europe to feed the demand for bio-fuels.

The ongoing forest clearing exacerbates the clearing of the past. According to the World Resources Institute, we only have half of the forest cover we had 300 years ago.[13] Indeed, scientists estimate that more than 40 per cent of the extra CO_2 that has accumulated in our atmosphere has come from deforestation in past centuries. Deforestation has a double impact on the climate – most of the carbon contained in logged trees goes into the atmosphere, and the destroyed trees no longer pull CO_2 from the air.

The Australian Government is aware of the problem and has announced a $200 million International Forest Carbon Initiative designed to help reduce emissions from deforestation and forest degradation in developing countries. Reducing

---

11   United Nations (1987), A/Res/42/187.
12   Gore (2009).
13   Australian Government (2010a).

emissions from deforestation and forest degradation goes by the acronym 'REDD'.[14] The international community agreed in December 2007 in Bali that action must be taken on REDD, to establish the necessary systems and financial mechanisms to ensure long-term emission reductions. The International Forest Carbon Initiative is Australia's contribution to this global effort.[15]

The Initiative aims to show that REDD can be part of an equitable and effective post-2012 global climate change agreement. A central element of the Initiative is taking practical action through collaborative Forest Carbon Partnerships with Indonesia and Papua New Guinea. Actions include assisting developing countries, particularly Indonesia and Papua New Guinea, to develop their own national forest accounting systems, and partnering with a consortium led by the Clinton Climate Initiative to use Australia's National Carbon Accounting System as a platform for a global forest carbon monitoring system.[16]

This is important work. But it needs to be said that the outcome for REDD, and for the whole question of getting agreement to protect tropical forests, is on a knife edge. At the moment, the United Nations Framework Convention on Climate Change working definition does not discriminate between native forests and plantations, such as the Indonesian palm oil plantations. Such an approach does nothing to protect any of the thousands of species I mentioned before which are under threat. Deforestation would not be monitored or detected. There is a proposed safeguard that REDD will not incentivize the conversion of native forests to plantations, but it is a matter of dispute.[17]

I believe we need to act to tackle climate change, and that that action needs to include protecting our tropical forests. I said at the outset I thought that concepts such as fiduciary duty and public trust are potentially useful in thinking about public policy, and this seems clear in this instance – the tropical forests are part of our global heritage, and we have a duty to protect them.

In the light of the failure of the Copenhagen Conference to make much progress, how are we going to adequately discharge the responsibility – fiduciary duty or not – that we have to protect our children and grandchildren from the effects of climate change? It is clear both that the Copenhagen Conference talks failed to achieve anything like the action that is needed to stop the Earth's temperature from rising to dangerous levels with unpredictable consequences, and that the dynamics at work which are preventing global agreement need to be shifted or else the crippling impasse at Copenhagen will continue indefinitely.

My proposals for breaking the present impasse are as follows. At present countries which want to do the right thing, or are trying to do the right thing,

---

14 Reducing Emissions from Deforestation and Forest Degradation (UN-REDD Programme).

15 See information on the International Forest Carbon Initiative at http://www. climatechange.gov.au/government/initiatives/international-forest-carbon-initiative.aspx

16 Australian Government (2010b).

17 UNEP (2010), p. 16.

are being undermined by countries who want to get away with doing as little as possible. It is time the latter countries were taught that nobody likes a 'bludger'.[18] Countries which are reducing their carbon emissions, or countries which have stabilized their carbon emissions, should give whatever notice is required by existing treaty obligations of their intention to impose green tariffs on imports from countries which are not cutting their carbon emissions.

The kind of formula which I think has merit is, if country A is cutting its carbon emissions by two per cent per annum, and country B is increasing its emissions by three per cent per annum, country A should impose a five per cent green tariff (the difference between country A's and country B's greenhouse performance) on goods coming from country B. The green tariff needs to be related to absolute emissions, not tricky substitutes like emissions intensity or emissions per capita. The climate is not interested in CO_2 intensity or emissions per capita.

This type of green tariff would achieve three things. First, it would punish bad conduct. At the moment there is no economic incentive for countries to behave well, and way too many countries are behaving selfishly and trying to get away with doing as little as possible. This has to change. Second, it protects good conduct. At the moment countries which bring in emissions trading or carbon taxes in order to reduce their carbon emissions run the risk of carbon leakage – companies moving their operations offshore. This risk if often exaggerated and used as an excuse by companies who are more interested in profits in the short term than the planet in the long term, but it is a risk that should not be there. If a country cuts its carbon emissions, its businesses should not face unfair competition from overseas businesses which have not done so. Action to cut carbon emissions should not be subverted by companies either going offshore or threatening such action. Third, it would provide a pool of money to be given to developing countries to help them down the renewable energy path, to decouple their economic growth from carbon emissions, and to help these countries deal with the adverse effects of climate change.

No doubt there will be objections to green tariffs, but I do not believe the objections are valid. The first objection will be that such an action constitutes an interference with trade liberalization, so called free trade. Indeed it does, but the climate crisis is urgent and desperate. Saving the planet is more important than free trade.

The second objection will be that this approach will hurt the poorer countries, because they are the ones who are going to need to increase their emissions to break out of their poverty. There are a number of flaws in this argument. This objection might have merit had the world accepted it, and come up with an agreed approach for the developing countries' emissions. The outcome of the Copenhagen Conference makes it clear that the world has not. This argument is a recipe for ongoing failure. It is simply not going to achieve the necessary carbon reductions. The ensuing stalemate damages many of the poorest countries most, because it

---

18   Australian slang for 'a hanger on' or 'a loafer' – a general term of abuse.

is they who are in the firing line from rising sea levels and increasing climate disasters such as hurricanes and floods, and have the least resources to adapt to climate change.

Third, putting green tariffs in place could produce money to help the poorer countries and reward those who are genuine about decoupling economic growth from carbon emissions. Of course it is desirable to close the gap between rich and poor, but I fear that trying to do this through the same mechanism as tackling climate change is trying to do more than we are humanly capable of.

Copenhagen failed because it had only a very small carrot, and no stick. In order to succeed we need a lot more of both carrot and stick. I think one of the impediments which have caused us to struggle with climate change solutions is that we have tended to focus on big, far-off targets, which can be an excuse for inaction on both counts. If the targets are big that can be paralyzing – we think that the mountain is too high to climb. And if the targets are long term, we think that we do not need to do anything now. We can act later, or even leave it to those who come after us.

I am attracted to thinking of the carbon reduction task in small bites, which are at once both more manageable and more demanding, because they require immediate action. Worryingly, Australia's carbon emissions are still continuing to rise. I think we should be setting a goal of stopping this in 2010 – stabilising our carbon emissions by the end of the year. Then, each year after, we should aim to cut our emissions by two per cent. It does not feel that impossible, taking it one step at a time. But if we could do it for the next 40 years, we would have cut our carbon emissions by 80 per cent by 2050 and if we could do it for the next 50 years, we would have made our country completely carbon neutral – 100 per cent reduction by 2060.

It is a good idea to put a price on carbon and put the market to work coming up with low-cost carbon abatement measures. But events have shown we need other measures as well. The Government has already reacted through lifting of the Renewable Energy Target[19] and funding a range of clean energy initiatives. It is desirable that there be more such action. Regulatory measures are possible, such as setting a date from which all new buildings must generate some percentage of their own electricity. Measures to generate your own electricity are a cost in the short run, but an advantage in the long run. Alternatively, the Government could use either a carrot or a stick to encourage a move to electric vehicles. Alternatively, it could act in the area of agriculture and forestry to promote the retention and replanting of native vegetation.

The Australian State and Territory Governments could also reconsider emissions trading schemes of their own. States either had introduced (for example, NSW), or were contemplating, emissions trading measures during the Howard Government years of inaction, but put their schemes on hold or wound them up in

---

19   Australian Government (2010b).

deference to the national Carbon Pollution Reduction Scheme. Given the present impasse, States should now seriously consider re-introducing these schemes.

Securing international co-operation to tackle greenhouse gas emissions has been problematic. The role of some rich countries that ought to know better – particularly the United States under President George W. Bush and Australia under Prime Minister John Howard – has been disgraceful.

No doubt, we need third world countries to be part of the climate change solution. I do not excuse for a moment rich countries using poor countries as an excuse for inaction, for example the excuse: 'We won't do anything until they do'. I do, however, agree absolutely that it is insufficient for rich countries to reduce their carbon emissions if this good work is undone by emissions from poor countries.

The science is telling us emissions must be cut – a figure with strong scientific backing is 60 per cent by 2050. It is not reasonable to say that all countries much share this load equally – the developed countries created the problem, and furthermore we cannot seriously expect the impoverished people of the third world to stay locked in that state of poverty indefinitely. But it will not work if the developing countries simply pick up the slack as developed countries reduce their emissions – this will not give us the cuts the scientists are clear are needed.

So what should happen concerning the developing countries?

Conclusion

I believe a three-part response is required. First, the developing countries must agree to move to stabilize their populations. There are many reasons for wanting population stabilisation, and carbon emission is a particularly important one.

Second, the developing countries must agree to halt deforestation. The world lost more than 70 million hectares of forests between 1990 and 2005. Most of the loss was in South America, Africa and the Caribbean. South America lost 64 million hectares of forests, some seven per cent of the world's total, and Africa lost eight million hectares of forest during this period. In South America, high food and fuel prices are driving continued forest clearance for production of livestock and crops for food, feed and bio-fuel.

Third, the rich countries must engage in technology transfer and basically bankroll the developing countries down renewable energy paths to development.

I believe these three steps would put the developing countries on a sustainable footing, and mean that the work of the rich countries in reducing emissions will not be in vain. We cannot sit by and allow the Copenhagen Conference impasse to drag on indefinitely. The damage both now and in prospect from climate change makes such a course of action utterly irresponsible.

References

Australian Government (2010a) 'Department of Climate Change and Energy Efficiency: External Scrutiny: Audits and Reviews', 29 June 2010, available at http://www.climatechange.gov.au/about/accountability/annual-reports/annual-report-0910/management-and-accountability/external-scrutiny.aspx.

Australian Government (2010b) 'Media Release, Enhanced Renewable Energy Target Scheme', 26 February.

J. Fischer, D. Lindenmayer, and R. Montague-Drake (2008) 'The role of landscape texture in conservation biogeography: a case study on birds in south-eastern Australia', 14(1) *Diversity and Distributions* 38.

A. Gore (2009) *Our Choice: A Plan to Solve the Climate Crisis*, Bloomsbury Publishing PLC.

International Union for Conservation of Nature (2009) 'Species', *e-Bulletin*, October, available at http://www.iucn.org/about/work/programmes/species/publications___technical_documents/publications/species_e/.

G. E. D. Pont, and D. R. C. Chalmers (2004) *Equity and trusts in Australia*, 3rd edn, Lawbook Co.

Lord Stern (2006) *Review on the Economics of Climate Change*, HM Treasury, available at http://webarchive.nationalarchives.gov.uk/+/http://www.hm-treasury.gov.uk/stern_review_report.htm.

J.E. Stiglitz, A. Sen and J.-P.Fitoussi (2009) *Report by the Commission on the Measurement of Economic Performance and Social Progress*, available at *www.stiglitz-sen-fitoussi.fr/ -*.

UN (1987) 'Report of the World Commission on Environment and Development', 96th Plenary Meeting of the General Assembly, A/Res/42/187, 11 December, available at http://www.un.org/documents/ga/res/42/ares42-187.htm.

UNEP (2010) *Convention on Biological Diversity*, 10th Annual Meeting, Nagoya, 18–29 October, UN EP/CBD/COP/10INF//20, available at *www.cbd.int/doc/meetings/cop/cop-10/information/cop-10-inf-20-en.doc*.

Chapter 12

A Ponzi Scheme on the Environment? Failures of Fiduciary Duty and the Challenges of Climate Governance

Fiona Haines[1]

Introduction

The recent financial crisis has raised the profile of the financial scam known as the Ponzi scheme, with analyses of Ponzi schemes finding their way beyond the confines of the business pages or the financial press. Many more of us are now familiar with the way Ponzi schemes are designed, with their central feature being that the money of new entrants to an 'investment opportunity' is used to pay off dividends to original investors. When the supply of new money dries up, the edifice collapses and the value of the scheme is revealed as a cruel hoax.

Yet, could the business practices of our current generation be understood more generally as a giant 'Ponzi scheme', as a hoax on our environment with current planetary resources being used at an unsustainable rate? This chapter takes seriously this idea of Ponzi scheme as a metaphor for our relationship with the environment, and explores the notion of a gross failure of fiduciary duty in the context of governance regimes aimed at combating climate change. Through an analysis of the more familiar Ponzi scheme, the chapter argues that fiduciary obligations in the financial sphere aimed at minimising financial misconduct by executives act to undermine governance regimes designed to combat climate change. Financial fiduciary obligations motivate business executives to look for loopholes in any emerging climate regime and to lobby to preserve shareholder value, in the short-term at least, in a manner that weakens emerging regulatory regimes aimed at institutionalising the necessary reduction in GHG emissions. Placing obligations to the environment within corporate law and more broadly within measures of economic well-being (that is, moving away from measures such as GDP) are needed to prevent obvious conflicts between fiduciary duties to

1 I would like to thank the organizers and participants of the workshop 'Fiduciary Duty, Public Trust and the Governance of Climate Change', and in particular the support of Donald Feaver and Ken Coghill, which led to the generation of the ideas presented in this chapter. This work draws on and extends my earlier work on political risk published in Haines (2011).

shareholders and investors and those to the environment. Without such changes, there remains a legal obligation and an economic imperative for companies and their executives to maximise investor value by sacrificing the planet.

At first glance, however, the notion of fiduciary duty is an inspiring one. An entry in a *Concise Desk Book of Business Finance* writes of a fiduciary as:

> ... a person, or organization such as a corporation, acting in the capacity of trust and special confidence so that he [*sic*] or it [*sic*] is obliged to act with the highest degree of good faith and to place ahead of his or its own interests the interest of those represented ... .[2]

Fiduciary duty highlights the obligations those in a position of trust have towards dependants. Yet, what has been prominent of late is a gross failure of fiduciary duty in the context of both the global financial and the environmental crises. Bernie Madoff has been paraded through the media and beyond as his investments were revealed as a massive Ponzi scheme with the inflow of billions of dollars of investor money used to fund a lavish lifestyle.[3] More broadly, the financial crisis itself has been likened to a global Ponzi scheme since the derivatives market relied on short-term borrowing to meet long-term commitments made through hedging transactions.[4]

This notion of using other peoples' money to fund lavish lifestyles and to shore up immediate obligations resonates with the crisis around climate change. Indeed, in this context our relationship to the earth has been characterised as a 'global Ponzi scheme'[5] as we deplete the natural resources of forests, fish and minerals and atmosphere at an unsustainable rate. The analogy of the Ponzi scheme as a metaphor for our relationship to the environment and particularly the climate is useful. It depicts graphically our problematic and unsustainable relationship to the environment – as a dereliction of our collective fiduciary duty. But it can do more than this. Characterising the problem as a Ponzi scheme mounted against future generations and non-human species, a massive failure of fiduciary duty, reveals some important insights for us in the development of a suitable regulatory framework, particularly that targeted at the business sector. The lessons in dealing with more traditional financial Ponzi schemes and financial fraud provide an insight into the challenge of climate governance aimed at reducing the GHG emissions of industry.

There are both problems as well as possibilities, with labelling business obligations to the environment as fiduciary in nature. Definitions of fiduciary obligations can be remarkably narrow. Indeed, the definition of fiduciary duties with the broad and honourable sentiment cited above comes down to earth once a

---

2   Moffat (1982).
3   Creswell and Thomas (2009).
4   Jabecki and Machaj (2009).
5   Climate Progress (2008).

legal text is consulted.[6] The concern within such a text is to define the conditions under which a claim for damages might arise where a paid agent fails to discharge their professional duty to the standard required. Fiduciary obligations are those owed by one party to another and are principally financial in nature. Legal obligations are owed by the trustee to the beneficiary with respect to the implied or written terms of the contract, no more and no less.

This narrow characterisation of fiduciary duties causes problems. These arise in at least two ways. First, narrow obligations invite gaming responses[7] where, in order to secure financial advantage, players abide by the rule but do not enter into the spirit of their obligations. Secondly, multiple regulatory regimes each based on narrow, but different, definitions of fiduciary duty arise. Diverse regulatory regimes create competing demands that require businesses to prioritise and deploy resources in disparate arenas.[8] Yet, and this is most important, there is an order to this regulatory complexity, an order where financial fiduciary duties sit at the apex of the pyramid of regulatory obligations.

Because of this ordering, developing an effective governance regime to tackle climate change requires change both in how we view the benefits arising out of business activity and how we measure it. Change is needed not only to how we formally and legally measure the financial health of a company but also the economic health of a country. The climate crisis means that the current calculations of financial and economic wealth of a country, which leave out its contributions to GHG emissions, cannot be allowed to dominate government policy or international debate around necessary governance regimes. This is no easy task since governments have considerable investment in maintaining public faith in the profitability of current environmentally unsustainable forms of industrial activity. The wealth generated by environmentally unsustainable businesses materially affects the health of government coffers and their capacity to fund necessary government expenditure. In addition, the perception of strong economic management also has an effect on a government's electoral prospects,[9] so incumbent governments have an incentive to place current forms of economic activity in the best light possible. A drastic change in perceptions of economic fortunes (however much this actually measures real wealth when the environment is taken into account) is likely to be resisted by incumbent governments.

---

6   See, for example, Vermeesch and Lindgren (1995).
7   Ayres and Braithwaite (1992).
8   Haines and Gurney (2003); Brimble et al. (2010).
9   Weatherford and Sergeyev (2000); Cohen (2004).

What Can We Learn from Attempts to Curb Financial 'Ponzi Schemes'? Understanding the Limits of the Law

The problem of narrow obligations inviting creative compliance is well illustrated by past attempts to deal with the more traditional forms of financial Ponzi schemes. Ponzi schemes and related frauds are a perennial problem in the financial arena. Successive reforms to financial regulation in Australia and beyond have attempted to curb speculative excess and outright fraud whilst maintaining the efficiency of financial markets.[10] In Australia, there has been a progressive tightening of regulations pertaining to capital adequacy requirements and to enhanced accounting and audit procedures in an attempt to reduce the incidence of financial fraud arising out of the collapse of HIH Insurance Group in 2001.[11]

However, problems remain. Reform of corporate law aimed at improving the integrity of business accounts are met with successive iterations of creative compliance.[12] Creative compliance ranges from business attempts to comply with the letter of the law whilst managing to avoid the spirit of the law to 'regulatory arbitrage', where tighter regulatory obligations are avoided by transferring funds to areas of lesser obligation.[13] So, companies develop techniques, from reinsurance contracts to hedging transactions, to reduce the amount of capital they are required by law to hold as a buffer against a call by investors or creditors. For example, funds are created and debts accrued outside of the regulator's gaze. As we have seen from the recent financial crisis, a large dose of creative compliance allowed the proliferation of debt to accrue in 'off balance sheet' accounts.[14]

It is important, however, to take a close look at what is involved in creative compliance and regulatory arbitrage. There are several elements to the puzzle. So, a company arranging its financial affairs to reduce its regulatory obligations may be a sensible strategy. Colander and his colleagues,[15] for example, argue that what ended up as 'regulatory arbitrage' in the context of the financial crisis actually began as a sensible method of spreading risk. Hedging was thought to reduce risk and so did not require banks to hold on to the same level of capital if they had not hedged their debts. For a number of complex reasons beyond the remit of this chapter, the models underlying the hedging process and the specific regulatory framework transformed the sensible spreading of risk into a recipe for enhancing risk and creating false impressions of the robustness of levels of debt and creditworthiness.

Further, inflexible rules by governments around financial reporting and corporate governance can create problems. In certain cases, 'creative compliance' may in fact be helpful to legitimate business prospects in the long term.

---

10   Black (2005); Acharaya and Richardson (2009); Clarke et al. (2003); Snider (2007).
11   ASIC (2009); Brown and Tarca (2005); Clarke and Dean (2005); Haines (2009).
12   Shah (1996); Clarke et al. (2003); McBarnet and Whelan (1999).
13   Stiglitz (2009); Colander et al. (2009).
14   Foster and Magdoff (2009); Krugman (2008).
15   Colander et al. (2009).

Regulations aimed at reducing the potential for various Ponzi schemes illustrate the problem of inflexible law. A prominent legal injunction acting as a backstop against the development of a Ponzi scheme is the prohibition against trading whilst insolvent (in Australia, Section 588G of the *Corporations Act 2001* (Cth)). Businesses are required to generate profit from the productive capacity of their operations. It is this core activity that must provide the means for payments to shareholders, other investors and creditors and for remuneration of executives and employees alike. In particular, new investor funds must not be used to line executive pockets or pay off debts to old investors as is the practice of a typical Ponzi arrangement. Yet, within Australia there is concern expressed by boards and their finance committees arguing that laws prohibiting trading whilst insolvent prevent directors save companies from extinction.[16] Directors and their boards are concerned that they may find themselves forced to declare bankruptcy, even if the company is salvageable, since they face prosecution if they continue to trade whilst technically insolvent. If the law is too inflexible directors may end up losing investors' money unnecessarily when, given more leeway, directors and boards could bring companies back into profit.[17]

This problem of inflexibility in the law, coupled with zealous enforcement, leads to calls for reform and lobbying by business to reform the law.[18] Complaints are made of the way multiple laws and regulations inhibit legitimate business which, in turn, leads to calls for reducing 'red tape'.[19] Ensuing reforms are then made in a manner more congenial to their aim of creating wealth. Businesses exert their 'regulatory authority'[20] and shape the regulations by which they are controlled. This process is critical to understanding the law's ineffectiveness in the control of those with malevolent intent. It is important to be clear what is happening here. Flexibility in law helps both the well intended and the malevolent. Regulatory regimes are designed to extract the harm (exploitation of investors for personal gain) from an otherwise beneficial activity (creating wealth). So, laws prohibiting trading whilst insolvent must be cognisant that they do not dampen desirable entrepreneurial activity whilst targeting those involved in 'Ponzi like' activities.

This problem of harm being tightly embedded in beneficial business behaviour reveals the difficulties of defining what is legal and what is not.[21] The business world provides many examples of enterprises that could earn the name of a Ponzi scheme in that their revenue comes not from their productive capacity but from investors – and where existing investors are paid from the pockets of the most recent investors. Prospecting for new mine sites, bio-technology ventures and a

---

16   Baxt (2008).
17   Baxt (2008); Grant (2009); Hahn (2008).
18   AICD (2009).
19   Banks (2005); Regulation Taskforce (2006).
20   Reichman (1998).
21   Sutton and Haines (2003).

myriad of start-up companies spend many years in debt and fail to turn a profit directly from their activities. Yet, these are seen as socially desirable activities that contribute to the vibrancy of the economy. But, before turning a profit from production they inhabit an ambiguous legal space. Indeed, their share price may rise and fall over many years before profitability is reached. If they fail, close scrutiny is likely to ensure that they were not engaging in problematic activity (see for example the fate of the mining company *Sons of Gwalia*),[22] yet when they succeed they are held up for acclaim and admiration. So, relevant laws need to be designed so that such entrepreneurial activity is not dampened whilst at the same time being adept at identifying directors and executives who play fast and loose with investor funds. Striking such a balance is not easy.

Creative Compliance and the Problem of Flexible Regulation in the Climate Arena

These problems of creative compliance, regulatory arbitrage and inflexibility all are present in the control of our impact on the climate. The emerging regime around a reduction of GHG emissions also risks developing the same chronic regulatory difficulties as those experienced in attempts to control financial Ponzi schemes. Creative compliance is likely to become a major concern in emerging emissions trading regimes.[23] For example, the capacity within some trading systems for offsetting has led to problems where companies can legally increase their emissions whilst buying offsets elsewhere.[24] The integrity of such offset regimes requires close attention, with some being less than robust in reducing the levels of greenhouse gases in the environment.[25] Indeed, as with Ponzi schemes, creative compliance may cross the line to develop into outright fraud with the falsification of data on emissions. Yet, this flexibility of emission trading regimes can work positively. At their best they can allow the most cost effective reduction in emissions,[26] for example by placing a monetary value on forests which without such support might rapidly disappear.[27] Flexibility in environmental initiatives can also work for both the well and not-so-well intentioned.

Lobbying by business to shape emerging regimes to suit their own interests is well recognised as highly problematic[28] and can directly affect the effectiveness of climate governance regimes. In the UK Emissions Trading Scheme (UK ETS) and European Union Greenhouse Gas Emission Trading System (EU ETS), the over-estimation of future emissions by market participants saw an excessive number of permits flood the trading floor as real levels of emissions fell well below permitted

---

22   Drummond (2005).
23   Hepburn (2007); Smith and Swierzbinski (2007); Haines (2008).
24   Oleschak and Springer (2007).
25   Wittman and Caron (2009).
26   Choi (2005); Brown et al. (2002).
27   Potvini et al. (2008).
28   Pearce (2007).

amounts.[29] Now, there is nothing illegal here, but this is hardly an outcome desirable in the pursuit of cutting emissions. The letter of the law has not been broken, merely shaped to suit dominant industry players.

Deconstructing 'Red Tape'

There is a further problem, however, with the development of a new governance regime designed to reduce our impact on the climate without a significant change in how governments relate to business. Regulation is primarily understood as a cost to business, one that needs to be kept in check.[30] Business lobby groups argue that they must be free to compete and that 'over-regulation' and 'red tape' must be minimised or eliminated.[31] Government agrees. On coming to office in 2007, the Australian Minister for Finance, Lindsay Tanner, took as a major policy platform deregulation arguing that the former Liberal-National Coalition Government had failed in its duty to deregulate. In a speech to the Sydney Institute, Tanner stated:

> (l)ike a smouldering fire the Liberals let the deregulation agenda in this country
> lie dormant for most of their eleven years in office.
> I intend to re-ignite it.[32]

Because regulation is seen as costly, governments also argue that it should be kept under tight control. As a result, new regulatory initiatives need to be tightly justified and their costs to industry kept as low as possible. Proponents of new regulatory regimes are required to define clearly and narrowly what 'the problem' or the harm is. A tight definition of the problem and a clear understanding of a causal pathway are seen as essential in the development of a regulatory regime,[33] one that will not unduly intrude on the competitive market place seen as so necessary to generate wealth. Business activity then is understood as predominantly productive (and therefore unproblematic) whilst regulation is seen as intrusive.

There are two inter-related elements at play here. The first is to understand the way regulation, when understood as an intrusion on economic activity, places a normative value on economic relationships and in particular market competition. That is, economic relationships centred on the competitive market are understood as separate from their social and environmental interdependencies. Hence, the ways in which economic relationships are necessarily embedded within a broader social context and certain environmental constraints are glossed over. Economic relationships are the norm by which other – social and environmental

---

29   Hepburn (2007).
30   Regulation Taskforce (2006).
31   AICD (2009).
32   Lindsay Tanner (2008) (Australian Labor Party), Minister for Finance and Deregulation, Speech to the Sydney Institute, 26 February.
33   Regulation Taskforce (2006); OECD (2004); Better Regulation Task Force (2005).

– relationships and dependencies are measured.[34] In this way, laws and regulatory regimes that promote social and environmental goals are understood as *intruding* on the market, rather than as generating the rules by which economic activity is allowed to take place.

The second element is that this narrow definition of 'the problem' to be regulated means that each new problem that arises requires a dedicated regime. Separate regimes develop, each concerned with a narrowly defined goal since governance regimes must be designed with a specific aim in mind. In the case of finance, regimes are designed to provide potential and current investors with a clear idea of the financial health of the enterprise into which they place their money. The regulator, Australian Securities and Investment Commission (ASIC), has no interest in the environmental bone fides of a company unless it will have a material effect on its profitability, despite ecological economists highlighting the unsustainable nature of much business activity.[35] So, if a company's profit is extracted at the expense of the environment ASIC has no concern. Even if the business strategy is one that is ecologically unsustainable, ASIC would defer to the jurisdiction of the various state environmental protection agencies.

The result of this tight coupling between separate regulatory regimes, each with identification of a specific (and separate) problem, is complexity and what business likes to call 'red tape'. In the academic arena, the development of multiple regimes is characterised as 'juridification', the proliferation of law each with its own tightly distinct instrumental orientation[36] and each with a limited capacity to communicate to the other. Yet, from an alternative perspective, the goal of regulation might be argued to be various attempts with differing levels of success to re-embed social relationships and environmental dependencies within economic relationships.[37]

Conflict in Law

Clearly, there are those in business who want to be part of the solution to climate change. However, those companies and directors with aspirations of becoming industry leaders in reducing their impact on the climate may be confronted by the inflexibility of law. The concern here is not with laws around environmental protection. Indeed, various environmental protection authorities would be only too glad to have such an 'industry leader' within their jurisdiction. Yet, if exemplary

---

34   There is a significant and extensive literature relevant here, much of it stemming from the work of Polanyi. See for example Polanyi 2001 and the introduction and commentary in that edition of *The Great Transformation* by Joseph Stiglitz and Fred Block. For an excellent analysis of the reductionist logic at play here in the US context, see Chen and Hanson (2004)

35   Constanza and O'Neill (1996); Goodland (1995); Holling (2001); O'Neill (1996).

36   Teubner (1998).

37   Cf. Polany (2001)

practice by such a company places them at a market disadvantage then they may find themselves experiencing not only the wrath of their shareholders but also legal authorities. Companies need to prioritise their various fiduciary obligations and may fall foul of corporate law if they risk shareholder returns by investment in green technology.

Responses to enhance the integrity of financial accounts emphasize the primary obligation of the director, understood as one owed to investor and shareholder funds. This narrow framing of financial regulatory duties may encourage 'creative compliance' with environmental obligations, particularly where such obligations are financially onerous. A primary responsibility to shareholders and investors may then encourage creative compliance with competing obligations, including those to the environment.

Resolving Conflict?

Nonetheless, there are ways that companies have tried to count their worth beyond a financial balance sheet. 'Triple bottom line' approaches and corporate social responsibility (CSR) schemes are ways that companies have tried to harmonise and maximise beneficial outcomes not only for shareholders but also the environment and the community.[38]

But these voluntary regimes do not overcome the problems of creative compliance and the inflexibility of law. In terms of conflict, a voluntary programme, even one with the credentials of a high profile CSR initiative, does not fare well if found in breach of legal obligations. Voluntary practices fail when they meet a contrary legal mandate. So, without tough legal environmental requirements, companies have limited room to go 'beyond compliance' in reducing their emissions.[39]

Further, this disjunction between legal mandate and voluntary aspiration may exacerbate the problem of creative compliance. Voluntary programmes can be vague and lack credibility, with accountability and transparency low.[40] CSR reports may be little more than 'greenwash', an attempt at image management rather than a sustained effort at creating change.[41]

So, without some sort of commensuration between legal obligations, the problems of conflicting obligations, juridification and creative compliance remain. The multiplication of tougher and tougher standards in a range of arenas such as environmental, corporate, safety and consumer protection law is necessary for increasing standards of corporate behaviour. But this also creates a complex array of competing fiduciary obligations.[42] In each case, the contract is drawn

---

38   McBarnet (2007).
39   Kagan et al. (2003); Thornton et al. (2008).
40   Parker (1999).
41   Chen and Bouvain (2008); Utting (2002).
42   Sunstein (2005).

up differently: between company and environment, company and shareholder, company and worker, company and consumer.

In the area of critical interest here, fiduciary obligations to the environment and climate, a potential way forward is to give environmental obligations the force of law in corporate legislation. That is to amend corporate law so that companies have an equal obligation to ensure they do not act as a Ponzi scheme on either their investors or future generations.

This would require a clear accounting regime dedicated to enumerating a business's impact on the environment. This is difficult, but not impossible.[43] Articulating the debt to the atmosphere within corporate law would sharpen the problematic nature of our relationship to the environment. It would also require placing a monetary value (or penalty) on GHG emissions, a step seen by some as essential to mitigating our impact on the climate.[44] Whether we like it or not, obligations that are essentially contractual and financial hold pride of place in a crowded regulatory pantheon. This suggests that serious attempts should be made to place a higher priority on our debt to the environment within standard business accounts. Debts to the environment need to be placed alongside the wealth accrued to business investors.[45]

Clearly, this is not a panacea. Methods of accounting for the environment also narrow conceptions of what our relationship to the environment should look like. The intrinsic value of our environment is transformed into exchange value.[46] The danger in measurement is that simplifying a complex system for the purposes of measurement and accountability may end up over-simplifying a complex problem.[47] Further, measurement for the purposes of audit and accountability can distort underlying relationships and change priorities so that only what is measured is considered important.[48]

Ultimately, however, constructing company accounts that do not include a debt to the climate (if not environmental sustainability more generally) is no longer acceptable. Without a clear sense of the financial impact of a company on the environment within standard company accounts 'decarbonising' economic activity is likely to remain at a lower priority than company profit.

---

43   However, measures of the impact on climate may be easier than broader measures of environmental sustainability, see Simnet and Nugent (2007).

44   Stern (2008); Goodland (1995); Goodland and Daly (1996).

45   Constanza and O'Neill (1996); O'Neill (1996); Goodland (1995).

46   Clark (1995); Cato (2009).

47   O'Neill (1996); Holling (2001)

48   Power (1997).

Governments and Political Risk

The problems of financial and environmental Ponzi behaviour, and the challenges of the regulatory response in generating sustainable businesses in both the financial and environmental sense, need to be placed within a political as well as an economic context. The principal incentive animating economic activity – and that motivate entrepreneur and Ponzi fraudster alike – is maximising profit. The beneficial side-effect of entrepreneurialism is the maximisation of the aggregate wealth of the country. This much is clear from the previous discussion. But the dominant logic of the market cannot be divorced from the needs of government.[49] I would argue that there is insufficient political motivation to develop laws to ameliorate Ponzi behaviour whether of the financial or climate variety. The reason for this is not hard to find. Governments, too, are dependent on entrepreneurial activity for their economic health. They remain sensitive to pressure by existing businesses that suggest that tighter restrictions on their business practices would harm future business investment. Indeed, governments can go out of their way to woo business and attempt to create an environment conducive to continuing investment.

Pressure on government to reform regulation follows a cyclical pattern. Ponzi schemes and their apparently more benign cousin, the unsustainable business model of a failing company, are most often revealed during times of recession. It is during times of economic hardship and in the face of significant company collapses that pressures for reform and tightening of laws arise. Yet, a cyclical downturn may not alter the number of companies engaging in 'excessive risk taking'. Rather, the receding tide simply makes them more visible. It is in the wake of business collapses and misdemeanours that government turns its attention from 'reducing red tape' to how to control such 'excessive risk taking'. Yet, as business reacts against tightening regulation and demands a more flexible regulatory regime, government attention returns once more to the problem of reducing 'red tape'.

The central problem here then is the problem of the addicted government. Government fortunes, both political and economic, are tightly related to their capacity to govern at a time of economic growth and rising prosperity. This is not just a problem of the 'hip pocket nerve'. Government depends for its own revenue on its capacity to 'skim off' funds from taxation and royalties.[50] These funds rely on robust business investment. Business threats of an 'investment strike' (even if rather unrealistic) are treated very seriously indeed.[51]

Make no mistake: government capacity to access resources affects us all. Their addiction to business investment is central to the funding of the full range

---

49   For an extensive discussion of political risk outlined in this section, see Haines (2011).

50   Habermas (1989).

51   Haines et al. (2008); Haines (2011).

of government activities, from education to health and beyond. Their addiction is our addiction.[52]

Yet, how economic sustainability and growth is actually calculated also needs close attention, particularly in light of the climate crisis. Arguably, a single-minded attention placed on GDP growth is the equivalent of a single-minded attention to corporate law as framing the legal context under which a sustainable business can thrive. Government capacity to deliver services to us depends upon GDP growth. Yet, just as profit to shareholders and good financial balance sheet is a poor indicator of an ecologically sustainable business, so too is GDP growth a poor indicator of an ecologically sustainable economy.[53] GDP growth is based on an economic model of well-being where impact on the environment, climate included, is entirely absent.

Clearly, developing a useful economic measure of our impact on our environment is a considerable task. Economic measures that are effective in measuring impact on the environment in a manner that can be easily integrated into national accounts are in their infancy. Certainly, there are a number of measures of environmental well-being that have been developed, including the World Bank's 'genuine saving measure'[54] and the easily identifiable Ecological Footprint.[55] Yet problems can arise. The 'genuine savings measure' purports to show that developed economies are the most sustainable. Yet, the measure takes scant account of a country's impact on the 'global commons' and so does little to reverse the Ponzi-like behaviour of developed countries with respect to the climate.

A central problem with some measures of ecological impact is their assumption (under so called 'weak' models of sustainability) that the various forms of capital – human, natural and economic – are perfectly able to substitute for one another. Hence, such calculations would purport to demonstrate that it is possible to be sustainable by substituting human capital (for example, higher levels of education) for natural capital (for example, forest cover). Many, if not most, would recognize this definition of sustainability as problematic.[56]

Arguably, a measure of ecological sustainability based on a country's 'ecological footprint' is attractive in that it is readily understood and easily remembered.[57] Measures of ecological footprint do not suffer from the problem of measuring only weak forms of sustainability. Indeed, the ecological footprint measure has been useful in demonstrating the limits to arguments that economic growth is good for the environment and hence will be effective in reducing climate change. Arguments about the positive impact of economic growth on the environment

---

52   Habermas (1989).

53   Asadourian (2008); Cavilinga-Harris et al. (2009); Constanza and O'Neill (1996); Goodland (1995); Goodland and Daly (1996); O'Neill (1996).

54   Goodland and Daly (1996).

55   Cato (2009).

56   Goodland and Daly (1996).

57   Cato (2009).

are often based on the idea of the 'Kuznets curve'[58] that attempts to show how increasing economic wealth improves environmental sustainability. There is some truth to the Kuznets curve if concern is focused on point source pollution and when complex interactions and ecosystems are taken out of consideration.[59] Yet, by mapping economic development alongside the ecological footprint, arguably a more comprehensive assessment of environmental impact, the relationship between economic development and environmental impact is reversed. Increasing wealth is associated with an increased impact on the environment. The negative impact of development and economic growth on GHG emissions is particularly marked.[60] Yet, even here, there are problems as the ecological footprint measure treats energy derived from nuclear power as if it were generated through burning fossil fuel – hardly helpful in the context of tackling greenhouse emissions.[61]

Conclusion

Effective governance regimes aimed to tackle climate change must take account of how companies and governments currently fulfil their fiduciary obligations. Both businesses and governments understand these obligations as largely financial and contractual. The virtue of the profit motive is embedded within legal structures governing the activities of business. Financially profitable businesses are also understood as critically important for the financial and political health of a government.

Because of this, environmental obligations of both companies and governments are relegated to a secondary importance. Strong incentives remain on both business and government to resist stronger environmental obligations and, where they do exist, to comply with the letter of any climate governance regime rather than entering into its spirit.

To change this, there is a need to reorganise the incentive structure underpinning both business and government activity. Notions of corporate social responsibility need to be enshrined in law – in particular in corporate law so that there are fiduciary obligations to enhancing the well-being of the environment and not just the financial well-being of investors. In this way fiduciary obligations can extend to include future generations and non-human species. At the broad level of the economy, governments need to be held to account under measures that take into consideration our impact on the climate. Current measures of GDP fail to do this and should be replaced.

Clearly, these are ambitious aims and considerable problems remain in how to operationalize measures of ecological sustainability at the level of both

---

58   Kuznets (1955)
59   Goodland and Daly (1996).
60   Cavilinga-Harris et al. (2009).
61   Cavilinga-Harris et al. (2009).

the company and the economy. Without any shifts in this direction, however, governance regimes designed to ameliorate our impact on the climate will continue to be dogged by narrow conceptions of fiduciary duty, contractual obligations principally involving the financial relationship between business and investor. Governments will continue to be measured by their economic performance – irrespective of their attempts to design regimes to reduce emissions and impact on the environment more generally. Laws will be made to combat climate change but are likely to be consistently weakened by entrenched business interests and where abiding by the letter of the law trumps enthusiasm and action designed to enter into its spirit.

Changing how we measure overall performance of both companies and the economy will not be a panacea. But, if successful, it will help change the underlying incentives that shape entrepreneurial and political motivation. With amenable incentives in place, dealing with the climate crisis through regulation may not be quite such a Sisyphean struggle.

References

V. V. Acharaya and M. Richardson (2009) 'Causes of the Financial Crisis', 21 (2–3) *Critical Review* 195.

AICD (2009) 'States fail the test on business friendliness, says AICD', Australian Institute of Company Directors, available at http://www.companydirectors. com.au/Media/Media+Releases/2009/States+fail+the+test+on+business+frie ndliness+says+AICD.htm.

E. Asadourian (2008) 'Global Economic Growth Continues at Expense of Ecological Systems', World Watch Institute, available at http://www. worldwatch.org/node/5456.

ASIC (2009) 'CLERP 9: Corporate reporting and disclosure laws ', Australian Securities and Investments Commission, available at http://www.asic.gov.au/ asic/asic.nsf/byheadline/CLERP+9?openDocument#how.

I. Ayres and J. Braithwaite (1992) *Responsive Regulation: Transcending the Deregulation Debate*, Oxford University Press.

G. Banks (2005) 'Regulation-making in Australia: Is it broke? How do we fix it?', *Public Lecture Series, Australian Centre of Regulatory Economics (ACORE)*, Faculty of Economics and Commerce, ANU, 7 July.

B. Baxt (2008) '"Encouraging" Entrepreneurialism: What Parts Do/Should the Courts Play?', 36(1) *Australian Business Law Review* 62.

Better Regulation Task Force (2005) 'Regulation – Less is More, Reducing Burdens, Improving Outcomes', available at http://archive.cabinetoffice.gov. uk/brc/upload/assets/www.brc.gov.uk/lessismore.pdf.

J. Black (2005) 'The Development of Risk-Based Regulation in Financial Services: Just "Modelling Through"?', in J. Black, M. Lodge and M. Thatcher (eds), *Regulatory Innovation: A Comparative Analysis*, Edward Elgar, pp. 156–80.

M. Brimble, J. Stewart and L. de Zwaan (2010) 'Climate Change and Financial Regulation', 19(1) *Griffith Law Review* 71.

P. Brown and A. Tarca (2005) '2005 – It's Here, Ready or Not: A Review of the Australian Financial Reporting Framework', 15(2) *Australian Accounting Review* 68.

S. Brown, I. R. Winglan, R. Hanbury-Tenison, G. T. Prance and N. Myers (2002) 'Changes in the Use and Management of Forests for Abating Carbon Emissions: Issues and Challenges under the Kyoto Protocol', 360 (1797) *Philosophical Transactions*: *Mathematical, Physical and Engineering Sciences* 1593.

M. S. Cato (2009) *Green Economics*: *An Introduction to Theory, Policy and Practice*, Earthscan.

J. L. Cavilinga-Harris, D. Chambers and J. R. Kahn (2009) 'Taking the "U" out of Kuznets: A comprehensive analysis of the EKC and environmental degradation', 68 *Ecological Economics* 1149.

S. Chen and P. Bouvain (2008) 'Is Corporate Responsibility Converging? A Comparison of Corporate Responsibility Reporting in the USA, UK, Australia and Germany', 87 *Journal of Business Ethics* 299.

I. Choi (2005) 'Global Climate Change and the Use of Economic Approaches: The Ideal Design Features of Domestic Greenhouse Gas Emissions Trading with an Analysis of the European Union's CO2 Emissions Trading Directive and Climate Stewardship Act', 45 *Natural Resources Journal* 865.

J. G. Clark (1995) 'Economic Development vs. Sustainable Societies: Reflections on the Players in a Crucial Contest', 26 *Annual Review of Ecology and Systematics* 225.

F. Clarke and G. Dean (2005) 'Corporate Governance: A Case of "Misplaced Concreteness?"', 11 *Advances in Public Interest Accounting* 15.

F. Clarke, G. Dean and K. Oliver (2003) *Corporate Collapse*: *Accounting, Regulatory and Ethical Failure*, Cambridge University Press.

Climate Progress (2008) 'Is the Global economy a Ponzi scheme?', 29 September, available at http://climateprogress.org/2009/03/08/ponzi-scheme-madoff-friedman-natural-capital-renewable-resources/.

J. E. Cohen (2004) 'Economic Perceptions and Executive Approval in Comparative Perspective', 26 (1) *Political Behaviour* 27.

D. Colander, M. Goldberg, A. Haas, K. Juselius, A. Kirman, T. Lux and B. Sloth (2009) 'The Financial Crisis and the Systematic Failure of the Economics Profession', 21(2–3) *Critical Review* 249.

R. Constanza and R. V. O'Neill (1996) 'Introduction: Ecological Economics and Sustainability', 6(4) *Ecological Applications* 975.

J. Creswell and L. Thomas (2009) 'The Talented Mr Madoff', *New York Times*, 25 June, available at http://www.nytimes.com/2009/01/25/business/25bernie.html.

M. Drummond (2005) 'How the sons of Lalor built then sank Sons of Gwalia', *Sydney Morning Herald*. 22 August, available at http://www.

smh.com.au/news/business/how-sons-of-lalor-built-then-sank-sons-of-gwalia/2005/08/21/1124562748347.html#.

J. B. Foster and F. Magdoff (2009) *The Great Financial Crisis: Causes and Consequences*, Monthly Review Press.

R. Goodland (1995) 'The Concept of Environmental Sustainability', 26 *Annual Review of Ecology and Systematics* 1.

R. Goodland and H. Daly (1996) 'Environmental Sustainability: Universal and Non-Negotiable', 6(4) *Ecological Applications* 1002.

B. Grant, B. (2009) 'Insolvent Trading Law Reform – Submission to the Hon Chris Bowen MP, Minister for Financial Services, Superannuation and Corporate Law', Law Council of Australia, 1 July available at http://www.lawcouncil. asn.au/shadomx/apps/fms/fmsdownload.cfm?file_uuid=3A035B95-1E4F-17FA-D292-DB5FC99438FA&siteName=lca.

J. Habermas, J. (1989) 'What does a crisis mean today? Legitimation problems in late capitalism', in S. Seidman (ed.), *Jurgen Habermas on Society and Politics: A Reader*, Beacon Press, pp. 266–83.

S. Hahn (2008) 'Insolvency Alert – Judicial Criticism Levelled at Insolvent Trading Laws', DibbsBarker, available at http://www.dibbsbarker.com/publication/Insolvency_Alert_-_Judicial_Criticism_Levelled_at_Insolvent_Trading_Laws.aspx.

F. Haines (2008) 'The Lure of the Market in Tackling Global Warming', refereed conference paper *TASA, The Australian Sociological Association* Annual Conference. Melbourne. 3–5 December, 2008.

F. Haines (2009) 'Regulatory Failures and Regulatory Solutions: A Characteristic Analysis of the Aftermath of Disasters', 34(1) *Law & Social Inquiry* 31.

F. Haines (2011) *The Paradox of Regulation: what regulation can achieve and what it cannot*, Edward Elgar.

F. Haines and D. Gurney (2003) 'The Shadows of the Law: Contemporary Approaches to Regulation and the Problem of Regulatory Conflict', 25(4) *Law & Policy* 353.

F. Haines, A. Sutton and C. Platania-Phung (2008) 'It's All About Risk Isn't it? Science, Politics, Public Opinion and Regulatory Reform', 10(3) *The Flinders Journal of Law Reform* 435.

C. Hepburn (2007) 'Carbon Trading: A Review of the Kyoto Mechanisms', 32 *Annual Review of Environment and Resources* 375.

C. S. Holling (2001) 'Understanding the Complexity of Economic, Ecological and Social Systems', 4(5) *Ecosystems*.

J. Jabecki and M. Machaj (2009) 'The Regulated Meltdown of 2008', 21(2–3) *Critical Review* 301.

R. A. Kagan, N. Gunningham and D. Thornton (2003) 'Explaining Corporate Environmental Performance: How Does Regulation Matter?', 37 *Law and Society Review* 51.

P. Krugman (2008) *The Return of Depression Economics and the Crisis of 2008*, Penguin.

D. McBarnet (2007) 'Corporate social responsibility beyond law, through law, for law: the new corporate accountability', in D. McBarnet, A. Voiculescu and T. Campbell (eds), *The New Corporate Accountability: Corporate Social Responsibility and the Law*, Cambridge University Press, pp. 9–56.

D. McBarnet and C. Whelan (1999) *Creative Accounting and the Cross-eyed Javelin Thrower*, J. Wiley.

D. W. Moffat (ed.) (1982) *Concise Desk Book of Business Finance*, Prentice-Hall Inc.

R. V. O'Neill (1996) 'Perspectives on Economics and Ecology', 6(4) *Ecological Applications* 1031.

OECD (2004) 'Regulatory Performance: Ex Post Evaluation of Regulation Tools and Institutions', available at http://www.oecd.org/dataoecd/32/52/34227774. pdf.

R. Oleschak and U. Springer (2007) 'Measuring host country risk in CDM and JI projects: a composite indicator', 7 *Climate Policy* 470.

C. Parker (1999) 'The Greenhouse Challenge: Trivial Pursuit?', 16(1) *Environmental and Planning Law Journal* 63.

G. Pearce (2007) *High & Dry: John Howard, climate change and the selling of Australia's future*, Viking.

C. Potvini, B. Guay and L. Pedroni (2008) 'Is reducing emissions from deforestation financially feasible? A Panamanian case study', 8 *Climate Policy* 23.

M. Power (1997) *The Audit Society: Rituals of Verification*, Oxford University Press.

Regulation Taskforce (2006) 'Rethinking Regulation: Report of the Taskforce on Reducing Regulatory Burdens on Business', *Report to the Prime Minister and Treasurer*, available at http://www.regulationtaskforce.gov.au/finalreport.

N. Reichman (1998) 'Moving Backstage: Uncovering the Role of Compliance Practices in Shaping Regulatory Policies', R. Baldwin, C. Scott and C. Hood (eds), *A Reader on Regulation*, Oxford University Press, pp. 325–46.

A. K. Shah (1996) 'Creative Accounting in Financial Reporting', 21(1) *Accounting, Organisations and Society* 23.

R. Simnet and M. Nugent (2007) 'Developing an Assurance Standard for Carbon Emissions Disclosures', 17(2) *Australian Accounting Review* 37.

S. Smith and J. Swierzbinski (2007) 'Assessing the performance of the UK Emissions Trading Scheme', 37 *Environment and Resource Economics* 131.

L. Snider (2007) '"This Time We Really Mean It!" Cracking Down on Stock Market Fraud', in H. N. Pontell and G. Geis (eds), *International Handbook of White-Collar and Corporate Crime*, Springer, pp. 627–47.

N. Stern (2008) 'The Economics of Climate Change', 98(2) *American Economic Review: Papers & Proceedings* 1.

J. Stiglitz (2009) 'The Anatomy of a Murder: Who Killed America's Economy?', 21(2–3) *Critical Review* 329.

C. Sunstein (2005) *Laws of Fear: Beyond the Precautionary Principle*, Cambridge University Press.

A. Sutton and F. Haines (2003) 'White Collar and Corporate Crime', in A. Goldsmith, M. Israel and K. Daly (eds), *Crime and Justice: An Australian Textbook in Criminology*, Lawbook Company, pp. 141–58.

G. Teubner (1998) 'Juridification: Concepts, Aspects, Limits, Solutions', in R. Baldwin, C. Scott and C. Hood (eds), *A Reader on Regulation*, Oxford University Press.

D. Thornton, R. A. Kagan and N. Gunningham (2008) 'Compliance costs, regulation, and environmental performance: Controlling truck emissions in the US', 2 *Regulation and Governance* 275.

P. Utting (2002) 'The Global Compact and Civil Society', 12(5) *Development in Practice* 644.

R. B. Vermeesch and K. E. Lindgren (1995) *Business Law of Australia*, Butterworths.

M. S. Weatherford and B. Sergeyev (2000) 'Thinking about Economic Interests: Class and recession in the New Deal', 22(4) *Political Behaviour* 311.

H. K. Wittman and C. Caron (2009) 'Carbon Offsets and Inequality: Social Costs and Co-Benefits in Guatemala and Sri Lanka', 22(8) *Society & Natural Resources* 710.

Legislation

Corporations Act 2001 (Cth)

Chapter 13

From Fiduciary States to Joint Trusteeship of the Atmosphere: The Right to a Healthy Environment through a Fiduciary Prism*

Evan Fox-Decent

Introduction

The present rate of humanity's economic production and consumption is unsustainable. A report by the United Nation's Department of Economic and Social Affairs estimates that humanity's ecological footprint (a comparison of humanity's ecological impacts with natural resources available to support key ecosystems) outstrips the earth's capacity to regenerate the biosphere by 30 per cent.[1] Five planets would be needed if the world's population adopted the average carbon-intensive lifestyle of North Americans.[2] By any reasonable measure, the world faces a grave environmental crisis.

The aim of this chapter is to show how a fiduciary theory of public authority can yield a human right to a healthy environment. By 'healthy environment' I mean a sustainable environment capable of supporting independent lives individuals have reason to value.[3] By 'sustainable' I mean an environment that can persist over generations without imposing unreasonable burdens on vulnerable persons, such as the burden of abandoning flooded homeland that results from global warming. By 'independent lives' I mean autonomous lives individuals can pursue within a legal regime of secure and equal freedom.

As we shall see in the next part of this chapter, the conventional understanding of human rights is ill-suited to address environmental concerns. I set out an alternative fiduciary theory under which human rights are conceived as norms arising from a fiduciary relationship that exists between states (or state-like actors)

* I am greatly indebted to Stefan Szpajda for invaluable research assistance and comments on an earlier draft. I owe a similar debt for comments to Evan Criddle and Jaye Ellis.

1 Department of Economic and Social Affairs (2010), p. 5: citing World Wildlife Fund (2008).

2 Department of Economic and Social Affairs (2010).

3 The definition proposed borrows from Amartya Sen's 'capabilities approach' to freedom; that is, freedom consists in substantive capabilities that enable individuals to pursue lives they have reason to value. See Sen (1999).

and the citizens and non-citizens subject to their power. In the following part, I explain how the fiduciary theory can supply a right to a healthy environment while overcoming the difficulties encountered under the conventional view.

Perhaps most significantly, the fiduciary theory explains why every state owes a cosmopolitan duty to extra-territorial non-citizens, a duty that is correlative to those persons' right to a healthy environment. States owe this duty because they owe human-rights-based duties to anyone, anywhere, whose legally cognizable interests are implicated by the possession or exercise of state sovereign power. In this case, the possession and exercise of sovereign powers over the emission of greenhouse gases engages the interest in a healthy environment held by every person in the world. So every state that emits greenhouse gases (for practical purposes, every state in the world) is under a cosmopolitan *erga omnes* obligation to respect the right of the world's inhabitants to a healthy environment. The content of the obligation will vary tremendously depending on contextual factors that might include the amount of greenhouse gas being presently emitted from the state, the state's historical record as a contributor to global warming, the state's level of economic development, and the state's population. But the critical point is that although determining the content of each state's obligation may be difficult, the fiduciary theory shows that all states are under an obligation to do their part to guarantee a healthy environment.

To take up the idea of an atmospheric trust announced in the title of this book, I focus on the international issue of global warming. We shall see that an implication of the fiduciary theory is that the world's states are joint trustees of the atmosphere, while the inhabitants of earth are their beneficiaries. I leave to another day an extension of the argument to regional issues such as acid rain, polluted rivers, and deforestation. Also left to another day is the relationship between the anthropocentric view of environmental rights defended here and eco- or biocentric approaches that regard the world's ecosystem as an intrinsic good.[4] Finally, the argument presented is mainly one of principle and as such foregoes detailed discussion of implementation and enforcement issues. But importantly, the argument is one of *legal* principle because the fiduciary concept on which it is based is a legal concept. As such, the argument is intended to resonate with judges and policy-makers entrusted to determine legal matters.

A Fiduciary Theory of Human Rights

Human rights are conventionally understood as timeless, natural, absolute rights that all humans possess simply by virtue of their shared humanity.[5] A. John Simmons provides an especially clear account of the conventional view:

---

4 Hayward (2007); Schroeder (1999).
5 See Donnelly (1998); Hart (1955); Simmons (2001).

[H]uman rights are possessed by all human beings simply in virtue of their humanity ... . Human rights are those natural rights that are innate and that cannot be lost (that is, that cannot be given away, forfeited, or taken away). Human rights, then, will have the properties of universality, independence (from social and legal recognition), naturalness, inalienability, non-forfeitability, and imprescriptability. Only so understood will an account of human rights capture the central idea of rights that can always be claimed by any human being.[6]

Simmons admits that this view of human rights is inconsistent with a wide range of rights recognized by international human rights law, since these rights require or presuppose legal institutions. Consider, for example, rights enshrined in the *International Covenant on Civil and Political Rights* to due process (arts. 9, 14, 15), unionization (art. 22), political participation (art. 25), and equality under the law (art. 26).[7] These rights cannot exist in a pre-institutional state of nature or in traditional societies that lack the institutions necessary for their realization. The point is not merely that institutions are necessary to enforce such rights, but rather that such rights cannot exist at all without institutions in place, since the rights themselves are designed to protect their bearers from threats posed by the institutions they regulate. Likewise, the modern idea of a right to a healthy environment achieves salience only in the context of developed societies with economic and political institutions that pose a threat to the environment.

Yet a further tension between the conventional view of human rights and global environmental protection is that human rights under the conventional view are considered absolute whereas global environmental concerns require tradeoffs, typically between development or industry on the one hand and reduced emissions on the other. If the right to a healthy environment is absolute, then in principle it cannot be infringed at all for the sake of socio-economic considerations. One could attempt to sidestep this implication by insisting that the standard set by 'healthy' is context-sensitive and tolerant of some environmental degradation. Yet even this concession may place the right to a healthy environment on too high a plane. In a severe crisis that posed a difficult choice between damaging the environment beyond 'healthy' limits and subjecting a population to imminent famine, other things being equal there is reason to sacrifice the environment. Indeed, there is an emerging 'right to development' that stands in tension with the (equally nascent) right to a healthy environment even in the absence of extraordinary circumstances.[8]

The conventional view of human rights, however, is not the only view on offer. In previous writings, Professor Evan Criddle and I have developed an alternative

---

6 Simmons (2001), p. 3, citing, *inter alia*, Gewirth (1986) (other citations omitted) (emphasis in original).

7 Others who adopt the conventional view also deny that the institution-dependent rights of international human rights law are properly characterized as human rights: see Cranston (1983); Finnis (1980), pp. 214–15.

8 See Atapattu (2002), p. 117.

fiduciary and relational account of human rights.[9] Specifically, we argue that human rights emanate from a fiduciary relationship between public institutions and persons subject to public powers. As fiduciaries, public institutions bear legal obligations to safeguard their subjects against domination-subjection to the threat of arbitrary state or private interference. The master-slave relationship is the classic example of domination. The master dominates the slave because he can treat the slave arbitrarily at any time, and this domination remains intact whether or not he in fact mistreats the slave (beyond keeping her in slavery). It is the *capacity* for arbitrary treatment that marks relations of domination, not mistreatment itself. Of course, public institutions under the fiduciary theory must also forgo mistreating or abusing the people subject to their power. They do so by satisfying the Kantian principle of non-instrumentalization. This principle requires the duty-bearer to treat persons always as ends-in-themselves and never as mere means. On the fiduciary theory of the state, human rights come into focus as institutionally grounded legal norms that arise from a state's assumption of sovereign powers and which reflect the demands of non-domination and non-instrumentalization.

We have argued further that one way to see the legal character of human rights is from the perspective of Immanuel Kant's theory of law. According to Kant, all persons have an innate right to as much freedom as can be reconciled with the freedom of everyone else.[10] The purpose of law on this account is to honour the dignity of all persons by enshrining legal rights within a regime of secure and equal freedom, such that no person can unilaterally impose terms of interaction on others. Within Kant's regime of secure and equal freedom, fiduciary obligations ensure that those who possess and exercise unilateral administrative powers over others' legal or practical interests are precluded from denying others' innate right to equal freedom through domination or instrumentalization.

Kant's conception of legal right supplies a sound philosophical justification for the republican conception of public officials and institutions as fiduciaries for their people.[11] Like other fiduciaries, the state's legislative, judicial, and executive branches all assume discretionary powers that are institutional, purposive, and other-regarding. Private parties are not legally entitled to exercise the state's powers and thus are particularly vulnerable to public authority, despite their ability within democracies to participate in democratic processes. The state's monopolization of public powers over its people can be understood therefore as a fiduciary

---

9   See Fox-Decent and Criddle (2010); Fox-Decent (2005); Criddle and Evan Fox-Decent (2009); Fox-Decent (2008). The summary in the text of the fiduciary theory of human rights is taken largely from a working paper: Criddle and Evan Fox-Decent, 'Human Rights, Emergencies and the Rule of Law', 34 *Human Rights Quarterly* (forthcoming, 2012).

10   Kant (1981), at 39–42.

11   See, for example, *Pa. Const. of 1776*, art. IV. ('[A]ll power being ... derived from the people: therefore all officers of government, whether legislative or executive, are their trustees and servants, and at all times accountable to them'); Locke (1690), at 87 (describing legislative power as 'a fiduciary power to act for certain ends').

relationship mediated by legality – rather than a relationship of domination or instrumentalization – only if a principle of legality prevents public institutions from exploiting their position to set unilaterally the terms of interaction with their people. The fiduciary principle is a principle of legality that does just this: it authorizes the state to exercise public powers for and on behalf of its people, but subject to strict legal norms that safeguard subjects' inherent dignity as free and equal co-beneficiaries of state action.

Human rights are one such kind of norm that governs the legitimate possession and exercise of state power. The fiduciary principle dictates that the state must forbear from adopting laws, policies, or practices that deliberately victimize or arbitrarily threaten persons subject to its powers. But the state bears a duty to adopt and enforce laws protecting persons from private domination and instrumentalization, such as may arise when non-state actors unjustifiably degrade the environment. Human rights are constitutive of state sovereignty on this account because they supply a normative framework within which the state can establish a secure order of equal freedom, an order marked by the absence of private or public domination and instrumentalization. In this sense, all public powers are constrained and constituted by the state's fiduciary duty to respect, protect, and fulfil human rights.

On the fiduciary theory, norms qualify as human rights if they satisfy certain substantive criteria. First, under a principle of *integrity*, human rights must have as their object the good of the people rather than the good of the state's institutions or officials. Second, a principle of *formal moral equality* requires fairness or even-handed treatment of persons subject to state power; human rights must regard individuals as equal co-beneficiaries of fiduciary states. Third, a principle of *solicitude* dictates that human rights must reflect proper solicitude toward the legitimate interests of a state's subjects. Collectively, these criteria support the fiduciary theory's thick substantive account of human rights as legal rights derived from the fiduciary obligations all states bear as sovereign actors, irrespective of whether they have consented to particular human rights conventions.

Distinguishing Jus Cogens *Norms from Other Human Rights*

The fiduciary theory also establishes a principled framework for distinguishing peremptory or *jus cogens* norms from other human rights: human rights qualify as peremptory norms if a state's compliance with these norms is always necessary to accomplish the state's fiduciary mission of guaranteeing secure and equal freedom. The prohibitions against slavery and racial discrimination qualify as *jus cogens* on this account because states cannot violate these prohibitions without undermining their own claim to treat all persons as equal co-beneficiaries of state action. Other state practices that exploit individuals as mere instruments of state policy are likewise inconsistent with states' basic fiduciary obligation to guarantee individuals' secure and equal freedom. For this reason, international norms that prohibit grave offences such as genocide, crimes against humanity, summary executions, forced disappearances, prolonged arbitrary detention, torture, and cruel, inhuman, and degrading treatment all qualify as *jus cogens*. States cannot violate these norms under any circumstances without forfeiting their claim to possess sovereign authority, because such practices always instrumentalize their victims and, as such, are never consistent with a regime of secure and equal freedom.

Of course, in international law not all human rights are peremptory. Some international norms such as the freedoms of expression, movement, and peaceable assembly are widely accepted as human rights, yet do not qualify as *jus cogens* because the fiduciary principle permits – and may even require – the state to restrict their exercise in certain contexts. A state's fiduciary duty to guarantee secure and equal freedom for its people arguably entitles the state to enact laws that require manufacturers to place warnings on their products that notify the public of possible health risks and other dangers. Although such laws restrict free expression, they do not violate human rights on the fiduciary theory because they are necessary and proportional means to guarantee the public's security from unilaterally imposed risks.

Although the schedules of peremptory and non-peremptory norms that appear in leading human rights conventions are generally consistent with the fiduciary theory, human rights do not derive their fundamental normative authority from state consent. Rather, the authority of human rights derives from their role as constitutive norms emanating from the state's institutional assumption of sovereign powers. States must honour human rights as a function of the fiduciary obligations that accompany their exercise of sovereignty – even if they have yet to ratify relevant human rights conventions. This is not to say that state consent and ratification of human rights conventions are irrelevant. Treaties signal the international community's best provisional estimate of the determinate content of particular human rights norms and the legal consequences of their breach. Consent also renders a state liable under the relevant treaty for a breach of the treaty's provisions, as well as generating liability for the remedial consequences stipulated in the treaty. But the basic normative authority of human rights remains traceable to the protection they afford against the threats of domination and instrumentalization engendered within the fiduciary relationship between public institutions and the persons subject to their powers.

The Right to a Healthy Environment under the Fiduciary Theory

Above, I suggested that the conventional view of human rights is hard pressed to accommodate the right to a healthy environment. Under the fiduciary theory, the right can readily be seen to satisfy its substantive qualifying criteria of integrity, formal moral equality, and solicitude. The right is intended to benefit the people subject to sovereign power and not state actors. The right regards each person as a morally equal right-holder. And the right expresses solicitude toward the legitimate interest everyone possesses in the enjoyment of a healthy environment. Below, we will see as well that the right emerges as a bulwark against threats of domination and instrumentalization posed by modern public institutions, which include private property and markets as well as the state's own institutions.[12]

There is, however, a more general difficulty that besets any attempt to justify the existence of a human right to a healthy environment. It is not clear that the purported right is ultimately distinct from rights to life, health, an adequate standard of living, and possibly other substantive human rights.[13] If we assume an anthropocentric perspective that eschews reliance on ecocentric views, as this chapter does, what is left for the right to a healthy environment to protect if human rights law already safeguards the major interests that depend on a healthy environment?

The principal answer supplied by the fiduciary theory is that human rights, properly understood, protect the fundamental means necessary for the enjoyment of an independent life (that is, a life one has reason to value, free of instrumentalization and domination) as well as physical life, health, and access to basic necessities. There is a good sense in which legal rights – equality before the law, due process, the presumption of innocence, and so on – are just means to enjoy an independent life. The same may be said for socio-economic rights, such as rights to housing and education, since the realization of these rights by the state ensures that individuals are not thrown on the mercy of private parties and thereby subject to domination (a domination that would persist even if the relevant private parties were beneficent and provided the needy with more than the state would otherwise provide).

I assume that for at least some people a healthy environment, like legal and socio-economic rights, is a fundamental means to the enjoyment of an independent life. This is especially true of peoples living in coastal regions threatened by rising water levels, others threatened by ever-more-frequent and violent storms, and yet others whose agricultural or hunter-gatherer ways of life are threatened by diminishing rains and other climatic changes.[14] As countless conflicts attest, many peoples have a deep and non-fungible attachment to what they perceive as *their* land and territory.[15] Implicit in this attachment is a commitment to the health of the relevant territory's environment.

If vulnerable peoples do not enjoy a right to a healthy environment, they are left to negotiate as best they can with those who are indifferent to the effects of

---

12   For discussion of rights as guarantees against 'standard threats', see Shue (1996); Pogge (2008) (arguing that human rights are 'moral claims on the organization of one's society').

13   Many authors ground an environmental human right in existing rights, such as rights to privacy, life, and health. See, for example, Atapattu (2002); Francioni (2010); Posner (2007); Nickel (1993).

14   See Frey et al. (2010); Arnell et al. (2002), p. 414: authors assess effects of climate change on 'natural vegetation, water resources, coastal flood risk, food supply and human health'; Burns (2006).

15   De Schutter (2010); McNamara and Gibson (2009); Kolmannskog and Myrstad (2009), p. 313 (addressing shortcomings in the refugee framework's capacity to handle persons displaced by environmental degradation).

environmental degradation. In practice, litigation through tort-based mechanisms has proven ineffective because complainants face virtually insurmountable problems with respect to causation and standing.[16] They also face a vexing procedural problem: degradation does not ground a *per se* claim, and there can be no public interest litigation in the absence of an individual injury.[17] The vulnerable are left to beg more powerful parties for compensation that is given as charity, if it is given at all, rather than awarded as a matter of right. In the absence of an effective right to a healthy environment, the powerful dominate those who are most vulnerable to climate change. If the less fortunate are wronged through an unjustifiable infringement of their right to a healthy environment so that others may enjoy greater prosperity, they are also treated as mere means of those others. They suffer instrumentalization as well as domination. Under the fiduciary theory, the right to a healthy environment emerges to protect them from these evils so as to enable them to live independent lives they have reason to value.

Conceptualizing the right to a healthy environment as a human right within the fiduciary model is attractive for other reasons, too. First, global environmental degradation is frequently justified as a consequence, however regrettable, of states exercising legitimate sovereign power in their peoples' best interests. On the fiduciary theory, sovereign power is constituted and constrained by human rights. So if the right to a healthy environment is a human right, then under the fiduciary theory it is a right that limits intrinsically the claims that can be made on grounds of sovereignty. In the result, recognizing the right to a healthy environment as a human right establishes it as a criterion for testing the validity of state action. The defence of sovereign immunity is rendered otiose because the fiduciary framework places squarely at issue the question of whether or not the impugned action is indeed a sovereign act. If the action falls below the standard set by the right to a healthy environment, then the impugned act is not a sovereign act that can benefit from the doctrine of sovereign immunity.

Secondly, the fiduciary model's reliance on Kant's legal theory allows it to trade fruitfully on the Kantian distinction between harms and wrongs. One of the challenges facing proponents of the right to a healthy environment is that the harm caused by climate change is gradual and non-linear in progression, severely affecting at present a relatively small portion of the earth's population.[18] Liability under Kant's theory, however, like liability generally as a matter of common law, tracks wrongs rather than harms (damages track harms, but only once liability is established). Some wrongs are not harms and some harms are not wrongs. If I open a tomato stand to compete with yours, I harm but do not wrong you. If I

---

16   See Posner and Sunstein (2008).

17   *Kyrtatos v Greece* (ECtHR Rep (2003-VI) 257) highlights these limitations. The illegal draining of a wetland was found not to injure the plaintiff, a nearby resident and thus the illegal action was not deemed a violation of Art. 8 of the *European Convention for the Protection of Human Rights and Fundamental Freedoms*; see Francioni (2010).

18   See Lind (1995); Posner and Sunstein (2008).

invade your privacy by reading your mail without permission, but the mail I read turns out to be meaningless junk, I have committed a wrong without harming. With this distinction in mind, we can see that if a state emits more greenhouse gases than it is entitled to emit, the harm to some may be vanishingly slight in any given year, but it is still a wrong because it violates the right to a healthy environment held by every person on the earth. The wrongfulness of the state's conduct is in no way excused by the fact that the harm caused appears slight, much less because many other states participate in similar wrongful conduct. Situating global environmental concerns within the fiduciary theory's Kantian framework shows that the 'no harm, no foul' argument sometimes essayed against environmentalists is just a category error.

The Status of the Right to a Healthy Environment

I suggested above that to treat the right to a healthy environment as an absolute or peremptory norm could place it on too high a plane because there may be circumstances in which intentional and significant environmental damage is warranted. Yet arguably there are some actions against which the right poses an absolute bar. These might include gratuitous or negligent environmental destruction that victimizes innocent parties by treating them as mere means, such as spilling petroleum in a heavily fished coastal area as a result of careless shipping or drilling operations. More dramatically still, an argument could be made that the use of nuclear weapons is enjoined absolutely given the environmental devastation their use implies. We might say, then, that there is a peremptory norm that bars gratuitous, negligent or severe environmental destruction that victimizes innocent parties, and that this norm applies with particular force when the destruction at issue has global consequences.[19] It bears emphasizing that while substantive rights to life, health, and other entitlements may also prohibit such action, only the right to a healthy environment protects comprehensively the right-bearer's interest in a sustainable environment capable of supporting an independent life.

In general, however, the right to a healthy environment, like most human rights, is non-peremptory, since circumstances may arise in which restrictions on the right are necessary to provide for individuals' secure and equal freedom. Suppose, for example, that a traditional community acquires half its sustenance through agriculture and the other half through fishing, but that a drought strikes one year leaving the community without crops to harvest. In the short term, to ward off famine, it may be justifiable for the community to take a catch from local fishing areas used by themselves and others that is unsustainable in the long term. The relevant legal standard would be proportionality, which is the standard

---

19 See Bridgeman (2003); Berat (1993); Uhlmann (1998), p. 135: although Uhlmann does not find a general peremptory norm, she writes that '[t]he prohibition of willful serious damage to the environment during armed conflicts is a *jus cogens* norm'.

usually applied when non-peremptory rights are infringed.[20] Inquiry under the proportionality standard would be guided by considerations of whether the right-infringing measure is rationally connected to the goal pursued, whether it minimally (or close to minimally) impairs the right, and whether on balance the deleterious consequences of the infringement outweigh the purported benefit flowing from it.

A Response to Global Warming

Perhaps the most difficult and pressing case is global warming caused by the emission of greenhouse gases produced mainly by the burning of fossil fuels. The case is pressing because reliable evidence indicates that if present trends continue, within the century sea levels are likely to rise precipitously, flooding low-lying coastal lands.[21] Moreover, droughts in already arid regions such as sub-Saharan Africa are likely to intensify, and severe weather is likely to become more common.[22] The case is difficult because reducing greenhouse gas emissions significantly will require global or at least massive multilateral cooperation. No normative theory of human rights can predict whether cooperative efforts will succeed, much less bring them about. But the fiduciary theory can supply useful arguments to guide public debate, policy-makers and possibly judges.

One advantage of the fiduciary theory is that it provides an immediate reply to the argument that states have duties to their citizens alone, or, more narrowly still, to the persons presently within their territory.[23] The predominant attitude of Western liberal democracies is that beneficence calls on them to provide charity to less-developed regions, but they are under no legal or binding moral obligation to do so. Under the fiduciary theory, public authorities owe relevant duties to anyone affected by their possession or exercise of public power, regardless of where the person happens to be and regardless of their civil or political status. So, for example, foreign terror suspects detained extra-territorially cannot be deprived of due process on grounds of their foreign nationality or on grounds that they are detained

---

20   In 2005 the Council of Europe adopted the Manual on Human Rights and the Environment, which summarized the principles of adjudicating environmental claims. Included among these principles is the right of national governments to balance individual and environmental rights. See Council of Europe Committee of Experts for the Development of Human Rights (2005–06); Francioni (2010).

21   See Burns (2006); Woodward, Hales and Weinstein (1998); Mimura (1999).

22   See Mirza (2003); Chhibber and Laajaj (2008).

23   See Horton (2007), pp. 11–12:

if I cease to be a British citizen … [m]y relationship to those who were my fellow citizens and what was once my government change. That government is no longer my government. … It will no longer be acting in my name. There are claims that I can no longer justifiably make against it, and demands that it can no longer legitimately make of me. Similarly, people who were formerly my fellow citizens are such no more; and various kinds of concern or lack of it towards their well-being, and that of my polity, take on a different ethical colouring.

extra-territorially, since they remain subject to the detaining state's power.[24] But for the constraint of due process and other safeguards, detainees would be subject to the arbitrary and unilateral will of their captors. They would be subject to mere coercive power rather than legal authority. On the fiduciary theory, public power is always a fiduciary power laced with duty so as to exist as authority.

With respect to global warming, in the normal case states have control over greenhouse gas emissions within their territory. They have this control because they claim a monopoly on the legitimate use of force within their territory, and typically have sufficient *de facto* power to enforce their claim. I assume that the preponderance of states, were they willing, could pass and enforce laws that would limit emissions, notwithstanding that such measures would likely impose costs on local industry and consumers. For convenience, let us call this power to control greenhouse gas emissions the state's environmental power.

Because no state at present can prevent greenhouse gases produced within its territory from spilling out into the global atmosphere, those gases erode a boundaryless atmosphere shared and relied upon by every inhabitant of earth, with potentially dire long-term consequences. For these reasons, every inhabitant of earth is subject to every state's possession and exercise of its environmental power. As a consequence, under the fiduciary theory, every state's environmental power is constrained by a duty to exercise that power in a manner consistent with the cosmopolitan right of every person to a healthy environment. Put slightly differently, the state's environmental power combines with spillover effects of greenhouse gas emissions to give rise to an extra-territorial duty owed to the world's inhabitants. The general content of this *erga omnes* duty is respect for the right to a healthy environment.

Implicit in this understanding of the cosmopolitan reach of the fiduciary theory is a commitment to the idea that states are essentially joint trustees of the earth's atmosphere.[25] The beneficiaries of the atmospheric trust are the earth's inhabitants. A signal advantage of the joint trusteeship model is that it provides a legal framework from within which to ground every state's duty to respect the right to a healthy environment. Taken seriously, and consistent with the fiduciary theory of human rights, the joint trusteeship model implies that states have more than a moral or charitable obligation to safeguard the world's environment. They have a legal duty to do so because the joint trust or fiduciary relationship is a legal relationship.

Judicial Review

It is well beyond the scope of this paper to suggest how the burden of emission reductions might be equitably shared. In practice these determinations will almost certainly have to be the result of international treaties that establish emissions standards or quotas. Assuming that the very difficult issue of emissions quotas has been settled through a treaty process, the fiduciary theory commends giving

---

24 See Fox-Decent (2005); Criddle and Fox-Decent (2009).
25 See Wood (2007).

standing before national and international tribunals to persons threatened by a state that refuses to do its fair share. Here the fiduciary theory shows its teeth. By establishing that the right to a healthy environment is a legal right arising from every state's joint trusteeship of the atmosphere, and under the assumption that an emissions treaty determines the specific content of each state's obligation, the fiduciary theory provides a juridical basis for the standing of non-state actors against wrongdoing states. The standing of non-state actors flows directly from their status as co-beneficiaries of the infringing state's joint trusteeship of the atmosphere. As is usually the case in international human rights law, complainants would be expected to exhaust national remedies before seeking review at the international level.[26]

Some might doubt that judges are institutionally competent to hear such cases, as they will invariably involve review of policy decisions with complex facts. Yet in principle this should be no bar to judicial review because judges have long adjudicated similar kinds of claims in other contexts. The case of riparian water rights is instructive as it provides a close analogy to the problem of destructive atmospheric spillover effects. Consider the finding of Chief Justice Latham in *H Jones & Co Pty Ltd v Kingborough Corporation*:[27]

> A right to use the water of a stream (and all the water thereof if that can lawfully be done) is illusory if the flow of the stream can be diminished at will by another person. A positive right in a landowner to the use of the water of a stream prima facie involves a right to prevent such interference with the stream as would prevent him from using the water.

The present claims of low-lying Pacific islanders threatened by rising sea levels are identical in kind to those made by downstream riparians. The islanders' rights to their lands are 'illusory' if the unilateral actions of other states are permitted to cause global warming and rising sea levels. The facts in the international case are no doubt much more complicated, but the basic principle is the same, and at this juncture we are assuming that a particular state has failed to meet a treaty-based target for emissions reduction.

Much more controversial is the idea that an aggrieved party could sue a state for failing to respect the right to a healthy environment in the absence of treaty-based emissions quotas. The idea is controversial in part because there is little agreement on what would constitute, for each state, a fair share of the burden of

---

26 See European Court of Human Rights (2006), p. 2: '[t]he exhaustion rule may be described as one that is golden rather than iron: the Commission and the Court have frequently underlined the need to apply the rule with some degree of flexibility and without excessive formalism, given the context of protecting human rights'; Udombana (2003), p. 2: the requirement that domestic remedies are exhausted is 'commonly applied in practice, its function is procedural – it is a question of admissibility and not of substance'.

27 [1950] HCA 11 at para. 18; (1950) 82 CLR 282 (16 May 1950)

carbon emissions reduction. But it is also controversial because, in the absence of a treaty or at least significant multilateral cooperation, a judicial ruling that prompted unilateral state action could reduce the incentives other states have to do their fair share (though presumably this would make a difference only if the ruling were truly dramatic, and a significant group of other states were particularly bloody-minded, and if the ruling's symbolic and precedential value for a healthy environment was limited). A further question is whether developed states have to compensate less-developed states for having produced excessive carbon emissions for so long. Yet even allowing for these and other complications, in some cases liability may be possible to establish.

If we assume that states must do something significant to curb global warming in order to respect the right to a healthy environment, and a particular state has done nothing meaningful in this regard, the *prima facie* case for liability is made. As is the case for common law riparians, liability can attach even if the precise standard is not set down in positive law. Such a finding might be justified if a state refused to enter multilateral negotiations toward an emissions treaty in which others were participating, or if the state's participation in such talks lacked good faith. In these circumstances, a declaratory or even injunctive remedy might be appropriate. Courts can review the issue of good faith treaty-making without dictating to the executive the substantive position it ought to adopt in negotiations of this kind.[28]

Of course, an appropriate compensatory remedy in cases where a state is flagrantly shirking its environmental duty is even more difficult to determine because the amount of damage caused by global warming is difficult to calculate. But medical malpractice cases are strewn with analogously difficult facts in which courts must determine the extent to which a negligently performed medical procedure caused an adverse outcome in the presence of a pre-existing condition.[29] Complicating facts such as these may reduce the damage award, but they do not extinguish liability. Moreover, if ordinary damages principles from private law were applied, a plaintiff representing the relevant affected class ought to be able to claim full damages from a single wrongdoing state under the doctrine of joint and several liability. It would then fall to the state to join other wrongdoers to the suit with hopes of distributing the damage award among them.

Conclusion

I have argued that the fiduciary theory of human rights provides a compelling basis on which to ground a right to a healthy environment. The theory explains why

---

28   See *Haida Nation v British Columbia (Minister of Forests)*, 2004 SCC 73 at para. 25, [2004] 3 S.C.R. 511 (requiring the Crown to negotiate in good faith with Aboriginal representatives).

29   See *Snell v Farrell*, [1990] 2 S.C.R. 311; *Athey v Leonati*, [1996] 3 S.C.R. 458.

extra-territorial non-citizens have legal claims against foreign states that violate this right. Perhaps the most interesting implication of the theory, however, is that states stand in the position of joint trustees of the earth's atmosphere. This is a somewhat surprising consequence, as one might think that the responsibilities of fiduciary states end at their borders. They do not end there, however, because every person in the world is subject to spillover effects under the control of every state's environmental power. And from this it follows that states are joint trustees of humanity's most important patrimony.

References

N. W. Arnell et al. (2002) 'The Consequences of CO_2 Stabilisation for the Impacts of Climate Change', 53 *Climate Change* 413.

Sumudu Atapattu (2002) 'The Right to a Healthy Life or the Right to Die Polluted?: The Emergence of a Human Right to a Healthy Environment Under International Law', 16 *Tulane Envir. L. J.* 65 at 117.

Ernest Barker (ed.) (1942) *Social Contract: Essays by Locke, Hume, and Rousseau,* OUP.

Lynn Berat (1993) 'Defending the Right to a Healthy Environment: Toward a Crime of Genocide in International Law', 11 *B.U. Int'l. L. J.* 327.

Natalie L. Bridgeman (2003) 'Human Rights Litigation under the ATCA as a Proxy for Environmental Claim', 6 *Yale Human Rights & Development L. J.* 1.

William C. G. Burns (2006) 'Potential Implications of Climate Change for the Coastal Resources of Pacific Island Developing Countries and Potential Legal and Policy Responses', 8 *Harvard Asia Pacific Review* 4.

Ajay Chhibber and Rachid Laajaj (2008) 'Disasters, Climate Change and Economic Development in Sub-Saharan Africa: Lessons and Directions', 17 *Journal of African Economies* ii7.

Council of Europe Committee of Experts for the Development of Human Rights, 'Final Activity Report on Human Rights and the Environment', 10 Nov. 2005, DH–DEV(2005)06 rev.

Maurice Cranston (1983) 'Are There Any Human Rights?', 112 *Daedalus* 1.

Evan J. Criddle and Evan Fox-Decent (forthcoming, 2012) 'Human Rights, Emergencies and the Rule of Law', 34 *Human Rights Quarterly.*

Olivier De Schutter (2010) 'The Emerging Human Right to Land', 12 *International Community Law Review* 303.

Department of Economic and Social Affairs (2010) 'Trends in Sustainable Development: Towards Sustainable Consumption and Production', Division for Sustainable Development.

Jack Donnelly (1998) *International Human Rights*, Westview Press.

European Court of Human Rights (2006) 'Key Case-law Issues: Exhaustion of Domestic Remedies', available at www.echr.coe.int.

Evan J. Criddle and Evan Fox-Decent (2009) 'A Fiduciary Theory of Jus Cogens', 34 *Yale J. Int'l. L.* 331.

John Finnis (1980) *Natural Law and Natural Rights*, Oxford University Press.

Evan Fox-Decent and Evan J. Criddle (2010) 'The Fiduciary Constitution of Human Rights', 15 *Legal Theory* 301.

Evan Fox-Decent (2005) 'The Fiduciary Nature of State Legal Authority', 31 *Queen's L.J.* 259.

Evan Fox-Decent (2008) 'Is the Rule of Law Really Indifferent to Human Rights?', 27 *Law & Phil.* 533.

Francesco Francioni (2010) 'International Human Rights in an Environmental Horizon', 21 *European Journal of International Law* 41.

Ashley E. Frey et al. (2010) 'Potential Impact of Climate Change on Hurricane Flooding Inundation, Population Affected and Property Damages in Corpus Christi', 46 *J.A.W.R.A.* 1049.

Alan Gewirth (1986) 'The Epistemology of Human Rights', in E. F. Paul, J. Paul and F. D. Miller, Jr (eds), *Human Rights* 1, 3.

Alan Gewirth (1982) *Human Rights: Essays on Justification and Applications*, University of Chicago Press.

H. L. A. Hart (1955) 'Are There Any Natural Rights?', 64 *Phil. Rev.* 175.

Tim Hayward (2007) 'Human Rights Versus Emissions Rights: Climate Justice and the Equitable Distribution of Ecological Space', 21 *Ethics and International Affairs* 431.

John Horton (2007) 'In Defence of Associative Political Obligations: Part Two', 55 *Political Studies* 1.

Immanuel Kant (1981) *Grounding For the Metaphysics of Morals*, James W. Ellington (trans.), Hackett Publishing.

Vikram Kolmannskog and Finn Myrstad (2009) 'Environmental Displacement in European Asylum Law', 11 *European Journal of Migration and Law* 313.

Robert C. Lind (1995) 'Intergenerational Equity, Discounting, and the Role of Cost-Benefit Analysis in Evaluating Global Climate Policy', 23 *Energy Policy* 379.

John Locke (1690) 'An Essay Concerning the True, Original Extent and End of Civil Government', in Sir Ernest Barker (ed.) (1942) *Social Contract: Essays by Locke, Hume, and Rousseau*, Wiley.

Karen Elizabeth McNamara and Chris Gibson (2009) '"We do not want to leave our land": Pacific Ambassadors at the United Nations Resist the Category of "Climate Refugees"', 40 *Geoforum* 475.

Nobuo Mimura (1999) 'Vulnerability of Island Countries in the South Pacific to Sea Level Rise and Climate Change', 12 *Climate Research* 137.

M. Monirul Qader Mirza (2003) 'Climate Change and Extreme Weather Events: Can Developing Countries Adapt?', 3 *Climate Policy* 233.

James W. Nickel (1993) 'The Human Right to a Safe Environment: Philosophical Perspectives on its Scope and Justification', 18 *Yale J. Int'l L.* 281.

Thomas Pogge (2008) *World Poverty and Human Rights: Cosmopolitan Responsibilities and Reforms*, 2nd edn, Polity.

Eric A. Posner and Cass Sunstein (2008) 'Justice and Climate Change', Discussion Paper 08–04, Harvard University Project on International Climate Agreements.
Eric A. Posner (2007) 'Climate Change and International Human Rights Litigation: A Critical Appraisal', 155 *U. Pa. L. Rev.* 1925.
Christopher H. Schroeder (1999) 'Third Way Environmentalism', 48 *U. Kan. L. Rev.* 801.
Amartya Sen (1999) *Development as Freedom*, Knopf.
Henry Shue (1996) *Basic Rights: Subsistence, Affluence, and US Foreign Policy*, 2nd edn, Princeton University Press.
A. John Simmons (2001) 'Human Rights and World Citizenship', *Justification and Legitimacy: Essays on Rights and Obligations*, Cambridge University Press.
Nsongurua J. Udombana (2003) 'So Far, So Fair: The Local Remedies Rule in the Jurisprudence of the African Commission on Human and Peoples' Rights', 97 *American Journal of International Law* 1.
Eva M. Kornicker Uhlmann (1998) 'State Community Interests, *Jus Cogens* and Protection of the Global Environment: Developing Criteria for Peremptory Norms', 11 *Geo. Int'l. Envtl. L. Rev.* 101.
Mary Christina Wood (2007) 'Nature's Trust: A Legal, Political and Moral Frame for Global Warming', 34 *B.C. Envt. Aff. L. Rev.* 577.
Alistair Woodward, Simon Hales and Philip Weinstein (1998) 'Climate Change and Human Health in the Asia Pacific Region: Who Will be Most Vulnerable?', 11 *Climate Research* 31.
World Wildlife Fund (2008) *Living Planet Report 2008*, Gland, Switzerland.

Cases

Athey v Leonati [1996] 3 S.C.R. 458
H Jones & Co Pty Ltd v Kingborough Corporation [1950] HCA 11 at para. 18; (1950) 82 CLR 282 (16 May 1950)
Haida Nation v British Columbia (Minister of Forests), 2004 SCC 73 at para. 25, [2004] 3 S.C.R. 511
Kyrtatos v Greece (ECtHR Rep (2003–VI) 257)
Snell v Farrell [1990] 2 S.C.R. 311

Legislation

International Covenant on Civil and Political Rights
European Convention for the Protection of Human Rights and Fundamental Freedoms
Pa. Const. of 1776, art. IV

Chapter 14

Conclusion

Ken Coghill, Charles Sampford and Tim Smith

Strongly contested views have been argued by the contributors to this volume. The authors have presented cases ranging from outright rejection of a perceived radical change to the established political order, to powerful advocacy of fiduciary duty as a necessary tool for addressing the direst threat to ever confront human civilization and an innovative theoretical basis for so doing.

The inspiration for the book and the workshop from which it derives arose from the profound frustration that many feel at the failure of legislatures and governments to implement well-known and understood solutions to this threat and a feeling that notions of public trust and fiduciary duty have the potential to guide policy-makers towards effective action. The nature of the threat and some suggestions on the potential role of fiduciary duty were introduced by McGoldrick, Feaver and Maver in chapter 2.

Finn's long-standing interest, developed in his academic career and continuing in his time on the Bench of the Australian Federal Court, has been reflected in chapter 3. In it, Finn points to the centuries-old origins of the concept of public trust and fiduciary duty. He argued for a general recognition of the fiduciary responsibility of government, deriving from the duty of persons holding positions of authority to conduct themselves not in their own interests but the interests of 'another' – in this case, the community.

In the peculiar case of Australia where it has not been formally recognized, he argues persuasively that, nonetheless, there are legal tools to achieve outcomes similar to those in other common law jurisdictions. The principles of interpretation, grounds of judicial review and the standards of 'fair play and fair dealing' practised by the state in the Australian federation could all be applied in a manner consistent with the historical pattern and methods of development of the law, to result in 'a more open acceptance' by the Legislature and the Executive 'of responsibility for the consequences of their decisions'. Finn foresees the potential for Australian courts to protect the public's interest in environmental sustainability in the same way as fundamental rights and interests are already protected from encroachment by legislation and executive action.

Sampford took issue in chapter 4 with Finn. He argued that constitutional developments dating from the eighteenth century have rendered fiduciary duty redundant insofar as it applies to the Legislature and the Executive. These developments are not complete and he argued that we still need to identify institutional means which make it more likely that power is used for the purpose for which it is given, rather than abused for personal or party political gain. He argued that the problem to which we now seek that solution is not the use of power by the Legislature or the Executive: it is the unsustainable vision of the good life developed by the West and increasingly sought by the rest. Rather than looking to trust doctrine developed in pre-democratic England, we should look to develop a global carbon integrity system inspired by the 'integrity systems' established to reduce the likelihood of corruption. Such integrity systems need to mobilize the long-term interests and values of individuals be they voters, shareholders or consumers, and make our institutions more responsive to those interests. The ethical 'pull' of a vision of a sustainable good life needs to be supplemented by an economic 'push' of an effective carbon tax (which he argues is far superior to carbon trading schemes). Sampford concluded his remarks with the observation that we are at a stage in which long-term interests need to be pursued if there is to be any long-term in which pursue our interests.

Glover surveyed a wide range of legal decisions and history in the United Kingdom, Canada and Australia and concluded that fiduciary duty has a very limited role in public (as opposed to private) law. In chapter 5, he distinguished the exclusivist nature of fiduciary duty and the inclusivist way in which public law addresses the use of power by the Executive and the Legislature. In this he was arguing that fiduciaries do not have obligations to the community but only to 'those who trust them *to the exclusion of all others*'. These obligations arise where the fiduciary owes loyalty to those to whose interests he is entrusted. Such cases are limited to private law or quasi-private law relationships. Glover went on to argue that invoking a fiduciary duty in cases where a member of the Executive was subject to statutory powers created by the Legislature would be to violate the sovereignty of the Legislature. He concluded by arguing against 'private fiduciary sanctions for abuse of power by public officials'.

Wood took a different approach. She developed the legal strategy known as Atmospheric Trust Litigation (ATL), based on principles of sovereign trust obligation. Her chapter 6 built on the 1992 United Nations Framework Convention on Climate Change (UNFCCC) which has been ratified and is now entered into force in most countries including Australia, Canada, UK and USA. It did so by building on the public trust doctrine which is recognized in jurisdictions as diverse as India, Philippines, South Africa and USA, but not established in Australian law.

Countries that ratified the UNFCCC accepted that they would be guided by principles including that they should 'protect the climate system for the benefit of present and future generations of humankind'. In so doing, they endorsed a responsibility to act to prevent degradation of the atmosphere, which is common to all, and implicitly accepted the public trust doctrine. Wood went on to argue that as all sovereign states are co-trustees of their common atmosphere, all share a binding fiduciary obligation to act to defend and protect it from further harm. She argued that the judicial branch has a role to compel the legislative and executive branches to act in accordance with their fiduciary obligations. This approach was further supported by the failure of political executives to agree on effective action at a global level and of most legislatures to pass the necessary laws. Wood set out in detail how ATL could operate at domestic and international levels, including the principles according to which responsibilities and remedies for harms to the atmosphere could be determined.

The nature of international organizations, their relationships to the nation states that have ceded powers to them and whether and, if so, to whom or what they have a fiduciary relationship was examined by Feaver in chapter 7. The chapter argued that powers transferred to international organizations can be conceptualized as political property rights and, as a consequence, the organizations 'stand in the position of a fiduciary' in exercising those powers. However, this obligation was distinguished from 'a positive fiduciary duty to act' to preserve the state of an asset such as the atmosphere, consistently with public trust doctrine. He argued that the public trust doctrine rests on the ancient *res communes* doctrine, discussed also in chapter 2, and human rights. The major difficulty and challenge at an international level is enforcement of a duty in the absence of institutional arrangements to enforce such action. Thus Feaver's argument was that the public trust doctrine cannot operate at the multilateral level because the sovereign powers to deal with the particular policy matters relevant to climate change have not been ceded through an intergovernmental agreement, rather than to suggest that it is at odds with legal principles. Indeed the argument concluded that an international organization assigned sufficient powers, independence and autonomy would likely be bound by the fiduciary principle to act to compel effective action 'to preserve the common property of mankind'.

He went on, however, to identify ways in which the public trust doctrine could be relied upon to bring about the necessary changes. He referred to the obligation of governments of nations, as sovereign trustees of trust assets harmed by atmospheric warming caused by other nation states (such as vulnerable coastal states), to protect those trust assets from damage[1] by taking proceedings in the International Court of Justice (ICJ) to protect the assets and to recover any loss. He explored the question whether the ICJ might recognize the public trust doctrine as a basis upon which to find in favour of those nation states harmed by atmospheric warming. He considered the long-standing norm in customary international law that States are obliged not to inflict damage on or violate the rights of other States as established in the arbitral decision in the Trail Smelter case[2] – the 'no-harm rule'. He discussed the elements of any such cause of action, and argued that establishing a breach of international law would be aided by the equitable rule that imposes a duty upon all nation states to preserve 'common property of mankind'. He drew attention to the fact that the legal status of the atmosphere as falling within the scope of 'common heritage property' was clarified by UN General Assembly Resolution 43/53 in 1988. Paragraph 1 of Resolution 43/53 provides that 'climate

---

1 *State v City of Bowling Green*, 313 N.E.2d 409, 411 (Ohio 1974).
2 Trail Smelter Arbitration of 1941, Reports of International Arbitration Awards (RIAA) III.

change is a common concern of mankind, since climate is an essential condition which sustains life on earth'.[3]

He suggested that a first step towards prompting the creation of adequate international institutional machinery to deal with issues relating to climate change related harm may be for those nation states that have been harmed to take the initiative and assert their rights under international law. At the very least, such action would provide an opportunity to clarify the status of the atmosphere and the international law relating to the common heritage of mankind.

Langford examined the history of fiduciary duty before the High Court of Australia, in particular 'whether climate change fiduciary duties are possible under Australian general law' (chapter 8). After reviewing a number of recent decisions in which the Court has had to determine the nature and extent of fiduciary duties, Langford found that although some aspects remain unclear and in flux, it is clear that they cannot be applied under Australian law to hold government accountable for protection of the atmosphere.

The debate was then taken up from three Australian political perspectives. Murray (chapter 9) concentrated comment on the Carbon Pollution Reduction Scheme – an emissions cap and trade scheme which was before the Australian Parliament at the time of the Workshop but was subsequently defeated in the Senate. He raised concerns about its environmental and economic viability – and was particularly concerned about the very large tax expenditures involved and the lack of any effective accounting for the benefits they are supposed to produce. Realpolitik means that the ALP needed an ally to share the blame if it went wrong – with the Liberals the only real possibility. He concluded that we should revisit the idea of a carbon tax, adding a number of arguments to those already raised by Sampford.

Clark's views argued in chapter 10 have particular significance. At the time of the Workshop he was a Victorian State Opposition spokesman but he became Attorney General and Minister for Finance with the change of Government in late 2010. Clark regards fiduciary duty as inappropriate for Westminster-style parliamentary systems as it would, he argued, undermine democratic institutions. He sees the solution as reform and strengthening of the powers and independence of agencies such as environment protection authorities through democratic processes rather than judicial intervention. Nonetheless, he endorsed use of 'the notion of fiduciary duty as a moral, political and/or ethical argument'.

Thomson, with experience as a member of both the Victorian State Parliament and the Australian House of Representatives and in Opposition and Government, drew attention in chapter 11 to the observed effects and projected risks of climate change and other factors affecting sustainability, particularly over-population. He argued that there is an obligation to protect and conserve the environment, summed up in Brundtland's definition of sustainability: 'meeting the needs of the present without compromising the ability of future generations to meet their needs'. That

---

3   General Assembly Resolution 43/53 (6 December 1988).

obligation should be seen by parliaments and governments as a fiduciary duty or public trust.

The concept of a Ponzi scheme was applied by Haines as she canvassed fiduciary obligations of both business enterprises and government in chapter 12. In doing so, she provided a wide-ranging analysis of the manner in which economic outputs are measured and the environmental impact is marginalized. She argued that comprehensive measurement would help re-orientate not only government but also the business world.

A fiduciary theory of human rights was advanced by Fox-Decent (chapter 13). In this analysis, 'the right to a healthy environment' is a human right. The theory implies that 'states (are) joint trustees of the earth's atmosphere' because the effects of every state's actions affecting the atmosphere are boundary-less and extend to affect every person on earth. As human rights are universal, sovereign power is constrained to respect the protection of the environment. The theory is strengthened by invoking the Kantian distinction between wrongs and harms. Thus, greenhouse gas pollution of the atmosphere is a wrong because of its potentially harmful effects even if it causes no immediate, discernible harm. Fox-Decent's fiduciary theory of human rights has some parallels to Finn's approach of protection of the public's environmental interests in the same way as fundamental rights. It complements and provides support for Wood's ATL strategy.

Our contributors have left us with ample information justifying urgent and far-reaching action by business, governments and individuals who are responsible for emissions, those with the responsibility for reducing them. In that latter category, we do not only include corporations, governments, agencies and watchdogs at all levels but all of us in our roles as consumers, voters and (as retirement benefit, pension and superannuation funds increasingly dominate markets) as the ultimate owners of capital. The authors have provided us with a rich selection of challenging ideas ranging from quite specific policy actions to a unifying theory.

The fiduciary theory of human rights provides a framework within which to locate other concepts and ideas discussed in this volume. Mankind's interest in the protection of the environment in the same way as fundamental rights, argued by Finn, clearly fits neatly within the framework.

Where court decisions have limited the potential of fiduciary duty to impose obligations on government, treatment of damage to the atmosphere as an infringement of the human right to a healthy environment may provide a fresh and innovative vehicle whereby the courts could reconsider the responsibilities of government to act effectively on environmental sustainability.

The framework also readily accommodates ATL, advocated by Wood, and, in so doing, may be said to be relevant to the argument that ATL is a response to failings of the system of government in the United States of America.

The fiduciary theory of human rights framework can be used by parliamentarians to argue much more strongly when seeking support for effective domestic and complementary international policy instruments. It provides a highly persuasive basis for arguing that the politician's own domestic jurisdiction has a responsibility to reduce and eliminate the wrong of diminishing the health of the citizens of the domestic polity and all other peoples who are immersed in and are unavoidably dependent on the boundaryless atmosphere. In the USA, where domestic political opposition has most constrained effective action, the framework has potential to greatly strengthen President Obama's argument in support of the Environment Protection Agency's regulation of emissions by fossil fuel power plants and petroleum refineries, announced in December 2010.[4]

In the global arena, it strengthens the hand of negotiators seeking international instruments which will bind all nation states to act in the common interest of all mankind.

There is, however, another crucial value in what has been argued. The behaviour of organizations, including political and judicial institutions, is not merely a product of their formal structures and codified rules. They are very much affected by the social environment in which they are embedded, shaped by the values adopted, applied and promoted by influential actors.[5] In this case, we are not concerned simply with whether concepts of fiduciary duty and public trust are or ought to be formally binding on government. We are concerned with the motivations felt by government actors. This volume has provided bases upon which key actors can argue persuasively for re-orientation of the behaviour of parliaments, other legislatures, courts and government towards protection of the rights to a healthy environment. Some may choose to do it through the development of legal arguments based on fiduciary duty and public trust. Some may see these as useful arguments to incorporate within the institutional ethics of political bodies, environmental agencies and corporations.

As humanity faces the gravest threat ever to civilization, our contributors have offered a way forward that invokes principles of fiduciary duty and public trust in a variety of ways – tempered with some cautions. It now remains for our politicians and citizens, corporate boards and shareholders, consumers and their suppliers – in short all those who hold power in this world – to consider how their powers will be used to address the existential threat posed by climate change and environmental degradation more generally.

References

Environment Protection Agency (USA) (2010) 'EPA to Set Modest Pace for Greenhouse Gas Standards / Agency stresses flexibility and public input in developing cost-effective and protective GHG standards for largest emitters', available at http://yosemite.epa.gov/opa/admpress.nsf/d0cf6618525a9efb85257359003fb69d/d2f038e9daed78de8525780200568bec!OpenDocument.

---

4   Environment Protection Agency (USA) (2010).
5   Meyer and Rowan (1977).

J. W. Meyer and B. Rowan (1977) 'Institutionalized Organisations: Formal Structures as Myth and Ceremony', 83(2) *American Journal of Sociology* 340.

United Nations Environment Programme Ozone Secretariat (2007) 'Montreal Protocol – Celebrating 20 years of progress in 2007', available at http://ozone.unep.org/.

Cases

State v City of Bowling Green, 313 N.E.2d 409, 411 (Ohio 1974).

Trail Smelter Arbitration of 1941, Reports of International Arbitration Awards (RIAA) III.

Legislation

General Assembly Resolution 43/53 (6 December 1988).

Index